Birding Brazil
A Check-list and Site Guide

by

Bruce C. Forrester

Contents :

Preface

The publication of this little book, with its modest title and unpretentious format, is an important event for birdwatchers interested in the New World Tropics, for it presents the most ample and detailed body of birdfinding information yet available for any country in South America. And - need it be said - such information is especially welcome for Brazil, so rich in birds and until recently at least, so poor in truly useful publications about them. But to call Bruce Forrester's guide useful would be an understatement - it is more exactly indispensable for anyone considering Brazil as a birdwatching destination or simply curious about the country's birds and their distribution.

It is an altogether admirable piece of work, resulting from an extraordinary degree of dedication: six successive summer trips and thousands of kilometers of overland travel in Brazil. Having been part of four of these trips, I can attest to the thoroughness with which the birdwatching sites were explored. Anyone who has travelled by car in Brazil and stared in perplexity at the branching of unmarked rural roads will recognize the contribution represented by the maps alone. But Birding Brazil is also a check-list of all the bird species known for the country and provides, as well, a bird list, as complete and accurate as possible, for each of the sites. The gathering and collating of such a volume of distributional information, much of it from unpublished sources, is itself a remarkable achievement.

It is to be hoped that Birding Brazil will inspire more ornithologists to visit the country and that as an ultimate result its birds, and in some cases their urgent need for protection, will be better known.

Davis W. Finch,
East Kingston,
NH, USA.
November 1992.

Long-tailed Tyrant

Introduction

Brazil is paradise to a curious mind, an ardent wanderer, a lover of our natural world. It has almost a fifth of the world's birds, a staggering 1,661 avian species, 177 of which are endemics.

This guide has been compiled primarily to encourage foreign visitors to travel around this beautiful and exciting country and to help them plan an itinerary. Comparatively few foreigners have come to birdwatch here despite the vast ornithological wealth of the country. The contents herein are intended for practical use in the field and for reference on specific species and sites. The guide has four main sections. Firstly, useful tips and background inform- ation on the country. Secondly, 42 sites are described, with a brief account of location, habitat, access and accommodation. Important species are noted and a simple map of the area is included. A coded reference, at the end of each site description, indicates what sources of information were used when collating the site list (this code is explained at the end of the guide). The third section of the guide contains site lists. These are in three groups of fourteen. Here, a full list is given of all species known to have occurred at each site. Lastly, there is a complete list of Brazil's birds, indicating on which of the three main site lists the species can be found.

To cover as many species as possible and in order to keep the list of sites to a manageable level, site sizes vary considerably. In a few cases, a site includes an entire state, *e.g.* Rondônia, which ideally should have been divided into perhaps four or five sites but due to lack of available information has been kept as one. Where possible, sites were chosen to include either very important National Parks and reserves or areas containing unique species. Sites are therefore dotted around the whole of the country. No visitor will attempt to visit all of them, on one trip, but the guide will help him choose a selection to meet his needs. I, myself, have not been able to visit every site, though I have been to 36 of them, many on a number of occasions.

Bruce C. Forrester,
Knockshinnoch Bungalow,
Rankinston,
Ayrshire.
KA6 7HL
U.K.

Slender Antbird

Acknowledgments

I have attempted to gather information from all possible sources: published material from various countries, unpublished material gathered from both foreign visitors and residents, and my own records from six trips 1987-1992. When I originally began this task, I had little information to hand but over the years have discovered a wealth of information, much of it in obscure locations and not readily available to the jet-setting birdwatcher who wants to have this at his fingertips in order to plan one trip before tackling another. I am indebted to all those individuals who have contributed their observations from the various sites. Without their help this guide would not have been possible. I would like here to single out a few of the many for contributing their time and help when supplying records - Walt Anderson, Mario Cohn-Haft, Nigel Collar, Davis Finch, Alan Greensmith, Ted Parker, Mark Pearman, Robert Ridgely, Derek Scott, Douglas Stotz, Dave Willis and Andrew Whittaker. A complete list of those who have supplied information forms part of the *literature consulted* section, on page 241. I am also grateful for those who were my fellow companions, Ian Puckrin (1987), Simon Cook (88), Davis Finch (88, 89, 91 & 92), Duncan Macdonald (89), Paul Coopmans (89), Dave Willis (90 & 92), and Eleanor Forrester (90, 91 & 92). Brazil needs the information collected by amateur ornithologists on vacation. Without their co-operation, these sightings would go unrecorded and as a result our knowledge of Brazil's fauna would be less complete.

A special thank you goes to Varig, Brazilian Airlines, for their kind support in this venture. Without their help none of this would have been possible. Hopefully, those who read this guide will be encouraged to seek new horizons and visit sites that otherwise would not have been considered, and when touring Brazil will travel Varig's extensive network both within the country and internationally.

My eternal gratitude goes to my wife, Eleanor, who has spent three years working on successive drafts of the guide. Thanks also to Davis Finch, Sandy Scott, David Clugston, Ronald Forrester and Robert and Chris Provan for their technical advice. I am indebted, as well, to BirdLife International, formerly the International Council for Bird Preservation (Cambridge), the Edward Grey Institute of Field Ornithology (Alexander Library, Oxford), the British Natural History Museum (Ornithology Library, Tring) and the Scottish Ornithologists' Club (Edinburgh) for access to their libraries.

General Advice

Upon arriving at Rio or São Paulo, visitors should buy a copy of *Guia Quatro Rodas* , an automobile guide with maps and information on hotels and other accommodation. Additionally, prior to departing from your home country, you may wish to obtain the *Rough Guide: Brazil* by D.Cleary, D. Jenkins, *et al.*, probably the most useful of the many general travel guides. The birdwatcher should also take with him a selection of books including Meyer de Schauensee's *A guide to the birds of South America*, the bible to the description of species; Johan Dalgas Frisch's *Aves Brasileiras* , very useful in southeast Brazil despite the seeming crudeness of its illustrations; John Dunning's *South American Birds: A Photographic Aid to Identification* , useful throughout the country. If visiting the Amazon, carry either Meyer de Schauensee and William Phelps's *A guide to the birds of Venezuela* or Hilty and Brown's *A guide to the birds of Colombia* . If concentrating on the south of the country, take Narosky and Yzurieta's *Birds of Argentina and Uruguay* and William Belton's *Birds of Rio Grande do Sul* . When visiting the Pantanal, take B. Dubs's *Birds of Southwestern Brazil* . For the library at home, Robert Ridgely and Guy Tudor's *The Birds of South America* will, when complete, form the new bible to South America. Dedicated enthusiasts will also acquire Helmut Sick's *Ornitologia Brasileira*.

To get around the country, it is advisable to make use of Varig's three week Airpass. While this was excellent value until 1990 (allowing unlimited use) it is no longer such a bargain. However, if you select carefully the five permitted flights, it remains good value. This Airpass now costs $440 (1993). Book all internal flights well in advance. This is particularly recommended for flights to Iguaçu and the Pantanal, especially during the months of July and August, as many Brazilians travel at this time of the year. I suggest that you hire cars from Localiza National which has proved over the years to be the best of the car hire firms in Brazil. They have offices at all the main airports. While car hire in general is expensive, Localiza normally has fairly reliable cars. Most Brazilian rental cars run on alcohol rather than gasoline. These have a distinct drawback since alcohol cars are very poor at starting under cold conditions, which can lead to many lost hours in the morning, particularly in the south of the country or at higher elevations. In such situations, hire a gasoline model where possible. Booking in advance will save considerable time on arrival at the airports. Be sure not to hire a car with a kilometre charge as this could be extremely expensive. It is best to pre-book through National Car Hire, which is affiliated with Localiza and may be found in your home country (Europcar in the U.K.), or alternatively write directly to Localiza in Belo Horizonte. In either case, ask for the International Dollar Tariff which will be charged at a set rate without a kilometre charge.

Visiting reserves can be a problem in Brazil. Commendably, the Brazilian authorities guard their reserves, often allowing the public to use only a small section, usually at the entrance to a National Park, for recreation. This undoubtedly helps protect the fauna of the area but it creates restrictions for the visiting birder and prevents the collection of data on the distribution and habits of that fauna. It is thus desirable to obtain permission to visit reserves. In most cases this *used* to be obtainable by writing directly to the IBAMA headquarters at Brasilia. However, since 1990 permission has to be obtained from the individual State headquarters, causing problems for the foreign visitor. Like many such administrative changes in Brazil, this could easily revert to the former procedure. National Parks where the public are welcome tend to be open from 8.00 until 17.00. With permission, it is possible to visit at any hour, and perhaps have basic accommodation on the reserve, with park personnel serving as your guides.

This guide has a strong bias towards July and August, the months of my visits. This may be the best time to tour Brazil since it is cool in the south and reasonably dry in the Amazon, where it might rain on alternate days for an hour or so. It is also dry in the Pantanal which, like the rest of Brazil, is almost free of insects during these months. At other times of the year, mosquitoes can be a real pest in wet areas. For most of Brazil, with the exception of the southeast, neither the temperature nor the weather changes markedly from season to season.

Red tape in Brazil changes from month to month and year to year, so it is impossible to predict and advise about almost anything relating to bureaucracy or banking practice. Some years, due to high inflation, it is advisable to change only a little money at a time (perhaps once per week). There may either be a large or a small difference between the official dollar rate and the unofficial or tourist rate. Thus, it may at certain times be advantageous to use a credit card and pay when you have returned home, perhaps making on the deal, but usually it is best to pay in cruzeiros or dollars. In 1990, it was more advantageous to change pounds sterling, rather than dollars, into cruzeiros, but this was unusual. Normally, the Banco do Brasil is the only bank willing to change travellers cheques and, usually, a favourable tourist rate is given but some years a better black market rate is offered by Cambios or Tourist offices. Therefore you must quickly ascertain the best method of changing money either upon arrival in Brazil, or if possible before leaving homo.

Brazil is notorious for its diseases and creatures with fatal bites. Thankfully, these are not normally encountered but they should not be totally ignored.

Mosquitoes, especially in the northern half of the country, can carry malaria, so malaria tablets are advisable. However, in the Amazon, due to the variety of strains, malaria tablets are not guaranteed to help. Yellow fever immunization, which lasts ten years, is a worthwhile precaution. Stomach upsets often occur when one arrives in the country, more due to a change in water and eating habits than to disease, but preventative tablets should be taken. Bottled water is widely available throughout the country, and in general most tap water, after an initial upset, has no lasting problem. Typhoid and hepatitis are both contractible and recently cholera has spread rapidly through Brazil. In the Amazon forest leishmaniasis may be encountered. This is transmitted by the bite of a particular fly, so it is advisable to wear long-sleeved shirts and trousers. Beware of staying in adobe shacks which may have beetle-like creatures carrying chagas' disease. Candura fish and piranhas can cause problems to swimmers in Amazonian rivers. Schistosomiasis or bilharzia can be found in stagnant water and is contractible through mere contact with it. Rabies is an added hazard. Though snakes are rarely seen, they are common throughout Brazil. Killer bees are also seldom encountered but have caused problems in some areas. This list is not intended to put anyone off going to Brazil but should be kept in mind while in the country. Prior to departure from your home country, it is advisable to consult your physician regarding immunization and protection from other health hazards.

Brazil has a very extensive police complement which is made up of various units. When driving in Brazil, you will probably be stopped from time to time, especially on departing from a town. Have car documentation handy plus an International Driver's Licence. I also carry my British Driver's Licence. British citizens do not require a visa to visit Brazil, though certain other nationalities do, including those from the USA.

Brazil has a good network of hotels. In almost every village, let alone town, there is a hotel of some description. In cheap hotels, food is usually very good value for money and often the largest quantities are served in the humblest of establishments which are honoured to have a foreign visitor. Since many birdwatching localities are in remote areas, it is fortunate that there is usually accommodation nearby. People choosing to stay at large expensive hotel complexes restrict themselves considerably, and miss out on many of the fantastic sights of the country. When visiting the Amazon, staying under the tree canopy is a pleasurable experience, though usually accommodation is available nearby.

Anyone not wishing to organise their own itinerary may wish to join an organised group. Doug Trent of Focus Tours runs general natural history tours and having an office in Belo Horizonte can readily make arrangements throughout the country. Both WINGS, Inc. and Field Guides Inc., based in the USA, have run trips to Brazil for many years and are very familiar with the locations and species. The addresses of these companies along with other useful addresses can be located at the end of this guide.

Silvery-cheeked Antshrike

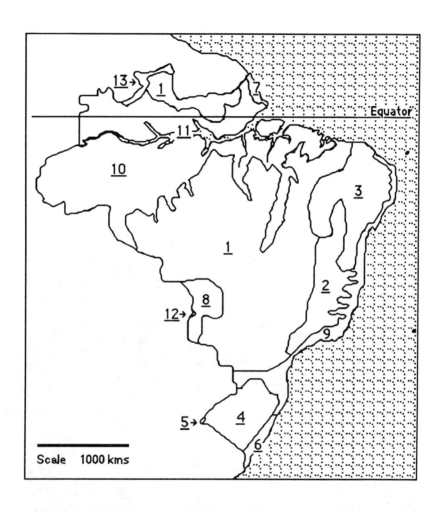

Equator

Scale 1000 kms

Habitats

Before going there, I, like many others, assumed that Brazil was nearly all steamy hot jungle. This, as will be seen, is not the case.

To make site lists more readable, they have been grouped in three geographic areas: Southern and Central Brazil, Eastern Brazil and lastly Amazonia and Atlantic Islands. The various habitats found in these areas are discussed below:

1. Cerrado which carpets the vast central plateau of Brazil, dominating the country, is savanna with scattered short trees. The savanna habitat in Roraima differs in being wetter.

2. Campo is similar to cerrado but is usually at higher elevation, with rocky outcrops, shorter grass and more scrub than trees.

3. Caatinga, extremely dry thorny scrub with small trees, is confined to Brazil's desert-like northeast.

4. Planalto and Campanha are characteristic of the south. These two habitats merge into each other and consist of short-grass plains with forest occurring in steep-sided valleys. From northern Rio Grande do Sul northward into Parana this area was once heavily forested in Araucaria, but following European immigration the forest was extensively felled, resulting in today's wide-open expanses.

5. Espinilho, a Uruguayan and Argentinian habitat, is found only as a fragment in southern Brazil. This is a forest complex of small leguminous trees.

6. Littoral here refers to low flat land occurring at scattered localities along Brazil's Atlantic coast, with marshes, lagoons and wasteland. The extreme southern coast is predominantly littoral.

7. Mangroves particularly in the north, occur at isolated points along Brazil's extensive coastline.

8. The Pantanal, the seasonally flooded region of western Brazil, is of world importance for water related birds.

9. The Atlantic Forest changes in composition from Araucaria in the south to deciduous farther north, becoming true rainforest at the northern part of the range. Its make-up also changes with elevation, the higher parts of the Serra do Mar being cloud forest.

10. Amazonian Forest is true rainforest. Amazonian trees have adapted to the poor soil, conserving nutrients through absorbing rainwater before it hits the ground and by having their roots confined near to the surface. Although given one name, Amazonian Forest is actually a wide range of different forest types. Buttress trunks are synonymous with Amazonian forest. Where the soil changes, so does the forest, including the animal and bird life. This is why one has to visit several Amazonian locations to find the widest variety of avian species.

Campina, dry forest on sandy soil, is confined to small patches in Amazonia.

11. Varzea Forest is found along the banks of the Amazon and is seasonally flooded.

Igapo Forest habitat is similar to Varzea but is almost continuously flooded and usually refers to blackwater rivers such as the Rio Negro.

Riparian areas are river banks that fluctuate drastically in width, creating damp open areas with low bushes or long grass.

12. Chaco Forest is very dry forest, characteristic of Paraguay, southern Bolivia and northern Argentine, but just creeps into Brazil along its western border of the Pantanal.

13. The Tepui region is located on the border with Venezuela and Guyana. This unique habitat is composed of flat-topped mountains with nearly vertical sides.

14. Oceanic - Brazil's best seawatching localities are in Rio Grande do Sul. Here, due to the drift migration of plankton from the south, large numbers of oceanic birds can be found.

Atlantic Islands - the two sites listed have had, and in the future will add new species to the Brazilian list due to their geographical locations.

Man has changed the make-up of Brazil, consequently few of these habitats can be viewed in their original state. In most cases, when in the field, two or more of these habitats will be encountered at the same time.

Although we all hear through the media of the destruction of Amazonian rainforest, Brazil's Atlantic Forest is much more seriously threatened. This habitat is no longer a continuous belt but is fragmented into a series of isolated forest islands forming oases for the remaining endemic bird species. Some of these are now confined to only one or two such island refuges. In this part of the world conservation is required NOW, not tomorrow. The Amazon is important for the world but not as important as the Atlantic Forest in Brazilian terms since Amazonian species can be found in Venezuelan Amazonia, Colombian Amazonia, Bolivian Amazonia and the Guianas. However, if the forest is to continue to be cut on the scale that it has been for the last twenty years then in a further ten years the situation will be very worrying. That is a problem to be faced after the successful preservation of the Atlantic Forest remnants. It is worth mentioning that the practice of reforestation with eucalyptus plantations, in recently deforested areas, is of no ecological benefit and are devoid of birds.

Brazil has many reserves, which are mostly well protected. Regrettably, the wardens are not always very knowledgable in matters of natural history, since their duties are principally law enforcement. Properly trained conservationists could make visiting Brazilian reserves a more educational experience for both international tourists and, more importantly, Brazilians.

White-eared Puffbird

Review of Brazilian Ornithology

Historically, Brazil attracted many pioneering ornithologists whose work and collections form the basis of knowledge about bird distribution in the country. These included G.K. Cherrie, W. Hoffmann, E. Holt, Emil Kaempfer, P.W. Lund, J. Natterer, O. Reiser, Clarence & Jessie Riker, H. H. Smith, Emilie & Heinrich Snethlage, J.B.Spix, plus many more. N.Gyldenstolpe's meticulous records of his Amazonian travels deserve special mention. Authors of scientific papers listing the collections of these explorers include J. A. Allen, H. Friedmann, C. E. Hellmayr, E.M.B. Naumburg and August von Pelzeln. More recently, Olivério M. de O. Pinto's many travels produced a mass of information for this guide. He must be considered one of the key figures of Brazilian ornithology. Much of the work of Helmut Sick and Augusto Ruschi has been of specific importance to me in preparing this guide. For endangered species, the recent publication *Threatened birds of the Americas : the ICBP-IUCN Red Data Book* , is the most concise and informative book available. Meyer de Schauensee's *The species of birds of South America and their distribution* remains the standard English work.

In recent years, important ornithological projects have been conducted in various parts of the country. The centres for ornithology are, naturally, the larger cities, Rio de Janeiro, São Paulo, Belém and Belo Horizonte. Much of the work is being carried out by university groups. Today, the Rio Bird Club (the national headquarters of the *Clube do Observadores de Aves*, which has a number of state chapters) is playing an important role by publishing recent sightings of rare species. The Sociedade Brasileira de Ornitologia has begun producing its own journal *"Ararajuba"* which is very informative, containing the results of recent bird surveys and leads the way to a new era in Brazilian ornithology.

Presently, active ornithologists in Brazil, include - M.A. Andrade, Paulo de Tarso Zuquim Antas, R. Bierregaard, R. B. Cavalcanti, Renato Cintra, Paulo S. Moreira da Fonseca, Luiz Gonzaga, J.B. Nacinovic, F.C. Novaes, J. Fernando Pacheco, David C. Oren, J.M.C. da Silva, Douglas F. Stotz, Dante M. Teixeira, J. Vielliard, W.A. Voss, A. Whittaker, Ed Willis and C. Yamashita. A warm thank you goes to these and many other Brazilian birdwatchers for their time and help with this guide.

Southern and Central Brazil

Under this heading are fourteen localities with habitats as varied as coastal Rio Grande do Sul, the fabulous Iguaçu Falls , the extensive grasslands of Brazil's tableland and the world-renowned wetlands of the Pantanal.

1 Taim Ecological Station.
2 The Mostardas Peninsula.
3 Aparados da Serra National Park.
4 Uruguaiana.
5 Foz do Iguaçu (Iguaçu National Park).
6 Campo Grande to Corumbá (Southern Pantanal).
7 Transpantaneira (Northern Pantanal).
8 Barranquinho.
9 Chapada dos Guimarães.
10 Emas National Park.
11 Serra da Canastra National Park.
12 Serra do Cipó National Park.
13 Brasilia National Park.
14 Araguaia National Park.

Although not listed here, other sites worth visiting within this region include the Esmeralda Ecological Station (Rio Grande do Sul), the São Joaquim National Park (Santa Catarina), the Chapada dos Veadeiros National Park (Goiás) and the Serra das Araras (Mato Grosso).

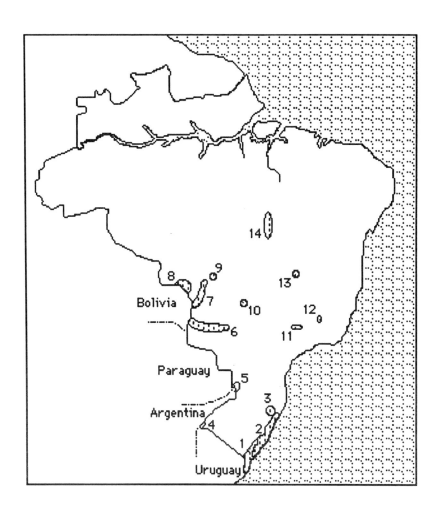

Bolivia

Paraguay

Argentina

Uruguay

8

9

7

6

14

13

10

12

11

5

3

2

4

1

Taim Ecological Station

The area covered under this heading includes the southeastern corner of Rio Grande do Sul from São Lourenço do Sul south to the Uruguay border at Chuí, the two most important areas being the marshes around the Taim Ecological Station and the seaside resort of Cassino.

Coastal Rio Grande do Sul is of world importance as a wintering area for both boreal and austral migrant shorebirds, according to season. The habitat is very low lying with large expanses of open water interspersed with marshy vegetation, ideal for rafts of White-faced Whistling-Ducks and masses of White-faced Ibises and many other aquatic species. In addition, Cassino and other spots along this coast are among the most important sites in Brazil for passage seabirds which come north following the great plankton drifts.

Unless you have permission to stay at the Ecological Station, you will have to stay at Cassino, Rio Grande or Pelotas, though bear in mind the 82km drive to Taim from Cassino (via Quinta). Cassino affords an opportunity to do some seawatching from the south jetty, at the mouth of the Lagoa dos Patos. To reach the jetty, drive down on to the hard-surfaced beach and drive six kilometres north. Off the end of the jetty, in winter, one can see Magellanic Penguin, Black-browed Albatross, Southern Giant-Petrel and White-chinned Petrel; there is always the possibility of a new or rare species for Brazil. The potential is certainly there for the hardy enthusiast willing to put in time and effort during rough seas. Hopefully, within the next few years many more new species will be added to the Brazilian list at Cassino as more birdwatchers take up this challenge.

Two-banded Plover

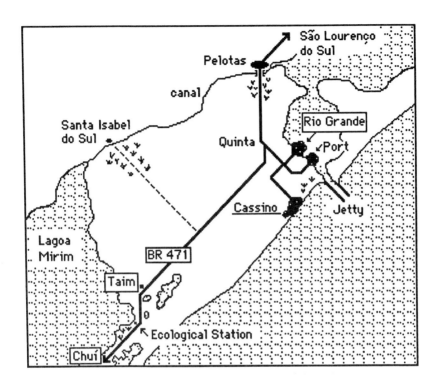

At Cassino, the dunes and the area behind them provide habitat for some species not so easily seen at Taim, such as, Common Miner, Freckle-breasted Thornbird and Bay-capped Wren-Spinetail.

This southern tip of Brazil is of course the northern limit of several Argentinian and Uruguayan species e.g. Curve-billed Reedhaunter, Stripe-crowned Spinetail, Wren-like Rushbird, Spectacled Tyrant, and in winter such species as Tawny-throated Dotterel, Rufous-chested Plover, Two-banded Plover, Bar-winged Cinclodes and Patagonian Negrito.

To reach this area fly into Porto Alegre and drive the 270 kms along a fairly good road to Pelotas. Alternatively, fly with Rio-Sul from Porto Alegre to Pelotas. Anyone spending time at this or the next three sites should consult William Belton's *Birds of Rio Grande do Sul, Brazil*, the most comprehensive work to any part of the country and a very impressive document. Finally, remember that the winter weather here is often damp and cold.

Ref: A10, B6, B7, C8, D6, F6, F10, M11, P7, P11, T3, T5, T10, V3, V4, W10.

The Mostardas Peninsula

The area covered in this site runs from the border with Santa Catarina south along the Rio Grande do Sul coastline as far west as the outskirts of Porto Alegre and south to the end of the 250 kilometre peninsula. The latter is not named on maps but in birdwatching circles, thanks to William Belton, has become known as the Mostardas Peninsula after the township halfway down its centre. Hotels can be found along the north coast and there is one at Mostardas.

The typical habitat, throughout the area, is *littoral*. The oceanic ingredient will also prove useful in producing an additional source of birds. The few rainforest species mentioned in the accompanying list refer to the remnant forest at the base of the escarpment, north of Capão da Canoa towards the border with Santa Catarina.

The road to Mostardas during the winter is undertaken with some trepidation since it is a 130 kilometre track of either rutted sand or deep slippery mud. If it rains while you are out on this isolated and little-travelled road you could be stuck for days. However, the birdwatching is excellent and you may see Chilean Flamingos on Lagoa do Peixe, now a National Park and a very important one for wintering shorebirds. An alternative route is to drive along the entire length of beach from Pinhal south to the end of the peninsula, which offers good driving conditions unless a south wind is blowing.

To reach the Lagoa do Peixe, drive just south of Tavares and look for a poor quality track (past the Jockey Club) leading to the lake shore. A neighbouring locality worth visiting is the Lagoa Capão da Fuma which can be reached by walking west from Mostardas. Here, huge flocks of ducks and ibises congregate. Look for Common Miners and Spotted Nothuras on the sandy roads.

Seawatching is good from near the Solidão Lighthouse: look for the sign on the Mostardas road marked *"Entrada Praia do Farol da Solidão"*(65 kms north of Mostardas). It should be noted that the 10 km track to the lighthouse is even worse than the main road.

Perhaps Brazil's best seawatching spot is the pier at Capão da Canoa. Both here and at Solidão, albatrosses, occasionally penguins, and other seabirds can be seen during the winter.

Ref: A9, B6, B7, D6, F1, F7, F10, M11, P11, P20, P38, R5, S5, W5.

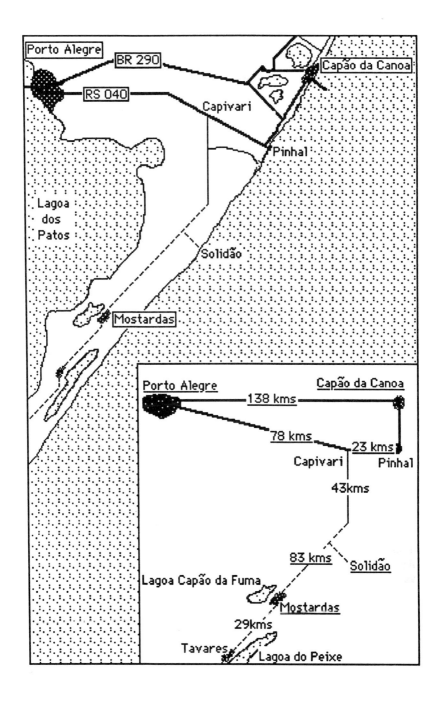

Aparados da Serra National Park

This site includes not only the National Park, situated on the edge of the Rio Grande do Sul escarpment at the Santa Catarina border, but also the road between São Francisco de Paula and the park, since most visitors to the park stay at the Hotel Veraneio Hampel (117 kms northeast of Porto Alegre and 70 kms southwest of the National Park).

To the west of the escarpment is a vast plateau of rolling grassland with isolated patches of forest. With the destruction of forest over most of the area, the Aparados da Serra National Park remains one of the best sites for the Paraná Pine (Araucaria). However, the park is more famous for the extremely deep and narrow Itaimbezinho Canyon. Along with the Araucaria forest, marshy pools with *eryngium* (a spiky plant favoured by Straight-billed Reedhaunter) provide an important habitat for some of the more interesting birds of the park.

Prior to 1989, lodging was available at the cafe within the park and there is basic accommodation in Cambará do Sul. If pleasant surroundings are important, then stay at the Hotel Veraneio Hampel which offers its own nature trail on private grounds. A rural housing development, between the hotel and the RS 020, has good numbers of Speckle-breasted Antpittas. The long drive between here and the park is worth spending a full day over, since it is excellent for grassland species such as Buff-necked Ibis, Yellow-billed Pintail, Red-legged Seriema, Black-and-white Monjita, Hellmayr's Pipit, Saffron-cowled Blackbird and Yellow-rumped Marshbird. The park itself offers such interesting species as Giant Snipe, Vinaceous-breasted Parrot, Long-tufted Screech-Owl, Biscutate Swift, Mottled Piculet, Araucaria Tit-Spinetail, Striolated Tit-Spinetail, Gray-bellied Spinetail, Long-tailed Cinclodes, Straight-billed Reedhaunter, Azure Jay, Lesser Grass-Finch, Black-and-rufous Warbling-Finch, and Long-tailed Reed-Finch.

Unlike most Brazilian National Parks there seem to be no restrictions on access: there are no locked gates, and camping is allowed, if you can stand the cold. During winter, beware the problems of bad weather such as heavy rain, sometimes snow and low temperatures especially at night. Access to the park from the north entrance is often impossible due to the high water at the ford. Also, the roads from São Francisco de Paula are dirt and can become very muddy.

Although Red-spectacled Parrot has been seen in this area, the main roosting site during the non-breeding season is at the Aracuri-Esmeralda Ecological Station north of Vacaria (250 kms to the northwest). However, since araucaria is the preferred habitat of these attractive parrots, it is probably worth looking out for a wandering party during the winter months.

Ref: B6, B7, C8, D6, F1, F7, F10, G1, P6, P11, P38, S6, W5, W10.

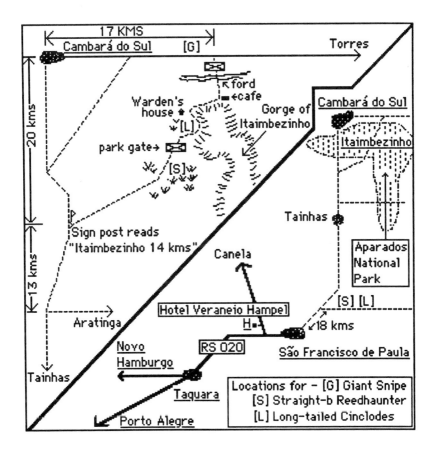

Uruguaiana

The city of Uruguaiana is situated in the southwest corner of Rio Grande do Sul, on the border with Argentina and close to Uruguay as well. About 62 kms southwest of here on the road to Barra do Quaraí is the remnant of espinilho parkland. This patch of important habitat is unique in the country and although certain species may be common across the border, in Brazil they occur only here. Species of particular note include Spot-winged Pigeon, Scimitar-billed Woodcreeper, Tufted Tit-Spinetail, Short-billed Canastero, Lark-like Brushrunner, Brown Cacholote and Black-capped Warbling-Finch.

When we visited this site in 1991, there was still a sufficient area of espinilho parkland to maintain its unique species. Other interesting birds seen in the area were Cinereous Harrier, Checkered Woodpecker, White-banded Mockingbird, Black-crowned Monjita, Chaco Suiriri, Golden-billed Saltator and a group of eight Yellow-billed Cardinals, a new record for Rio Grande do Sul.

Rio-Sul flies to Uruguaiana from Porto Alegre. There is a hotel in Barra do Quaraí and a number of them in Uruguaiana.

Ref: B6, B7, F10, G6, P28, P31, T14.

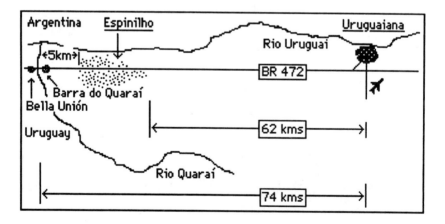

Foz do Iguaçu (Iguaçu National Park)

Species in parenthesis on the main list have been found on the Argentine side, though some of them could easily occur in similar habitat on the Brazilian side.

The reserve is located on the border with Argentina and very close to Paraguay, in the western corner of Paraná. Between the Brazilian National Park and the Argentine (Iguazú National Park) lie the world famous Iguaçu Falls, perhaps the largest in the world. They extend almost three kms in width and form a natural amphitheatre containing 275 separate falls, the Devil's Throat in the centre dropping 90 metres. Heavy rain (2,000 mm annually) and constant spray from the Falls keep the tropical forest wet and rich in lianas and epiphytes. The volume of water passing over the falls is greatest in January and February. The park is a wonderland for the world's twitching lepidopterists. Although the park has an impressive list of birds, due to the dense undergrowth it is difficult to penetrate this forest. The ornithological highlights are Helmeted Woodpecker and Spotted Bamboowren. Access to the Argentine side is fairly easy as visas are not required for short trips. The Argentine side also has more extensive forest with a good trail system which allows better birdwatching.

There are many good, if somewhat expensive, hotels between the airport and the park, the most convenient being the Hotel das Cataratas, one of the Tropical Hotel chain. It is located above the Falls and is the only hotel within the park. Hotels in nearby Foz do Iguaçu are much cheaper and a good bus service runs between the park and town, though this does not begin until well after dawn. From Foz do Iguaçu it is easy to cross over into Paraguay or Argentina. If going to the the Argentine National Park stay at the Hotel Internacional Iguazú.

On the Brazilian side, there are few paths into the forest. There are a few near the entrance and museum, where the forest is a little more open. The Poço Preto Trail, located two and a half kilometres from the gate, behind an obvious house, is decidedly the best trail on the Brazilian side. It has a locked gate, so permission to enter should be sought from the director of the park. A short trail, called Macuco, going down to the river about halfway between the entrance and the Falls is used by a local guide who offers safaris and boat trips but permits walkers on the trail. Around the Hotel das Cataratas there are a few minute paths, most of which close in quickly. Unlike most of Brazil's National Parks, Iguaçu (along the main road to the Falls) is open to the public and written permission is not required, for this section of the park. A walk along the road can be productive, though during the tourist season is very busy with passing coaches. If staying at the Hotel das Cataratas, book in advance.

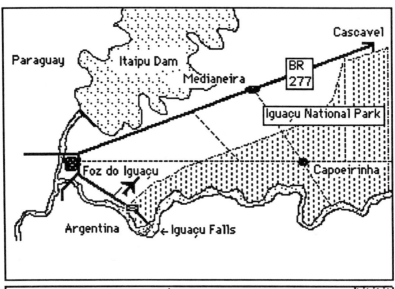

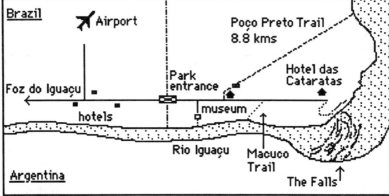

Likewise, book flights to and from Foz do Iguaçu, well in advance to avoid delays, as there are only a few flights each day.

The main reason for visiting this reserve is undoubtedly the Falls. Anyone failing to obtain permission to enter, either the *real* National Park or the Poço Preto trail, will find this site ornithologically disappointing. In such a situation the Argentine side would be more productive and could occupy four or five days. If a permit is granted, the main part of the park, located on several by-roads leading south from the Foz do Iguaçu to Cascavel road, should prove productive. This section of park is off limits to the public, with the roads within this area closed and certain villages abandoned.

Iguaçu is famous for its huge flock of Great Dusky Swifts that swarm into the cliffs along the Falls in the evening. It is also a good locality for Ochre-collared Piculet, Russet-winged Spadebill, Southern Bristle-Tyrant, Bay-ringed Tyrannulet, São Paulo Tyrannulet, Cream-bellied Gnatcatcher, Green-chinned Euphonia and Red-ruffed Fruitcrow, the largest passerine in South America. It is also the only place where I have seen Jaguar.

Ref: A2, A3, A7, B1, B9, C7, C8, C9, D6, F1, F3, F6, F10, F12, G1, G9, I3, I7, M5, M10, P6, P12, P38, R6, R11, R16, R18, S1, S5, S16, W3, W5, W10.

Campo Grande to Corumbá (Southern Pantanal)

This site covers the northwest corner of Mato Grosso do Sul. The Pantanal is a seasonally flooded swampland which at times can become the largest marshland in the world. It is caused by the high water in rivers to the north engulfing the Rio Paraguai, causing it to overflow its banks and flood the neighbouring land for hundreds of miles. The wet season runs from October to April, with January and February having the heaviest rains. These fall mostly in the north, and at first, the southern pantanal is drier. Later, the water flows into the southern half and floods this area and the north begins to dry out. When the water is lowest, huge flocks of birds congregate in the shallower waters. Of the two areas, the northern half appears generally better, at least in July, though in the south certain interesting species are easier to see.

The area described here contains three principal habitats. Firstly, between Campo Grande and Miranda, there are rolling hills with cerrado, good for hummingbirds and parrots, the latter showing well when making their frenzied evening excursions. Secondly, between Miranda and the Rio Paraguai is the low, flat, wet pantanal, famous for its huge numbers of ducks and herons. A third habitat, only represented by a small area within our site, is dry *chaco forest*, on the west side of the Rio Paraguai north to Corumbá and extending westward into Paraguay and Bolivia. Most of the *historical* records refer to Urucum, the locality for certain species not otherwise found in Brazil.

Some of the more noteworthy species of the southern Pantanal include Hyacinth Macaw, Blaze-winged Parakeet, Green-cheeked Parakeet, Black-hooded Parakeet, White-fronted Woodpecker, Great Rufous Woodcreeper, Mato Grosso Antbird, Cinereous Tyrant, Bearded Tachuri, Fawn-breasted Wren, Black-backed Grosbeak, Black-and-tawny Seedeater, Tawny-bellied Seedeater, Dark-throated Seedeater, Marsh Seedeater, Rufous-rumped Seedeater and Chestnut Seedeater. Most of these birds can be seen within 100 kms of Corumbá, though the drier and higher ground east of the true Pantanal (*ie.* east of Miranda) is worth exploring.

There are a number of basic hotels in the border town of Corumbá and a few ranches in the Pantanal where one can stay, including the Fazenda Santa Clara 98 kms east of Corumbá. This is a favourite among birdwatchers because Hyacinth Macaw and many of the other birds mentioned are fairly easily found nearby. Corumbá can be reached by plane though not by Varig, and there is also a railway from Campo Grande, though this is very slow. A good road (Pantanal section dirt) linking the two cities allows the 403 kms to be driven comfortably in two days, perhaps with a stop at Aquidauana (131 kms from Campo Grande). The hotel here is very basic. Alternatively, the drive can be made between Campo Grande and Corumbá in one day, though with little birdwatching. As in many towns on the border with Bolivia, on leaving Corumbá you can expect to be scrutinized thoroughly by the police, who are looking for drug trafficking. This may involve having your car almost taken apart.

In the Pantanal, it is difficult to drive anywhere during the wet season as the road may be several feet under water, leaving a boat the only means of access. Even during the dry season, the vast majority of the terrain is essentially inaccessible.

The area is very popular with fisherman and in recent years foreign visitors, who come to see the thousands of waterbirds and caimans, plus Capybaras, the world`s largest rodent, Black Howler Monkeys, Giant and River Otters and anacondas. The area is a photographer`s paradise. During July there are surprisingly few mosquitoes, but at other times of the year these can be unbearable. Anyone visiting this, or the following three sites, for long should obtain the *Birds of Southwestern Brazil* by B. Dubs.

Ref: A4, A7, C8, C9, D1, D7, D8, F6, F12, M5, N2, N3, P19, P26, P28, P30, P37, R8, R11, R13, R17, R19, S16, W2.

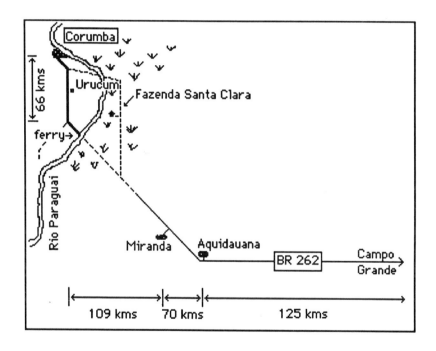

Transpantaneira (Northern Pantanal)

This site, in southwest Mato Grosso, is principally the main road from Cuiabá to Porto Jofre. Nearby lies the Pantanal Matogrossense National Park, located at the meeting of the Rio Cuiabá and the Rio Paraguai, southwest and down river from Porto Jofre. Access to the park could be arranged through IBAMA. Housing is available for scientists, but is not intended for casual birdwatchers. However, nearly all the wildlife of the area can be seen from the Transpantaneira and its 126 bridges along the 148 kms from Poconé to Porto Jofre.

The habitat from Cuiabá to Poconé is rolling cerrado, gradually descending into the bowl at the heart of South America. From Poconé to Porto Jofre you are in the true Pantanal. Here, the Rio Paraguai and its affluents slowly wind the 2,000 kms to the sea off Buenos Aires. Unlike the southern Pantanal, the Transpantaneira traverses a considerable amount of forest. Personally, I found this area to be better in July than the southern site, with far greater numbers of waterbirds but fewer parrots. However, I have not visited either site at other times of the year.

The Transpantaneira is a dead-end road. At one time it was planned to continue the highway to Corumbá but, fortunately for the wildlife, this did not come about. There are various hotels along the dusty road, and due to the increase in tourism and fishing parties there is a high demand for accommodation, so booking in advance is advisable. Also, there are only two gasoline stations south of Poconé and they often run short, so fill up before leaving Poconé. Fortunately, the hotel owners appreciate the importance of Hyacinth Macaws as a tourist attraction and go out of their way to preserve these wonderful birds, the world's largest parrot. Seeing these magnificent birds flying in to roost may be the most memorable part of your trip to this site. The best bases for spending a few days in this area are Pixaim and Porto Jofre. The two hotels at Pixaim are rather small. The Hotel Santa Rosa at Porto Jofre is larger and more expensive, but there is also a camp site here. The temperature in this part of Brazil is fairly high and evenings can be a little unpleasant, with high humidity. Mosquitoes can be a pest for much of the year, but not in July. October to April can be very wet and the hotels can be cut off from the road due to flooding. The road surface and many of the bridges have deteriorated in recent years so care should be taken when driving, especially at night.

Nearly all species can be viewed from the road, but for some of the passerines you will require to walk a short distance off the road into dry forest. There are several trails leading off the main road near Pixaim and at Porto Jofre. Birds to look for are Buff-bellied Hermit, Mato Grosso Antbird,

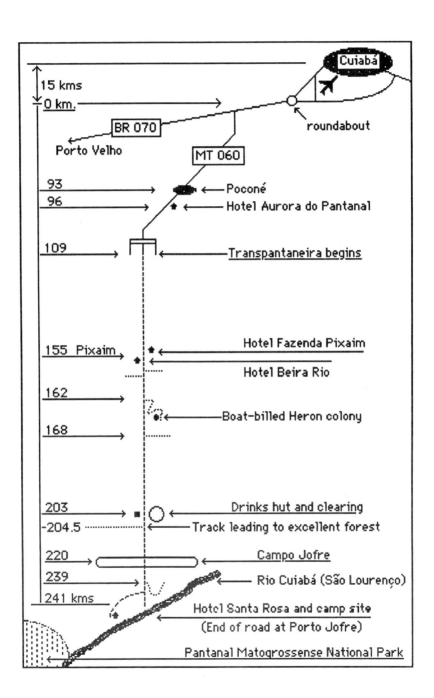

Cuiabá

15 kms
0 km.

roundabout

BR 070

Porto Velho

MT 060

93 ← Poconé
96 ← Hotel Aurora do Pantanal

109 ← Transpantaneira begins

Hotel Fazenda Pixaim
155 Pixaim
Hotel Beira Rio

162
← Boat-billed Heron colony
168

203 ← Drinks hut and clearing
-204.5 ← Track leading to excellent forest

220 ← Campo Jofre
239 ← Rio Cuiabá (São Lourenço)
241 kms

Hotel Santa Rosa and camp site
(End of road at Porto Jofre)

Pantanal Matogrossense National Park

Band-tailed Antbird, White-naped Xenopsaris, White-eyed Attila, Bearded Tachuri, Plain Tyrannulet and Black-backed Grosbeak. However, it is the herons, ibises, Jabirus, Sunbitterns, Chaco Chachalacas and other large birds that birdwatchers particularly associate with this wonderland. Probably more species can be seen here in a day than at any other site in Brazil. Caiman, Capybaras, turtles, iguanas, Black Howler Monkeys and other large animals add to the vivid memories that you will take away with you, along with photographs of the most colourful sunsets.

Especially impressive in late afternoon and evening are Campo Jofre, the vast open area 20 kms north of Porto Jofre, and the area just north of Pixaim. Hyacinth Macaws can be seen at any hotel along the Transpantaneira, with the drinks hut overlooking a clearing with palm trees 38 kms north of Porto Jofre being particularly good, even during the daytime. The area between the Hotel Santa Rosa and the camp site at Porto Jofre is however perhaps the most reliable.

Ref: A7, B1, B9, C3, C8, D6, D7, D8, F1, F3, F7, F9, G1, H4, H6, I4, M5, M8, M10, N2, N3, P2, P11, P19, P28, P30, P37, R6, R11, R13, R16, R17, S1, S5, S16, S27, T15, W7, W10, W24, W29.

Hyacinth Macaw

Barranquinho

Located on the northern edge of the Pantanal in southwestern Mato Grosso, the Hotel Fazenda Barranquinho, 60 kms from Cáceres, offers a pleasant way to see many of the birds of the Pantanal in much more peaceful surroundings than the Transpantaneira.

The site list includes birds seen in the Cáceres area and Descalvados, (located, by boat, 80 kms to the south, on the west side of the Rio Paraguai). Descalvados had many of the historical records for this area. The habitat of this site ranges from cerrado, on the edge of Brazil's tableland, dropping into the Pantanal. Differing in habitat and latitude from the neighbouring Transpantaneira site, the Barranquinho area has attracted certain species which have normally a more Bolivian or Amazonian distribution. The most unusual are Agami Heron, Andean Condor (the only record for Brazil), Dot-eared Coquette, Many-spotted Hummingbird and White-bellied Hummingbird. Obviously, this is a good place for rare hummingbirds. If time permits it may be worth visiting the nearby Serra das Araras Ecological Station, located to the northeast of Cáceres, where Blue-eyed Ground-Dove has occurred.

From the Fazenda, it is possible to arrange a boat trip down the Rio Jauru which will allow good views of typical Pantanal species, including Sunbittern. This site offers a combination of birds associated with the previous site (Transpantaneira) and birds that can be seen at the following site (Chapada).

Ref: C8, D7, D8, I4, M5, N2, P19, P26, P31, P37, P38, R11, R19, S16, S27, S30, W24.

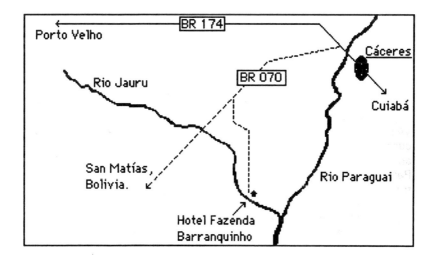

Chapada dos Guimarães

Chapada dos Guimarães is located on the edge of Brazil's central plateau 68 kms northeast of Cuiabá (Mato Grosso), in campo. This site complements a visit to the Transpantaneira since the higher and drier habitat supports a different avifauna. The area is now a National Park attracting many tourists who come to see Chapada's impressive monolithic rock formations and high waterfalls.

Chapada dos Guimarães offers the birdwatcher a very small, concise habitat in which to work. It is only an hour's drive from Cuiabá and has several reasonable hotels such as the Hotel e Restaurante Turismo in the village or the Pousada da Chapada 2 kms to the west. The species characteristic of this area are limited in number, but there is the opportunity to see birds from the plateau region plus the chance of Amazonian vagrants. Over the years, a surprisingly large number of species associated with the Amazon have been seen here, since, at Chapada, you are at the southern edge of the Amazonian watershed: water running off the hills to the north of Chapada flows into the affluents of the Rio Xingu and another stream nearby eventually enters the Rio Tapajós. Although the open hills are certainly not Amazonian in character, the gallery forest hugging the sides of those streams and rivers is still basically tropical. Chapada may not have the huge variety of birds that the Pantanal has, but it offers species near the southern edge of their range and therefore of importance to birdwatchers not intending to visit the Amazon.

The area can occupy an interested birdwatcher at least three days, though in recent years most birders have spent only a day on the way to or from the Pantanal. The first stop should be some 52 kms from Cuiabá, at the Portão do Inferno, with its superb views over the plateau escarpment. At certain times of year, Red-and-green Macaws and to a lesser degree Blue-winged Macaws can be seen nesting on the cliffs. The next stop, 5 kms farther east should be the Véu de Noiva. Here, Biscutate and Great Dusky Swifts have been seen and Crested Black-Tyrants are regularly found along the top of the cliffs. A steep trail starting at the right of the cafe leads over the cliff to the bottom of the waterfall, 61 metres below. This is the most extensive forest in the area. Thirdly, stop at the waterfall just a little farther up the main road, upstream from the Véu de Noiva. Here, small paths lead through the gallery forest, though during weekends these can be busy with tourists. Watch for Brown Jacamar and Helmeted Manakin. Next drive the 7 kms of dirt track passing the Casa de Pedra. Two kms farther east, on the Chapada road, is an excellent asphalt road leading six and a half kms to a radar station. This *private* road has good forest. However, the forest behind the Portão da Fe is perhaps the best of all the areas to be visited. Cinnamon-throated Hermit, Pale-crested Woodpecker and Band-tailed Manakin have regularly been seen here and Fiery-capped

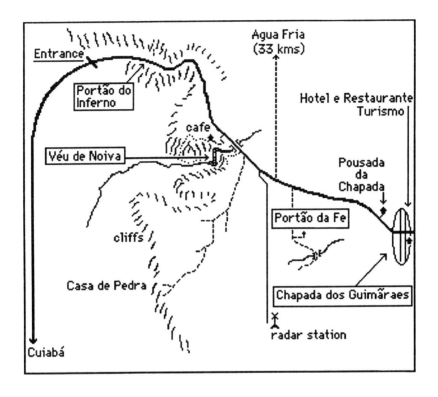

Manakin and Dot-eared Coquette have also been encountered less regularly. The entrance road to this religious centre, 6 kms from Chapada, is also good for campo birds such as Horned Sungem, White-banded Tanager and others. Lastly, several dirt roads leading east from Chapada will take you into a much drier habitat. Alternatively, perhaps try the dirt track leading to Agua Fria, 33 kms to the north, since Point-tailed Palmcreeper has been seen here in recent years.

There has been some confusion over the years with Chapada. At the end of the last century, H.H. Smith visited a site which in J.A. Allen's account is referred to as "Chapada", and Allen gives a description of the site as 30 miles southwest of Cuiabá at an altitude of 2500 feet. I suspect that he meant to write 30 miles northeast of Cuiabá. There are several towns in Brazil with part of their name being Chapada but today if you refer to a place merely as "Chapada" then it is presumed you are referring to Chapada dos Guimarães. In the Chapada bird list, all species refer definitely to Chapada dos Guimarães unless given an "H" symbol. These I presume to refer to our Chapada and most of these are Smith's records.

Ref:A4,B1,C3,D5,D6,D7,D8,F1,F3,F4,F7,F9,G1,H4,H6,I4,M5,M8,P7,P11, P19,R6,R16, S1, S16, S27, T15, W7, W10, W23, W24.

Red-legged Seriema

Emas National Park

Located in southwestern Goiás, the Emas National Park is the most famous and probably the best cerrado park in Brazil. Unfortunately, it is a long drive of 550 kms from Goiânia. Alternatively, charter a flight from Goiânia or Campo Grande.

With written permission one can stay at the park headquarters, and, as in most reserves, prior permission is required to visit anyway. Since this is a remote locality, you should bring your own food, which can be prepared for you by the park staff. It is suggested that you bring food for them as well.

The park is just about the best place in Brazil to see Giant Anteater as well as Maned Wolf and, with nocturnal effort Giant Armadillo. Puma, Brazilian Tapir, Pampas Deer, Marsh Deer, Capybara, Giant Otter and many other mammals make this one of the best places to view large game in Brazil. It also contains most of Brazil's cerrado birds, such as Greater Rhea, for which the park is named, Red-legged Seriema, Red-winged Tinamou, Campo Miner, Collared Crescentchest, Cock-tailed Tyrant, Sharp-tailed Tyrant, White-banded Tanager, White-rumped Tanager, Black-throated Saltator, Coal-crested Finch and Black-masked Finch. In addition, it contains some rare species only found at two or three places in Brazil, such as White-winged Nightjar, Russet-mantled Foliage-gleaner, Marsh Seedeater, Rufous-rumped Seedeater, Chestnut Seedeater and Cinereous Warbling-Finch. For raptor enthusiasts, Emas offers the possibility of seeing Crowned Eagle.

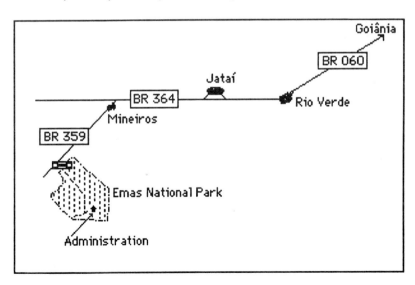

My sincere thanks to Walt Anderson for supplying information on this site.

Ref: A7, B9, C7, C8,D6, I7, P2, R11, R13, R17, S1, S23.

Giant Anteater

Serra da Canastra National Park

Serra da Canastra National Park is located in southwestern Minas Gerais near the border with São Paulo, 330 kms west of Belo Horizonte. Like Emas, visiting Canastra involves a long drive. Also, like Emas, Canastra is basically a cerrado reserve but its location on top of a steep walled escarpment makes the habitat tend towards campo. The reserve includes the headwaters of the Rio São Francisco which flows northward for over 2,500 kms before entering the Atlantic, the longest river totally within Brazil.

Besides Giant Anteater, Giant Armadillo and Maned Wolf, mammals for which this reserve is famous, there are two ornithological specialties, Brazilian Merganser and Brasilia Tapaculo. The merganser is becoming very scarce due to habitat loss and its need for privacy. The tapaculo seems scarce due to its very secretive habits. The merganser inhabits the small hill streams on the plateau and I have photographed them with young at the top of the Casca D'Anta waterfall. In recent years they have also been regularly found below the waterfall on the Rio São Francisco. At one time or another, the mergansers have been seen on virtually every watercourse in the reserve but they are very difficult to pin down. July and August have proved good months to see these elusive birds since at this time they are on the rivers with their newly fledged young.

The tapaculo has been seen along densely forested streams at locations (C) and (G) on the map. I have seen them in undergrowth that allowed only a few metres visibility bordering a stream only a metre wide (G). Patience is required to catch a glimpse of these mouse-like birds that skulk only up to a few inches above the ground.

Finding these species requires two very different forms of birdwatching. For the mergansers, you have to be prepared to walk watercourses for several days. To see the tapaculo, you have to be willing to sit for long periods in dense undergrowth with every kind of insect investigating you, as well as the odd snake!

One highlight of a trip to Canastra is the chance to observe the Giant Anteaters that freely roam the plateau and can be picked out even from several kilometres by their distinctive "two boulder" silhouette. At such a distance, they look remarkably like two large rocks joined together. Giant Anteaters have very poor eyesight but their sense of smell is very acute, so approach these peculiar animals down wind and you may be rewarded with exceptionally close views.

Other cerrado species occurring here are Greater Rhea, Black-chested Buzzard-Eagle, Red-legged Seriema, Giant Snipe, Campo Miner, Cock-tailed Tyrant, Sharp-tailed Tyrant, Hellmayr's Pipit, White-striped Warbler and Stripe-tailed Yellow-Finch.

Just outside the park along the Rio São Francisco, patches of mature forest offer a wider variety of species, including Golden-capped Parakeet, Stripe-breasted Starthroat, Red-ruffed Fruitcrow and Gilt-edged Tanager.

The park is officially closed for much of the winter but with written permission, entry and even accommodation on the reserve is possible. Canastra has some of the most professional wardens of any National Park, truly involved in the preservation of their wildlife and proud of their custodial responsibilities. Hotel accommodation can be found in nearby São Roque de Minas (7 kms). The Hotel Faria, although basic, was very friendly and helpful and as good an establishment as can be expected in rural Brazil.

Ref: C7, C8, D6, F7, F9, G1, P11, P12, R6, R16, S6, S16, W10.

Brazilian Mergansers

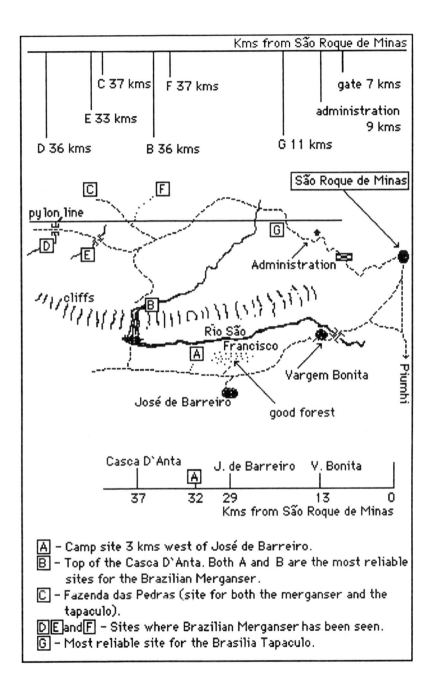

Kms from São Roque de Minas

C 37 kms F 37 kms

gate 7 kms

E 33 kms

administration
9 kms

D 36 kms B 36 kms G 11 kms

São Roque de Minas

C F

pylon line

G

D

Administration

E

cliffs

B

Rio São

Francisco

A

José de Barreiro

good forest

Vargem Bonita

Piumhi

Casca D'Anta   J. de Barreiro   V. Bonita

A

37   32 29   13   0

Kms from São Roque de Minas

A – Camp site 3 kms west of José de Barreiro.
B – Top of the Casca D'Anta. Both A and B are the most reliable
   sites for the Brazilian Merganser.
C – Fazenda das Pedras (site for both the merganser and the
   tapaculo).
D E and F – Sites where Brazilian Merganser has been seen.
G – Most reliable site for the Brasilia Tapaculo.

Serra do Cipó National Park

Serra do Cipó National Park is located 90 kms northeast of Belo Horizonte airport in Minas Gerais. From the airport, take the main road southward towards Belo Horizonte, watching for the Lagoa Santa turn off. On reaching Lagoa Santa continue northward following signs to the park. The site list includes the area south to Lagoa Santa which was studied thoroughly many years ago: most species listed with historical status were collected by P.W. Lund.

The reserve is situated at the southern edge of the Serra do Espinhaco, a chain of mountains that runs north into Bahia. The higher parts of the reserve are true campo, while the lower parts are a mixture of cerrado and gallery forest. Its spectacular waterfalls and gorges have made the lower section a favourite tourist attraction, but the higher elevations, with more interesting birds, are much less frequented.

Stay either at the very pleasant Hotel Veraneio, which offers splendid accommodation at the park entrance, or at the very basic and much cheaper Chapéu de Sol Hotel 8.7 kms farther up the mountain track.

Four endemic birds are fairly readily found in the higher part of the reserve and close to the road: Hyacinth Visorbearer, Gray-backed Tachuri, Buff-throated Pampa-Finch and Cipó Canastero. The latter is a recent discovery to science. In 1988, while looking for the first three species mentioned, Simon Cook and I found some odd brown birds skulking in the rocks. They looked like Short-billed Canasteros, a species occurring principally in Argentina and Paraguay but found just into Brazil 1,700 kms to the southwest in Rio Grande do Sul. I suspected at the time that they might be a new species. Surprisingly, within a month of our sighting, Mark Pearman and John Hurrell made the same discovery. However, it has since come to light that Frederico Lencioni Neto actually discovered them first, in 1985 (see *Uma nova especie de Asthenes da serra do Cipo, Minas Gerais, Brasil*, by J. Vielliard, Ararajuba 1). A number of British birdwatchers, including myself, have returned to see them and all found them to be skulkers but not impossible to see. They do not appear threatened in any way and may well have a wider distribution since their habitat continues, fragmentary, for hundreds of kilometres northward along the Serra do Espinhaco.

Since the canastero will be of interest to all who visit Cipó, and there is little information about it available, I include here a brief description of the species:

About 15 cms in length. In shape and colour very similar to Short-billed Canastero. Upperparts generally an earthy greyish-brown. Apart from the tail, the cap is the brightest part of the bird. This is a warm brown, darker

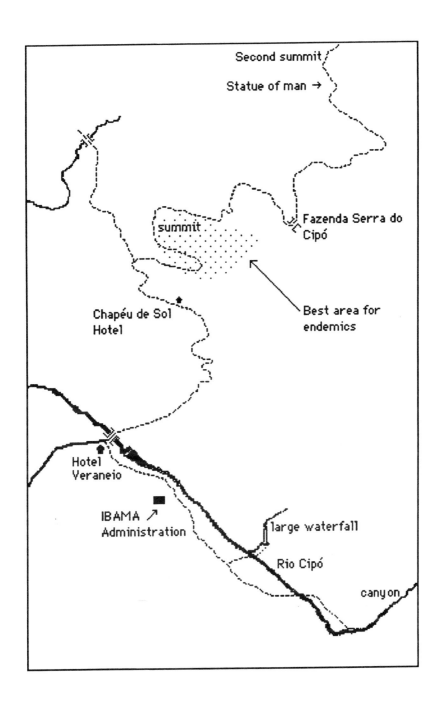

Second summit

Statue of man →

summit

Fazenda Serra do
Cipó

Best area for
endemics

Chapéu de Sol
Hotel

Hotel
Veraneio

IBAMA ↗
Administration

large waterfall

Rio Cipó

canyon

than the nape and mantle which are greyish-brown, mantle with a hint of olive. Wings a dull brown. Primaries and secondaries a darker brown. Tertials with dark brown centres and pale grey edges. Alula appears to be a pale cream. Uppertail with dark brown central feathers, and outer two or three rufous. Tail graduated, with dull-pointed tips giving a ragged appearance. Undertail entirely rufous. Underparts a pale grey, sometimes appearing a sooty grey. The grey tone of the underparts extends up the side of the neck onto the nape. Cheeks a darker grey, faintly streaked. Supercilium obvious and greyish white, fairly long and narrow, beginning just in front of the eye and curving down fractionally behind the ear coverts. Thin dark eye stripe. Throat streaked black and often appearing as a dark throat patch from a distance. Bill and legs grey. Bill about three quarters the length of the head. Eye dark.

The birds skulk among the rocky outcrops. When agitated, they will perch on top of an obvious vantage point, usually a rock but sometimes a bush, and deliver a loud descending trill of about ten notes. For discussion of call see Mark Pearman: "Behaviour and vocalizations of an undescribed Canastero *Asthenes sp.* from Brazil." B.O.C. 110 (3). Usually holds tail horizontal but, if agitated, cocks tail overhead like a wren. Behaviour indicates that it is very territorial.

Permission to enter the lower parts of the reserve appears to be fairly easily obtained. Patches of good forest can be found along the dirt track leading past the Administration Building towards the canyon 12 kms away. The upper part of the reserve appears to have no restrictions and it is doubtful that you would come across anybody away from the road. There is certainly no habitat which could be damaged by walking over the terrain. The most reliable site for the canastero is at the first summit, 13.7 kms from the Hotel Veraneiro, and the other specialties can be seen here as well. Various stops are advised between here and the second summit, which is 13 kms farther. The most productive areas are where gallery forest crosses the road. A very rough track leading off to the northwest 3 kms past the Chapéu de Sol Hotel, brings one to an iron bridge some 5 kms away. Here, Giant Snipe and Cinereous Warbling-Finch have been seen.

Other species of note that have been recorded in the area are Lesser Nothura, Long-trained Nightjar, Horned Sungem, Checkered Woodpecker, Serra Antwren, Collared Crescentchest, White-winged Black-Tyrant and White-naped Jay. Cipó, I am sure, will in the future be put firmly on the map as a site of great ornithological interest.

Cipó Canastero

Ref: B8, C8, D6, F7, F9, G1, H4, K5, N2, N3, P1, P11, P12, P13, P19, P28, P31, P32, P37, R11, R13, R16, S16, V2, W7.

Brasilia National Park

Of the cerrado sites listed in this guide, Brasilia is the easiest to reach and one of the richest in cerrado species. One can easily be on the reserve from dawn until dusk and still have a comfortable hotel only ten minutes away. The National Park is located on the northern edge of Brasilia some 9 kms from the city centre. At one time, most of central Brazil was cerrado, or long grass with scattered trees, but today most of this part of Brazil has been agriculturalised. Sadly cerrado, as with Brazil's rainforests, is shrinking in size. This national park preserves intact a significant tract of cerrado.

Brasilia is a modern city designed and built in the 1950s and, for the motorist, is well laid out. The streets are very wide, free of normal city traffic, and road signs are very clear. Hotels are located together right in the centre. The park is only a few minutes away by taxi and buses are also possible. It is perhaps the most typical of all the National Parks in that the public is encouraged to visit the entrance which has leisure facilities such as a swimming pool, but the rest of the reserve is closed to them. This does allow protection of the wildlife but I feel the Brazilian public have to be educated about their heritage and encouraged to take an interest in the fauna of their countryside, which means allowing them more access to the park so that they will come to appreciate the plants, mammals and birds that it holds. Otherwise, the great outdoors means nothing to the majority and they will not see the importance of preserving those vanishing habitats.

Written permission is definitely required for admission to the true reserve. A vehicle is essential to tour the 28,000 hectares via a complex network of tracks. The lake in the centre of the reserve, Barragem Santa Maria, can be worth a visit; this is the only known site for Rufous-faced Crake in Brazil and Southern Pochard has often occurred here. Despite the number of pleasure seekers to the swimming pool, the entrance area is actually one of the best spots on the reserve, as once the crowds have departed for the evening, the large trees surrounding the pool become a haven for birds and monkeys that often come out to feed on picnic scraps. Even Bare-faced Curassows come out for the evening feast. For the remaining hour of daylight, this is the best place to be for species such as Small-billed Tinamou, Green Ibis, Toco Toucan and Curl-crested Jay, each vying for your attention.

This cerrado habitat holds the usual grassland species, the most unusual ones being:- Lesser Nothura, Dwarf Tinamou, Ocellated Crake, Yellow-faced Parrot, Sickle-winged Nightjar, Horned Sungem, Checkered Woodpecker, Campo Miner, Russet-mantled Foliage-gleaner, Brasilia Tapaculo, Collared Crescentchest, Cock-tailed Tyrant, Black-bellied Seedeater and Coal-crested Finch. Chestnut-capped Foliage-gleaner is said to occur at the city zoo.

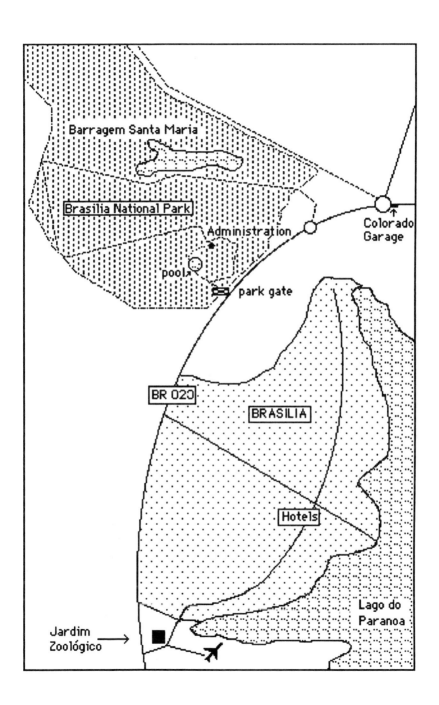

Barragem Santa Maria

Brasilia National Park

Administration

pool

park gate

Colorado Garage

BR 020

BRASILIA

Hotels

Jardim Zoológico →

Lago do Paranoa

A special thank you goes to Paulo de Tarso Zuquim Antas for his help with this site.

Ref: A7, A8, C7, C8,D6, F7, F8, G1, G6, I7, K2, L4, N4, P11, P12, R11, R13, R17, R19, R20,S5, S13, S16, S34, T2, W29.

Araguaia National Park

This cerrado reserve is located 750 kms northwest of Brasilia on the Ilha do Bananal in the new State of Tocantins, formerly part of Goiás. In fact, the reserve is a transitional zone between tropical Amazonia and the dry central plateau, and, as such, has a mixture of scattered trees in grassland plus gallery forest. The Ilha do Bananal is the largest fluvial island in the world; part of it floods regularly, creating a superb habitat for water related birds. Thus the park area offers much variety in both habitat and bird life.

This remote location, almost in the centre of Brazil, is not easy to reach. Flights to the small town of Santa Terezinha operate from Brasilia daily, except Sundays. It should be possible either to stay in the town or cross the Rio Araguaia to the reserve and, with permission, stay in the park headquarters. In the past, the ideal way to view much of this habitat was by a riverboat hotel. Unfortunately, this is no longer possible, though Christoph Hrdina may be able to arrange access for parties of interested naturalists.

Birds to look for along the river are Orinoco Goose, Sunbittern, Giant Wood-Rail, Hoatzin, Green-throated Mango, Glossy Antshrike, White-naped Xenopsaris, Riverside Tyrant, Pale-tipped Tyrannulet, Crimson-fronted Cardinal plus Scarlet-throated Tanager. In the gallery forest can be found Chestnut-bellied

Guan, Long-billed Woodcreeper, Bananal Antbird, Band-tailed Manakin and Rose-breasted Chat. When birdwatching in the dry higher elevation woodland look for the secretive Brasilian Tinamou, Rufous Nightjar, Cinnamon-throated Hermit and Gray-headed Spinetail. Finally, in cerrado look for Long-tailed Ground-Dove, Hyacinth Macaw and Yellow-faced Parrot. The once endemic "*Bananal Tyrannulet* " was apparently a misidentified Gray Elaenia.

As will be apparent from the list, an impressive array of birds occur in this area, many of them represent species that are normally associated with the Amazon, the northeast or the central belt. Here, at the crossroads of central Brazil, all these faunas come together to create a mosaic of interesting sightings.

Ref: C7, C8, G6, I1, I7, M5, R7, R11, R17, R19, S5, S13, S16, S26.

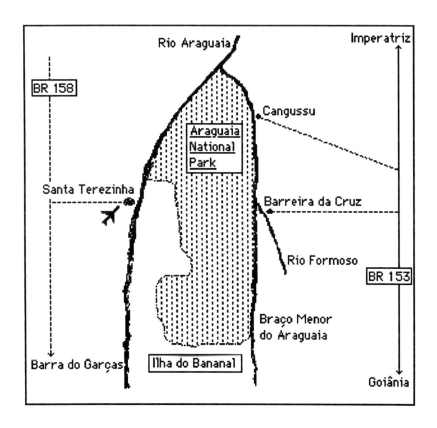

-49-

Eastern Brazil

The fourteen localities listed here contain a high percentage of Brazil's endemic bird species. Most are forested areas in very picturesque settings. The northernmost sites are in much drier habitat which has its own brand of endemism. First-time visitors to Brazil will most likely visit a selection of the following sites:

1 Serra do Mar.
2 Itatiaia National Park.
3 Serra dos Orgãos.
4 Rio de Janeiro.
5 Parque Natural do Caraça.
6 Nova Lombardia Biological Reserve.
7 Fazenda Montes Claros.
8 Rio Doce State Park.
9 Sooretama Biological Reserve.
10 Southern Bahia.
11 Salvador.
12 Boa Nova.
13 Northeastern Bahia.
14 Maceió.

Other areas worth visiting, if time permits, are the Paranaguá area, including Marumbí State Park (Paraná), Ilha do Cardoso State Park, run by C.E.P.A.R.N.I.C. (São Paulo), Caparão National Park (Minas Gerais), Chapada da Diamantina National Park (Bahia), the Rio São Francisco Valley between Januária (Minas Gerais) and Barra (Bahia), the Serra da Ibiapaba area, including Seven Cities National Park (Piauí), Serra da Capivara National Park (Piauí) and Ubajara National Park (Ceará).

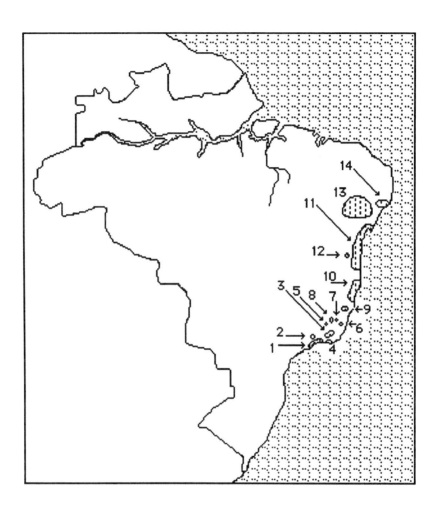

Serra do Mar

This site covers the coastal mountain range, known as the Serra do Mar, from eastern Sao Paulo east into the state of Rio de Janeiro as far as Angra dos Reis. Within this area lie the Serra da Bocaina National Park, Ubatuba and to the west the Boraceia Forest Reserve. The habitat thus ranges from rocky coast and mangroves to mountain peak with a mixture of Atlantic Forest, plantations and agricultural land. This is a very expanded site which would perhaps be better treated as four separate ones.

The coast from Angra dos Reis (151 kms from Rio) to Paratí (100 kms from Angra) is known for its mangroves and waterbirds. This area contains the microscopically limited range of the recently re-discovered Black-hooded Antwren, a cosmic amongst antbirds with its striking plumage (black head and body with brilliant rusty back). Unicolored Antwren, Red-eyed Thornbird and Black-legged Dacnis are found here as well. There are a number of moderately priced hotels in both towns and many buses to and from Rio.

The Serra da Bocaina National Park is an important site since it embraces as much as 100,000 hectares of the Serra do Mar and contains many species not found farther inland. Surprisingly little information is available for this site, but there will surely be great reward for anyone spending time here. A recommended base is the Pousada do Vale dos Veados, in the northern corner of the Park, reached by taking the SP 66 to São José do Barreiro and continuing south on a minor road.

The Ubatuba Experimental Station is 5 kms from the Ubatuba roundabout (229 kms from São Paulo and 323 kms from Rio), on the road towards Taubaté. Again, there are many moderately priced hotels in this coastal town. The Fazenda Capricornio, owned up to now by Hans Scavenius, and his neighbour's property to the east are both excellent sites for seeing southeastern Brazilian specialties. These include Spotted Bamboowren and Fork-tailed Tody-Tyrant which are both common in the large tracts of bamboo found on the latter fazenda. The Ubatuba area also has Saw-billed Hermit, Helmeted Woodpecker, Salvadori's Antwren, Squamate Antbird, Slaty Bristlefront, Buff-throated Purpletuft, Eye-ringed Tody-Tyrant, Bay-ringed Tyrannulet, Oustalet's Tyrannulet and many other interesting species. The Buff-throated Purpletuft is regularly seen in the clearing, behind Hans`s lodging, in isolated cecropia trees. Although Hans is returning to Denmark in the near future, he hopes to set up a development which, while utilizing the lower section for housing, will protect the upper section and offer, within a conference centre, accommodation for researchers into natural history. Fazenda Capricornio is located north of the BR101, just a short way east of kilometre post 45, on an obvious bend. Having turned off the main road, continue straight ahead for Hans`s fazenda or turn immediately right for his neighbour`s fazenda.

Lastly, to the west lies the Boraceia Forest Reserve, about a three hour drive east of São Paulo. Take the SP 088 to Biritiba Mirim and then follow a minor road south to Casa Grande and then the reserve. Permission to stay here is required and should be sought from Dra. Francesca do Val, Museu de Zoologia, Avenida Nazare 481, São Paulo, Caixa Postal 7172 - 01051 São Paulo. This site has produced Purple-winged Ground-Dove, Blue-bellied Parrot, Canebrake Groundcreeper and Black-backed Tanager.

Together these sites form an impressive package which should occupy a birdwatching visitor for at least a week.

Ref: A6, C8, F3, F5, F9, F10, F12, G1, G5, I7, M1, M2, M5, M9, N2, N3, P1, P2, P7, P19, P28, P29, P37, R10, R11, S14, S16, S25, S32, T13, W7, W18, W20, W21, W22, W27.

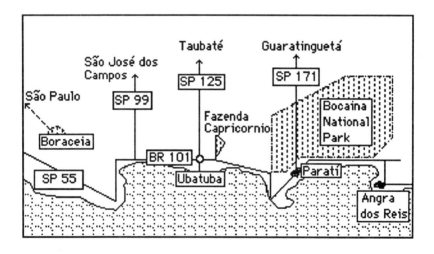

Itatiaia National Park

Located on the border between the states of Rio de Janeiro and Minas Gerais, and almost halfway between the cities of Rio (174 kms) and São Paulo (257 kms), the Itatiaia National Park lies just north of the Dutra Highway (BR 116) in the Serra da Mantiqueira.

Itatiaia was the first national park established in Brazil, and is probably the one most visited by birdwatchers. It presents a fine example of Atlantic Forest rising to grassy, treeless summits including Agulhas Negras which, at 2787 metres is the second-highest mountain wholly within Brazil. If one includes the species occurring in the surrounding lowlands, the park list is impressively long.

Within the park lie several good hotels. Of particularly high quality is the Hotel do Ypê, 13 kms above the village of Itatiaia and situated in superb habitat, the finest birdwatching usually within a few hundred metres of the hotel itself. Almost equally good are the Hotel Simon, a little farther down the mountain and the Hotel Repouso, lower still. Cheaper accommodation can be found in the village. However, the Hotel do Ypê is perhaps the best hotel in Brazil for birdwatching and the food is excellent and abundant even by Brazilian standards. It is the most superb location to relax after several weeks of touring or to become acquainted with the birds of south-eastern Brazil. Either way, it can be highly recommended. At each of these hotels it is worth checking the hummingbird feeders and bird tables which attract many species and, occasionally, some unusual migrants. Note that the hotels within the park are often full on weekends.

Itatiaia is a place to relax over several days where birdwatching is done on foot between the many meals. However, there are two all-day walks that are always rewarding, and the best of these is the "jeep trail" leading up the valley and, ultimately, to the top of Agulhas Negras. This trail would take several days if done at the pace necessary for serious birdwatching, but half the distance can be easily undertaken in a day and affords the opportunity to see many good birds. Technically, permission is required for access to this trail but in practice very few people are questioned about their presence. The second long walk, usually very productive for birds, is the Tres Picos Trail which begins behind the Hotel Simon. This path is narrower and can at times be slippery. One day should also be allocated for the high tops. This involves driving down to the Dutra Highway and heading west before coming off at the Caxambú exit (BR 354). Continue for 26 kms before taking a very rough track to the summit. The specialties of this section are found at or above tree-line. A small marsh, 6 kms from the BR 116, is worth a stop for grassland species. The hotels can supply transport for a group if this is not personally

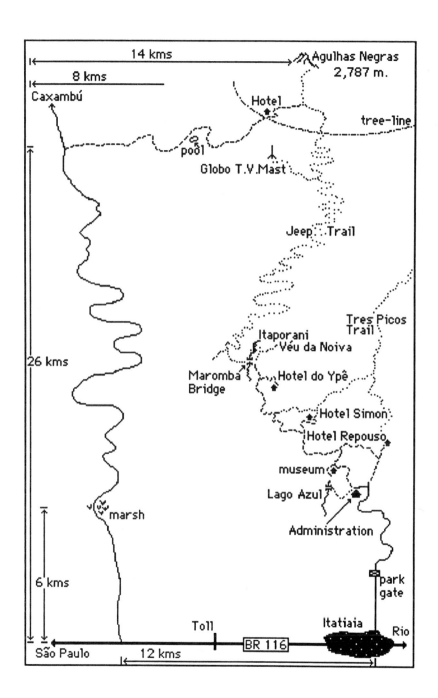

14 kms

8 kms

Caxambú

Agulhas Negras
2,787 m.

Hotel

tree-line

pool

Globo T.V.Mast

Jeep Trail

26 kms

Tres Picos
Trail

Itaporani
Véu da Noiva

Maromba
Bridge

Hotel do Ypê

Hotel Simon

Hotel Repouso

museum

Lago Azul

Administration

marsh

6 kms

park
gate

Toll

Itatiaia

Rio

São Paulo

BR 116

12 kms

available. The trip allows one to see some of the high altitude species not encountered near the main hotels, *e.g.* Itatiaia Spinetail, Serra do Mar Tyrannulet and Long-tailed Reed-Finch. The Alsene Hotel, near the Agulhas Negras summit, offers very basic accommodation.

Itatiaia is an excellent place to see Giant Antshrike, especially at or near the Hotel do Ypê. Three elusive species; White-bearded Antshrike (which shows a preference for bamboo), Swallow-tailed Cotinga and Shrike-like Cotinga have all been seen in the lower sections of the park. Black-and-gold Cotinga and Black-capped Manakin should be looked for at higher altitudes. The latter behave more like Tanagers. Itatiaia is a very good site for such endemic tanagers as Gilt-edged, Brassy-breasted, Brown and Olive-green. Recently, Rio de Janeiro Antbird has been located, a little to the east of the park, near Penedo.

Some species can cause confusion. At Itatiaia, the common large wood-creeper is Planalto though White-throated does occur. Similarly, the common Trogon is the eastern Brazilian race of Surucua, though White-tailed is also here. White-browed Woodpecker often causes confusion since its brow is actually cream. Although both Brown-breasted and Drab-breasted Bamboo-Tyrants occur, the former is usually found at higher elevations, while the latter can be found as low as the hotels.

Ref: A3, B1, B9, C7, C8, D6, F1, F3, F6, F7, F9, F10, G1, G9, H8, I4, I7, K2, M1, M2, M9, M10, P6, P11, P25, P28, P34, P37, P38, R6, R16, S1, S6, S16, S22, S27, S32, W3, W5, W7, W10, W22, W27, W29.

Black Jacobins

Serra dos Orgãos

This site covers the Serra dos Orgãos stretching from Tinguá and Santo Aleixo in the south to Cantagalo in the north. While the majority of species known for this area can be found in the Serra dos Orgãos National Park, a few records do refer to these other sites, particularly the historical records which usually refer to the Nova Friburgo and Cantagalo region. Taken as a whole, this area has more endemic bird species than any other site in Brazil and so might be considered the centre of Brazilian endemism.

The Serra dos Orgãos National Park is only 55 kms north of Rio de Janeiro. It has habitat very similar to that of Itatiaia but has the advantage of being closer to Rio. Two sections of the park are open to the public during normal park hours.

The town of Teresópolis is located nearby and the Hotel Teresópolis is only a short walk from the main entrance. Accommodation is available within the park with prior permission.

Orgãos is a very scenic part of Brazil, often appearing in travel brochures. Its barren rock pinnacles, especially the famous "Finger of God", are very photogenic. The park has not been visited by foreign birdwatchers to the same extent as Itatiaia, yet it has almost identical birds. It is also the home of the elusive Gray-winged Cotinga which has only been seen on a handful of occasions. This species resembles the female Black-and-gold Cotinga and is found at even higher elevations. After many attempts, we finally caught up with it in 1991. Having driven through the main entrance and up to the dam, follow the mountain trail which winds up the forested mountain slope. It is about a five-hour walk to the very top. We located the *Cotinga* just below the first demolished habitation high up on the mountain trail, at an altitude of 6,020 ft , well below the known elevation for this species on this side of the mountain, which had previously been thought to occur only in elfin cloud-forest. The fruit-laden trees were 45 feet high and took three hours of walking to reach. The species has also been seen at the second demolished habitation used by campers which is a further hour`s walk.

A male bird called sparingly, giving a one-second call at intervals of fifteen minutes. This, though shorter than the song of its cousin, the Black-and-gold Cotinga, had the same eerie undulating whine on three pitches, that from a distance seems to flatten out to one note. In contrast, the song of the Black-and-gold Cotinga lasts four seconds, but, due to the rota system of calling by several birds in a loose group, an unabating whine is produced.

The path was closed to the public for many years but was reopened in 1990 and is now being heavily used by hillwalkers. While this gives access to the high tops and allows the study of its rare species, it may lead to their disturbance. This is a typical example of how things in Brazil can change totally and abruptly, as only a year earlier I had written of the severe deterioration of the "closed" path which had become overgrown with bamboo. Now the situation has been completely reversed. It is impossible to predict events in this country, as Brazil is like a vast sand dune that continually changes shape.

The road between Teresópolis and Petrópolis has been deforested but a few open-country birds can be seen, including Swallow-tailed Cotinga which seems to favour such scrub habitat. The lower section of the park is also worth a look as it provides a variety of species not encountered at the higher levels. This section, called the Subsede do Parque da Serra dos Orgãos, is about 14 kms toward Rio from the main park entrance. Lower still is Santo Aleixo, the site of the recently discovered Rio de Janeiro Antwren which unfortunately has not been relocated. The habitat north of Nova Friburgo changes with diminishing altitude, becoming rolling hills with patches of forest confined to the tops. In this habitat, in the far north of the site, Three-toed Jacamar and Rio de Janeiro Antbird should be looked for.

Ref: C7, C8, C10, D6, F1, F6, F7, F10, G1, G2, G4, G5, H4, I2, I4, I7, K2, M9, N2, N3, P1, P11, P19, P28, P37, P38, R10, S4, S14, S16, S32, W4, W5, W10, W13, W22, W27.

Swallow-tailed Cotinga

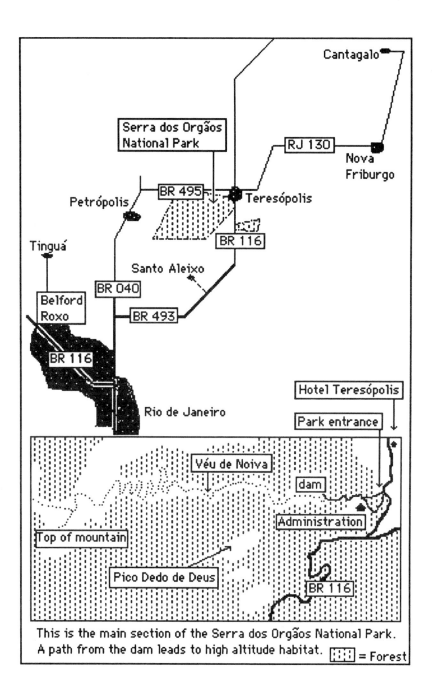

Cantagalo

Serra dos Orgãos National Park

RJ 130

Nova Friburgo

BR 495

Petrópolis

Teresópolis

BR 116

Tinguá

Santo Aleixo

BR 040

Belford Roxo

BR 493

BR 116

Rio de Janeiro

Hotel Teresópolis

Park entrance

Véu de Noiva

dam

Administration

Top of mountain

Pico Dedo de Deus

BR 116

This is the main section of the Serra dos Orgãos National Park.
A path from the dam leads to high altitude habitat. [::::] = Forest

Rio de Janeiro

This site includes the Tijuca National Park in the centre of Rio east to Lagoa de Saquarema 90 kms from Rio. The site includes such diverse habitats as montane forest, scrub, marshes, lakes and coast.

Rio de Janeiro is one of the most spectacular cities in the world. Few places can boast of having a national park in the heart of the city with buildings clinging to the steep slopes below the mountain, like a fortress protecting the mother mountain. Below lie the world's most beautiful beaches, and behind are a multitude of rounded rock peaks that gradually recede into the distant mountain ranges to the north. While overhead glide Magnificent Frigatebirds in all their glory, the guardians of this city.

Most people visiting Brazil will fly into Rio. It is the main gateway to the country and one of Brazil's most popular tourist attractions, the Christ Statue on Corcovado being the centre point of all photographs. Unfortunately, this paradisiacal scene has been marred by numerous muggings of tourists in recent years and I do not recommend that foreigners stay in Rio for long. Being such a breathtaking city, it is particularly sad that this is the case. I would recommend that visitors should leave Rio as quickly as possible but allow a full day to see the city at the end of the trip, and even then take very few valuables with them. Once away from Rio, I have never felt any concern for my safety and, generally speaking, Brazilians are very friendly when approached, though initially wary of strangers in remote areas. On arrival, hotel accommodation can be arranged through the airport tourist office. Just advise what category of hotel you require and something in your price range will be booked for you.

Tijuca National Park and the Botanical Gardens are good places to become acquainted with Brazilian birds, though all but a very few of them occur also in the more peaceful surroundings of Itatiaia or Orgãos.

Tijuca has a sprawling, irregular shape and can be approached from various points. The main entrance, at Alto da Boa Vista, off the Estrada das Furnas, can be reached by taking a No. 233 or No. 234 bus from outside the Rodoviaria (bus station) or by taxi. Having entered the park, follow signs to Bom Retiro, passing the Mayrink Chapel. This is a walk of about an hour and a half. At Bom Retiro, the main path ends, but a small path continues to the summit of Pico da Tijuca. Take the path to the right of the drinking fountain. It will require a further hour to walk this trail. An alternative walk is the Estrada do Redentor between Tijuca and Corcovado which can be interesting for birds and can be driven on weekdays. Watch for a Dusky-throated Hermit lek, halfway between the Corcovado toll gate and Corcovado itself. Other noteworthy birds to look for in Tijuca include Black-cheeked Gnateater, Pin-tailed Manakin and Eye-ringed Tody-Tyrant.

If time permits, it may be worthwhile to visit the Lagoa Piratininga for Southern Pochards or the Reserva Biológica da Barra for Crested Doradito and Rio de Janeiro Seedeater (a pallid race of Capped Seedeater). The Lagoa de Jacarepaguá, may also be worth a look. The beach-scrub from Praia de Jacone, near Saquarema and farther east contain the restricted population of Restinga Antwren. Probably the best seawatching north of Santa Catarina is at Cabo Frio, though birds for this site have not been included on this guide's Rio list because of Cabo Frio's geographical location 168 kms east of the city.

Ref: A3, A7, B1, C8, D6, F1, F3, F6, F7, F10, G4, G5, G9, H4, I4, I5, L4, M1, M2, M7, M9, P6, P11, P28, P37, R16, R17, S1, S5, S6, S11, S14, S16, S27, S32, T13, W5, W10, W14.

Parque Natural do Caraça

In Minas Gerais, 123 kms east of Belo Horizonte, this beautiful mountain region is one of the most idyllic settings in Brazil. I recommend that anyone visiting here stay at the Religious Sanctuary itself, though this has to be booked in advance and is usually full at weekends. Alternatively, accommodation is available at Santa Bárbara, 24 kms away. To book accommodation at Caraça, write to Superior do Colegio do Caraça, 35960 - Santa Barbara, M.G.

This picturesque site is famous for its monastery and religious college set in the high "Alps" of Minas Gerais. Today it is a popular tourist resort among walkers. Recently, ornithologists have turned their attention to the mountain fauna and have discovered many species that previously were considered specialties of the Serra do Espinhaco to the north. Caraça is also world famous for the Maned Wolf. This is the only location where Maned Wolf is almost guaranteed, since a pair has been coming every evening for many years to the steps of the monastery, where food is put out for them. These elegant creatures are truly one of the most majestic and beautiful of Brazil's mammals, and a chance of seeing them at a range of only a couple of feet is not to be missed. The scene can be a little like a circus, but this drawback is minor when one considers the alternative localities for the species where a distant glimpse is not even likely unless you are in the right habitat for weeks. It is well worth coming here for the Wolf alone, but Caraça is also exceedingly good for birds, so the ornithologist will not be disappointed: Hyacinth Visor-bearer, Gray-breasted Sabrewing, Serra Antwren, Shrike-like Cotinga, Red-ruffed Fruitcrow, Gray-backed Tachuri and Buff-throated Pampa-Finch are all possibilities. Both Brasilia and White-breasted Tapaculos have been claimed for this site, and may imply the occurrence of an undescribed *scytalopus* (see reference C8). The best woodland habitat is on the path leading to Tanque Grande and the picnic sites below the monastery. For drier habitat take the paths below the monastery leading to Cascatona and Cruzeiro. If more time is available, a multitude of paths can be be explored. This is an ideal location for anyone wishing to combine hillwalking with birdwatching.

I am sure the habitat would occupy anyone at least four days if the site were to be done justice. Nearby, the Estação Ambiental de Peti reserve would be worth visiting after Caraça, though the habitat is not as good. Peti is 15 kms from Santa Bárbara. Go south through town, passing the Rodoviaria and continue about 1 km to edge of town. Here, take a left turn along a dirt track, which after 10 kms forks. Continue on the major left track, passing a dammed river and eventually you will come to the park which contains the dam itself. Written permission is required to enter this reserve. Write to C.E.M.I.G., Belo Horizonte (see address at rear of guide). Red-eyed Thornbird is the highlight in this area, and has a preference for scrubby marsh. If entry is not allowed into Peti, try the marshy habitat, on both sides, farther along the road.

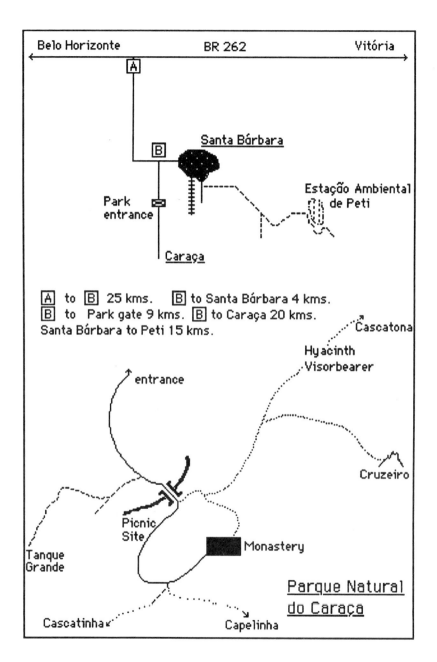

Belo Horizonte — BR 262 — Vitória

A

Santa Bárbara

B

Park entrance

Estação Ambiental de Peti

Caraça

A to B 25 kms.   B to Santa Bárbara 4 kms.
B to Park gate 9 kms. B to Caraça 20 kms.
Santa Bárbara to Peti 15 kms.

entrance

Cascatona

Hyacinth
Visorbearer

Cruzeiro

Picnic
Site

Tanque
Grande

Monastery

Parque Natural
do Caraça

Cascatinha

Capelinha

Ref: C4, C8, F9, G1, K2, P12, R13, S16, W5, W7.

Nova Lombardia Biological Reserve

The Nova Lombardia Biological Reserve is in Espirito Santo about 9 kms north of Santa Teresa, in the hills 90 kms northwest of Vitória. Although a public dirt road runs through the reserve, permission to birdwatch or stay must be previously obtained from the IBAMA offices in Vitória. The reserve is a fine example of upland Atlantic Forest and contains many species associated with the northern Serra do Mar region. Some of the more interesting species to occur here include Salvadori's Antwren, White-bibbed Antbird, Shrike-like Cotinga, Cinnamon-vented Piha, Russet-winged Spadebill, Oustalet's Tyran-nulet, Rufous-brown Solitaire, Sooty Grassquit, Dubois's Seedeater and Buffy-fronted Seedeater. Nova Lombardia is the home of the notorious Black-billed Hermit which is generally thought to be a variant of Scale-throated Hermit.

To reach the Nova Lombardia Reserve take the Colatina road out of town and immediately look for an unmarked dirt track up a very steep hill on the right at the edge of town. After about 6 kms, the main track swings left and the track to the reserve goes off on your right, again twisting up a hill. A gatehouse will be seen at the top of the hill and the warden's house is over the hill just before the main forest. To view forest birds, I recommend walking the two roads in the reserve. Alternatively, a good path leads east from near the warden's abode and a shorter one runs west from the rear of the house. An open area, 1 km south of the reserve entrance, can be good for grassland birds.

This site includes the small hummingbird reserve at the home of the late Augusto Ruschi, situated at the east end of the town of Santa Teresa. It is well worth visiting and is the only place in Brazil where I have seen dozens of hummingbird feeders attracting hundreds of hummingbirds. Up to twenty species occur here, Black Jacobins being especially abundant. This reserve is only open on Saturdays from 9 a.m. to 5 p.m. Santa Teresa should be reached from Vitória via the BR 101 to Fundão and then the ES 259. Alternative routes that appear shorter on the map take twice as long to drive. The Hotel Pierazzo, in the one-way street system of Santa Teresa, is very friendly and comfortable.

Ref: A1, C8, D6, F6, G1, G3, G4, G9, H2, I5, I7, K2, M2, P2, P6, P11, P28, P37, R16, R19, R20, S1, S16, S27, T15, W3, W5, W13, W27.

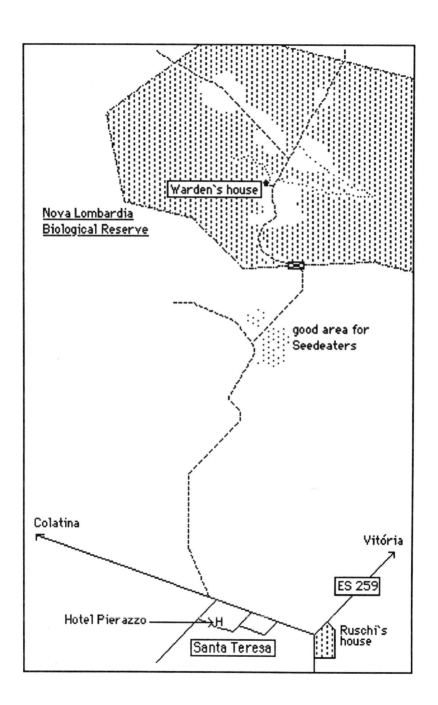

Nova Lombardia
Biological Reserve

Warden's house

good area for
Seedeaters

Colatina

Vitória

ES 259

Hotel Pierazzo →H

Santa Teresa

Ruschi's
house

Fazenda Montes Claros

This privately owned reserve in eastern Minas Gerais, halfway between Belo Horizonte and Vitória, is of world importance for the preservation of the Muriquí or Woolly Spider Monkey. This is the last stronghold for this very rare species, the largest primate in South America. It is to be hoped that the reserve will be made more permanent to ensure the survival of this species.

The property also has several other primates including the Brown Howler Monkey, Tufted-ear Marmoset and Brown Capuchin Monkey. Unfortunately, this primate haven is very small and represents only a fragment of what was once an extensive upland forest similar in nature to Nova Lombardia. Today, most of the area is agricultural land with forest confined almost totally to the hilltops. Regrettably, tree felling is still continuing.

The owner, Feliciano Abdalla, is pleased to have such rare species on his land and has allowed a research station to be set up, primarily to study the remaining groups of Muriquí. Accommodation and food is available at the station at a cost of about $20 per day. Write to the Director, Eduardo Veado, Estação Biológica de Caratinga, CP.82 36 950 Ipanema, M.G.

Access to the reserve is easier via Ipanema, though a rough road does exist from Caratinga. The research station is about 2 kms from the Fazenda. The reserve has a maze of footpaths through the forest for the benefit of the researchers but these prove equally useful to the birdwatcher too.

Highlights of this reserve, apart from the Muriquí are Yellow-legged Tinamou, Golden-capped Parakeet (with fiery red caps here), Ochre-marked Parakeet, Vinaceous-breasted Parrot, Tawny-browed Owl, Plumbeous Antvireo and Serra Antwren. Three-toed Jacamars were found here until comparatively recent, but have almost certainly disappeared from the actual site though may still occur in suitable habitat in the vicinity.

Ref: B8, B9, C8, F7, R13, T15.

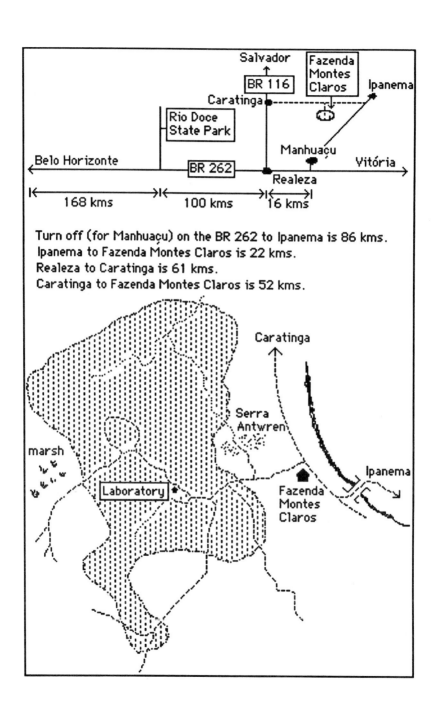

Turn off (for Manhuaçu) on the BR 262 to Ipanema is 86 kms.
Ipanema to Fazenda Montes Claros is 22 kms.
Realeza to Caratinga is 61 kms.
Caratinga to Fazenda Montes Claros is 52 kms.

Rio Doce State Park

Located 210 kms east of Belo Horizonte in eastern Minas Gerais, the Rio Doce State Park, despite being little-known outside of the country, has perhaps the largest single block of forest in southeastern Brazil. As such, it is of world importance.

Although the park is open to the public, when we visited in July 1988, it was officially closed for the winter. So it is advisable to obtain a permit beforehand, obtainable from the I.E.F. in Belo Horizonte (refer to rear of guide for address). The Hotel Pousada was under construction in 1988 and so is presumably now open. We were permitted to use the Casa de Tabua lodge, within the park, and this was extremely comfortable. Food was available only from the Pensão Araouso in Baixa Verde. Although this establishment did not even have a sign outside, it offered the largest meal of our trip for the lowest price: as much as you could eat for about $1.

The park has impressive forest but with, unfortunately, few access points. Within the forest a multitude of lakes, several hundred in all, prevent access to nearly the entire reserve: but obviously help to protect the fauna. There are two fairly short footpaths: one through primary forest located on the eastern half of the reserve, the other through secondary forest 2.3 kms from the entrance or 4 kms before the lodge. The two pools marked on the map are also worth visiting. The main road bisecting the park diagonally permits good views of the forest ideal for birdwatching. However, it is heavily travelled and heightens the danger of a disastrous fire.

Look for Forbes's Blackbird in the more open administration area. Other noteworthy birds include Red-billed Curassow, Rufous-vented Ground-Cuckoo, Black-banded Owl, Rusty-barred Owl, Striped Owl, Minute Hermit and Red-ruffed Fruitcrow. Capybara and Brazilian Tapir also occur here. Rio Doce has a high percentage of the birds known for Sooretama. Since much of the reserve is impenetrable, I suspect that many species have yet to be discovered at this unique site, and that Rio Doce will soon be recognised as one of the most important sites in Brazil.

Ref: C1, C7, C8, F7, F12, R19, S14, S16, W10.

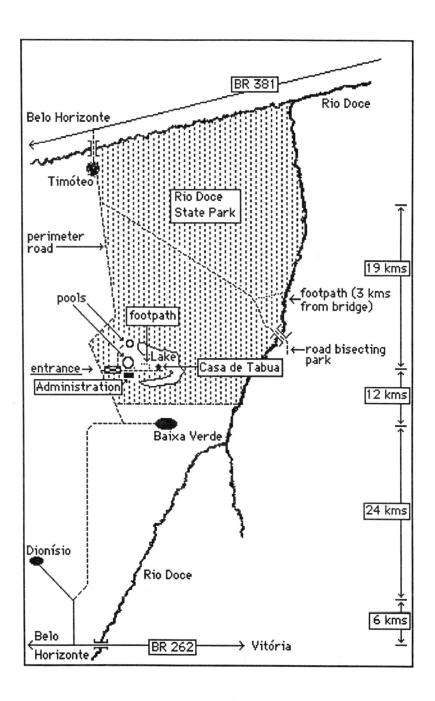

Sooretama Biological Reserve

Very similar to the previous site but at a much lower elevation, Sooretama is about 50 kms north of Linhares in Espirito Santo. The site includes the privately owned Linhares Forest Reserve (CVRD) on the east side of the BR 101, which has similar habitat. The combined reserves, though slightly disjunct, constitute the largest patch of Atlantic Forest in Brazil, though because of its low elevation, it seems almost Amazonian. For permission to visit or stay at Sooretama contact IBAMA at Vitória. The Linhares Forest Reserve is owned by the mining company Companhia Vale do Rio Doce (CVRD).

Ideally, one should seek permission to stay at the park headquarters, otherwise there are basic hotels at São Mateus and Linhares where the Hotel Linhatur is good. Sooretama lies halfway between the two.

The minor dirt road bisecting the reserve northwest of the headquarters is very good for birds, especially the open area around the marsh. After it has rained, Red-billed Curassows often feed near the road. The best footpath is the 6 km Estrada Quiranão, though a small path, leading to a drinking pool, located behind the administration area, is also good. Two tracks on the perimeter of the park along the north and south edges can be driven with care, if permission has been obtained, but are better walked. Another vantage point is the main road (BR 101), although constant traffic can be annoying.

White-tailed Trogon

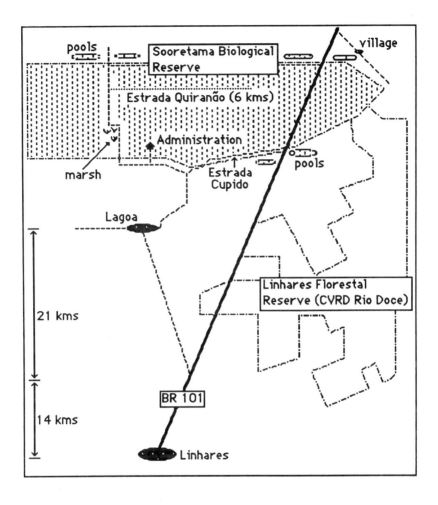

Sooretama has an impressive list of birds including Red-browed Parrot, Rufous-vented Ground-Cuckoo, Hook-billed Hermit, Minute Hermit, Striated Softtail, Scalloped Antbird, Banded Cotinga and White-winged Cotinga. The notorious *Black Barbthroat* , which previously appeared on many Brazilian lists, has since been defrocked, being now recognised as a variant of Rufous-breasted Hermit, if not, merely its immature plumage.

Ref: C5, C8, D6, F1, F6, G1, G9, H7, I7, K2, M1, M2, P6, P11, R16, S1, S4, S5, S14, S16, S32, W3, W5, W10, W13.

Southern Bahia

This site covers the coastal area from the Rio Jequitinhonha south to the border with Espirito Santo and includes Monte Pascoal National Park and Porto Seguro.

Monte Pascoal National Park, 348 kms south of Ilhéus, includes such diverse habitats as mangroves, sand dunes, coastal tropical forest and Monte Pascoal itself, a steep isolated hill with forested sides. Part of the reserve has been returned to the Pataxos Indians. Despite the variety of habitats, most visitors will only visit Monte Pascoal, since the rest of the reserve is not open to the public, and even if permission were obtained, the impressive coastal forest has few access points. Hotels are available at Itamaraju, 32 kms south of the reserve.

The park has many birds typical of northern Atlantic Forest, such as White-necked Hawk, Black-fronted Piping-Guan, Red-billed Currasow, Ochre-marked Parakeet, Red-browed Parrot, Racket-tailed Coquette, Long-tailed Woodnymph, Cream-colored Woodpecker, Salvadori's Antwren, Band-tailed Antwren and Striped Manakin. The habitat of this site also attracts various large birds of prey including Harpy Eagle, Black-and-white Hawk-Eagle, Ornate Hawk-Eagle and Black Hawk-Eagle, so it is worth keeping an eye open overhead at this site. The visitors' centre offers an ideal viewing area over the flat forest below and good views of Monte Pascoal itself. Bare-throated Bellbirds are often heard here, their distinctive call ringing out over the forest.

Porto Seguro, Brazil's oldest town, is a popular holiday resort with a variety of small hotels. The coastal setting is very pleasant and the mangroves in the delta would be worth further investigating, since hundreds of parrots and parakeets swarm into them to roost in the evening.

The Pau Brasil Ecological Station and the Porto Seguro Reserve are owned by the Companhia Vale do Rio Doce and are adjacent to each other about 14 kms west of Porto Seguro and 322 kms south of Ilhéus. Much of the land east of the main road BR 101 in southern Bahia has interesting forest, though much of it is being felled, including parts of both reserves. However, they still contain many of the good birds associated with Monte Pascoal or Sooretama to the south or the Una Biological Station and the Ilhéus area to the north. Porto Seguro is well known for its Cotingas: Black-headed Berryeater, Banded Cotinga and White-winged Cotinga all occurring here.

Ref: C6, C7, C8, C10, F6, F12, G1, G9, H4, I7, L4, M5, P11, P26, P37, R19, S12, S14, S16, W4, W10.

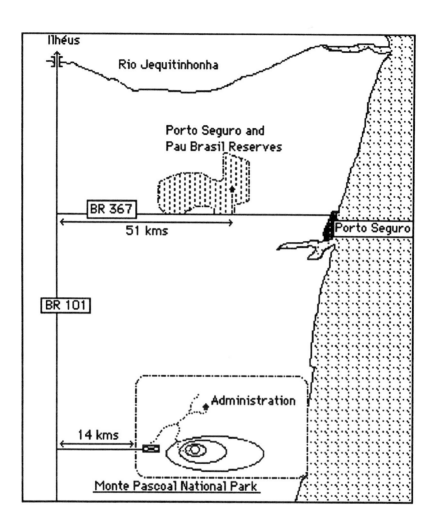

Ilhéus

Rio Jequitinhonha

Porto Seguro and
Pau Brasil Reserves

BR 367

51 kms

BR 101

Porto Seguro

Administration

14 kms

Monte Pascoal National Park

Salvador

The site list includes the Bahia coast from Salvador and Santo Amaro south to Canavieiras, including the Una Biological Reserve 65 kms south of Ilhéus.

Many of the birds listed for this site are historical records, mostly because few ornithological studies have been made here in recent times. This site is predominantly coastal Atlantic Forest though here, as elsewhere, this is gradually being converted to farmland and sugar cane.

Coastal mangroves form another important habitat. In 1990, we visited a very impressive area at the mouth of the Rio Traripe. Take the BA 026 west towards Santo Amaro and turn south after 6 kms towards São Francisco do Conde. After 8 kms you will arrive at a crossroads. The right road leads to S.F. do Conde, left to Candeias. Go straight ahead onto a minor road. This leads to a network of small roads leading down into the mangroves. At low tide, Little Wood-Rails were easily seen foraging on the banks of the river about 1 km from its mouth. Also present were Yellow-crowned Night-Heron, Little Blue Heron and Bicolored Conebill, all mangrove specialties.

The Una Biological Reserve has many of the birds occurring at the previous site (Southern Bahia) and must be worth inspecting, though both Porto Seguro and Monte Pascoal have had better coverage and are known as reliable sites. The area north of Una is undoubtedly also good. While driving through Bahia in 1990, I was very impressed by the amount of forest remaining along the road between Ubaitaba and the Valença turn-off on the BR 101. This at least gives hope for the safe preservation of many species.

Rare birds recorded from this area are Spotted Piculet, Fringe-backed Fire-eye (Santo Amaro), Bahia Tapaculo (Valença and Ilhéus) and Stresemann's Bristlefront (Salvador and Ilhéus). This is a very impressive selection when one considers the small total bird list for the area. It is also worth looking for the rare Gold-and-black Lion Tamarin, which is endemic in this part of coastal Bahia.

One locality, said to be good for the Fire-eye, is approximately 8 kms west of Santo Amaro, on the south side of the BR 026, at the base of a scrub covered escarpment. The Bahia Tapaculo record, for the Valença vicinity, was located in flooded dense vegetation, in lowland forest.

Ref: C6, C7, C8, C10, D2, F8, F9, G6, M11, N6, P18, P19, P26, P28, P37, R10, R19, S14, S16, S25, T12.

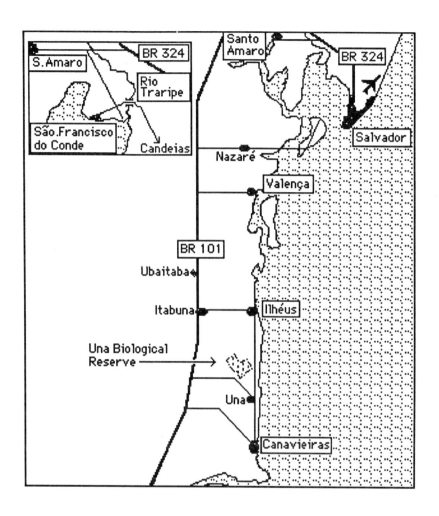

BR 324

S.Amaro

Rio
Traripe

São.Francisco
do Conde

Candeias

Santo
Amaro

BR 324

Salvador

Nazaré

Valença

BR 101

Ubaitaba

Itabuna

Ilhéus

Una Biological
Reserve

Una

Canavieiras

Boa Nova

Boa Nova is directly inland from Ilhéus in central Bahia. The area covered by this site encompasses more endemics per square kilometre than any other site in Brazil. This is a little surprising, since one tends to think of the centre of Brazilian endemism as farther south. The reason is the site's varied habitat. To the west of the village of Boa Nova is *caatinga*, very dry thorny scrubland confined to northeastern Brazil. Surrounding the village is rolling grassland with the remnants of dry forest now confined to the hill tops. To the east of the village, hugging the steep hillsides, are the remains of tropical forest that once stretched uninterrupted from Alagoas south to Rio Grande do Sul, but today stands isolated as an oasis, like so many others in eastern Brazil. Historical records for this site often refer to the nearby Rio Gongoji between Dario Meira and Gongoji. Only the Serra do Mar and the Serra dos Orgãos have more endemics and these sites cover immense areas by comparison.

The habitat surrounding Boa Nova today is under great threat of total destruction and is one of the most fragile and forgotten corners of Brazil. It is in urgent need of governmental protection in the form of an official reserve to safeguard the remaining fauna.

To reach the village of Boa Nova, drive from Ilhéus via Ibicarai and Poções for 230 kms. The dirt road via Dario Meira may be shorter but it takes much longer, though it leads through several patches of good forest, the best at a point 5 kms north of Almadina.

Species known for the Boa Nova area include Broad-tipped Hermit, Spotted Piculet, Ochre-cheeked Spinetail, Gray-headed Spinetail, Striated Softtail, Silvery-cheeked Antshrike, Rufous-winged Antshrike, Rio de Janeiro Antbird, White-bibbed Antbird and Gilt-edged Tanager. For many species, Boa Nova is the most northerly point of their range. In addition, the few remaining Slender Antbirds that inhabit the dry forest here make the site particularly important. This scarce species is confined to a very precise habitat, dry forest hilltops containing enormous terrestrial bromeliads. Within the same habitat can be found Pileated Antwren, Narrow-billed Antwren and White-browed Antpitta.

In 1989, Davis Finch, Duncan Macdonald and I found a tapaculo looking like Mouse-colored Tapaculo, but its song was slower and more drawn-out. For the present, we consider it to be Mouse-colored Tapaculo, somewhat north of its previous known limits, though further study is required. This bird was seen again in 1990 at the same locality, the "good rainforest trail". During both years, we also found Rio de Janeiro Antbird at the start of this trail.

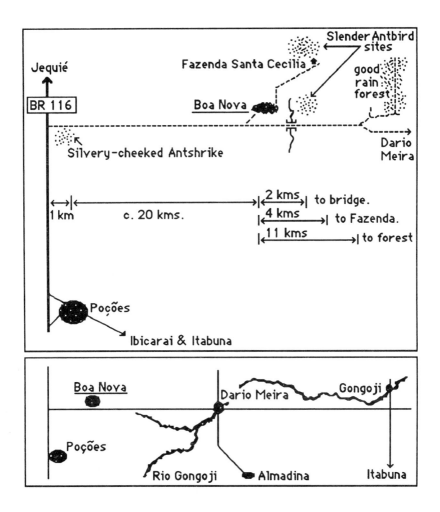

Ref: C8, F8, F9, G1, N3, P12, P19, P26, P28, P37, R10, R19, S16, T12, T14, W4, W13, W19.

Northeastern Bahia

This site covers the area from Itaberaba north to Paulo Afonso and Juazeiro, and includes the famous Raso da Catarina.

The Raso da Catarina was once a name that brought to mind California's Death Valley. It was a place inspiring awe; a place appearing on maps as only a white area indicating there was nothing there; no water, no forest, no town, no roads, just emptiness. This is the infamous *sertão*, the site of Brazil's bloody rebellion in the late 1800's.

Ornithologically, the Raso da Catarina was put on the map when in 1978 Helmut Sick tracked down the enigmatic Indigo Macaw to this locality. For over a century, the Indigo Macaw had been known only from collections, while its range and habitat remained a mystery. After many weeks of searching, Sick ultimately found the species nesting in a canyon on the edge of the Raso, near Cocorobo. In 1989, the population was believed to number about 64 birds, occupying two more or less adjacent canyons. Seeing these parrots flying in to roost in the setting sun over beautiful red rock canyons was the highlight of my 1989 trip and remains one of the most memorable of all my Brazilian experiences. This very rare parrot, a small version of the Hyacinth Macaw, has obviously always been restricted in range and appears to be slowly disappearing for good, as did its cousin, the Glaucous Macaw, earlier this century.

Regrettably, another macaw endemic to this region is in dire straits. The Little Blue Macaw has had a sparce distribution since its discovery last century. At present, the only surviving wild bird is in caraiba woodland, near Curaçá, where it was found in 1990 apparently mated with a Blue-winged Macaw. The outlook for this noble species is particularly bleak.

Reaching the village of Cocorobo, also known as Canudos, involves under-taking a mini-expedition. It lies 320 kms of rough road northwest of Aracaju, and 177 kms of this can at times be more mud than road. If conditions are good, the journey can be completed in one full day, though if birdwatching is done on the way, it will take longer. About 21 kms west of Jeremoabo is proper forest where Pectoral Antwren and Scarlet-throated Tanager can be found. There are hotels at Cocorobo and Jeremoabo. At Cocorobo, SEMA (Secretaria Especial do Meio Ambiente) wardens who guard the macaw roosts can take you to the site. During the day, the macaws feed on the Raso (plateau) and only return to the cliffs as darkness falls. Reaching the canyons involves a long walk which should only be undertaken in the company of wardens.

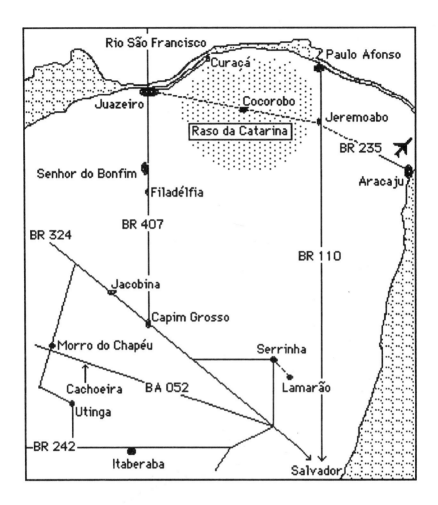

The Raso da Catarina has good caatinga with White-browed Guan, Caatinga Parakeet, Broad-tipped Hermit, Red-shouldered Spinetail, Black-bellied Antwren, Lesser Wagtail-Tyrant and White-naped Jay. Caatinga is also found around Morro do Chapéu and a small patch remains surrounding a rubbish dump along the BR 407 near Filadelfia. Surprisingly, Moustached Woodcreeper has not been recorded within this site, but has occurred close to it and should be looked for in suitable caatinga habitat. Recently, Great Xenops has been found near Serrinha, a locality which historically had various caatinga species.

At Senhor do Bonfim, 384 kms from Salvador, Pheasant Cuckoo and Pileated Antwren have been recorded at least historically, and the latter species has also been seen historically at Lamarão. It seems to be confined to caatinga while its cousin the widespread Black-capped Antwren, although preferring larger trees, can be found in a variety of habitat within this site.

In the southwestern corner of this area lies Morro do Chapéu which has become known as one of the main sites for Hooded Visorbearer. These can be found at the "Cachoeira", or waterfall, 19 kms east of town on the BA 052. Like the Hyacinth Visorbearer, this species has a preference for low vegetation, only a metre or so high. The road south of town towards Utinga has good forest habitat containing Spotted Piculet. Morro do Chapéu also has Buff-throated Pampa-Finch, Caatinga Parakeet and Gray-backed Tachuri. The Palace Hotel in the centre of town is very pleasant.

I was surprised to find many waterfowl in the region during my 1990 trip, when on almost every pond, we encountered a few ducks. These included Southern Pochard, a species normally associated, in Brazil at least, with large reservoirs, but here occurring on several reedy ponds between Capim Grosso and Senhor do Bonfim, and also near Campo Formoso west of Bonfim. A large flock of Comb Ducks, along with a few White-cheeked Pintails, hundreds of Brazilian Ducks and White-faced Whistling-Ducks were found between Jacobina and Capim Grosso, on the BR 324. Bahia has not been known for its waterfowl, but should perhaps be considered an important area for them during the austral winter.

Ref: A1, C8, F8, F9, G1, G6, H4, J3, P12, P19, P26, P28, P31, P37, S15, S16, S25, T14, V1, V5, W4, W13, W14, Y1, Z3.

White-naped Jay

Maceió

The two main areas included in this site, in the State of Alagoas, are the Parque Estadual da Pedra Talhada, 160 kms west of Maceio, and Pedra Branca, 70 kms northwest of Maceio. I found that the first offered the easiest access and probably has the largest intact forest.

The Parque Estadual da Pedra Talhada lies on the border with Pernambuco. The reserve was set up with the enthusiasm of Anita Studer who, having made a study of Forbes's Blackbird here, realised that there was no point in studying a species if its habitat was disappearing. For the protection of this species and others endemic to northeastern Brazil, a reserve was essential. In collaboration with the local people, *Associação Nordeste* was born. When we visited in 1989, the continued existence of the reserve was in doubt, since it was only made a temporary reserve, until more influential bodies came up with the financing to secure the forest for the future. IBAMA is now involved in the running of the reserve.

Anita Studer has involved the people of Quebrângulo and neighbouring areas, and has introduced a programme for regenerating the forest. This project, known as "Arco Iris", is highly commendable and a model for other regions. It gives hope for the future of Brazil, its aim being to link the remaining forest of Pedra Talhada with other smaller forest refuges in the surrounding area. The next stage, astonishingly ambitious, is to create forest bridges to the north, following the riverbanks or hilltops, and, year by year, gradually work to link up the remaining forest refuges between Alagoas and the Amazonian forest of Brazil's north coast. Such a project needs the world's support. It is to be hoped that, by the turn of the century, the fruits of the scheme will be obvious. Brazil needs such initiative.

Pedra Talhada holds a large percentage of the northeastern endemics. Some only recently discovered are in fact quite common here. Pinto's Spinetail appeared to be common on the forest edges and the Alagoas Antwren was numerous though extremely difficult to see. Since they kept to the tops of the highest trees, it took days of neck-breaking efforts before we had good views. The males are like small Black-capped Antwrens, which are also found here, while the females have rusty underparts. Alagoas Tyrannulet was also common. Other species recorded at Talhada are Pygmy Nightjar, Tawny Piculet, Long-tailed Woodnymph, Bearded Bellbird, Seven-colored Tanager and Yellow-faced Siskin.

Alagoas in recent years has seen vast areas of its forest removed for the planting of sugar cane. While we were at Talhada, the signs of deforestation were very obvious. The previous week a flash flood had washed away many buildings in the nearby village of Quebrângulo. It was as if an earthquake had demolished parts of the town. While this is a tragedy for the community, perhaps it will make the people of Alagoas consider the disastrous effects of deforestation in the future. Regrettably, Fazenda Pedra Branca, that famous location of so many new species, has largely transformed into a wasteland as far as forest birds are concerned. In a little over fifteen years, 7000 hectares of pristine forest has been felded, which reiterates the fagility of the mere exist-ence of these birds. Thankfully, at present, they can still be located on the neighbouring Fazenda Bananeira. The best area to visit, at the time of writing, is Usina Bititinga.   About 10 kms north of the Murici junction on the BR 101 lies a very rough track, known as *Usina Bititinga II,* though perhaps only drivable during the dry months from October to February. This track leads west towards Murici, and after 10 kms one reaches the village of Bititinga. From here a really rough track leads 11 kms north, into excellent forest, though a tractor or horse are required for this second leg. This is probably the main site of interest to birdwatchers at Pedra Branca since it is the site of the Alagoas Foliage-gleaner and the Bititinga Antwren, formerly considered a race of Unicolored Antwren. White-collared Kite and Alagoas Purpletuft have also occurred here. These, like many other northeastern endemics, should eventually be found at Pedra Talhada. At Quebrângulo and União dos Palmeres, there are basic hotels.

Other sites of interest are Palmeiras dos Indios which has a drier, more caatinga-like habitat, and São Miguel dos Campos east to the sea, which has the remnant population of Alagoas Curassows (if they still exist) and has had White-collared Kite.   However, this once-forested area has been largely cleared for sugar cane farms, and finding trees is as hard as finding the curassows.

Interesting birds known for this general area include White-bellied Nothura, Southern Pochard, Paint-billed Crake, Klabin-Farm Long-tailed Hermit, Ash-throated Casiornis, Smoky-fronted Tody-Flycatcher, Buff-breasted Tody-Tyrant and Greater Wagtail-Tyrant.

As will be appreciated, this whole area is of major importance for the preser-vation of some of Brazil's endemic species. It is particularly worrisome that this area has been almost totally converted to sugar cane farming. Only through the snowballing enthusiasm for forest regeneration, nurtured by Anita, can there be any hope for the survival of certain species, including some only very recently discovered.

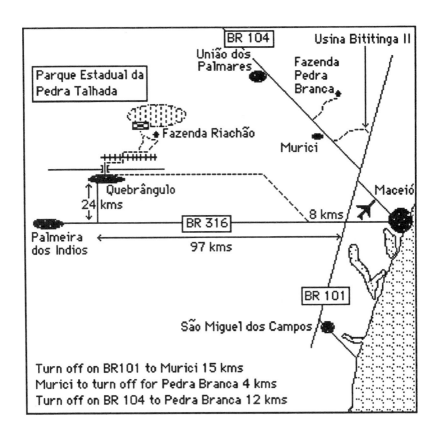

Parque Estadual da Pedra Talhada

BR 104
União dos Palmares
Usina Bititinga II

Fazenda Pedra Branca

Fazenda Riachão

Murici

Quebrângulo
24 kms

Maceió

Palmeira dos Indios

BR 316

8 kms

97 kms

BR 101

São Miguel dos Campos

Turn off on BR101 to Murici 15 kms
Murici to turn off for Pedra Branca 4 kms
Turn off on BR 104 to Pedra Branca 12 kms

Ref: C7, C8, C10, F8, G1, G9, H2, H7, M5, P12, P19, P36, P37, R11, S5, S25, S38, T4, T6, T7, T8, T9, T11.

Amazonia and Atlantic Islands

Sites in this section are twelve in northern Brazil and two groups of Atlantic islands.

On a trip to the Amazon, it is as well to select various sites along its length, since Amazonian forest offers diverse faunas within the overall ecosystem. Millions of years ago, the Amazon area contracted, leaving only isolated forest refuges similar to what has happened in recent times with the Atlantic Forest, though here the causes were different. Species became separated and subsequently diversified. When in due course the Amazon region became united again, these isolated populations had become sufficiently distinct as not to interbreed, and it is this repeated process of speciation that accounts for Amazonia's remarkable avian diversity. Today, the ground upon which the forest grows varies according to the particular river system nearby. Consequently the make-up of the forest is also different, although this is not always apparent to the layman. In addition the Amazon and its larger tributaries constitute barriers to bird dispersal, with the result that different species often occupy opposite banks.

The temperature at the Amazonian sites does not fluctuate significantly throughout the year but most have heavy rain between October and May, Belém and Porto Velho being the wettest.

1   São Luís.
2   Belém.
3   Amapá.
4   Santarém.
5   Amazônia National Park.
6   Manaus.
7   Tefé.
8   Tabatinga and Benjamin Constant.
9   Pico da Neblina.
10   Northern Roraima.
11   Rio Branco.
12   Rondônia.
13   Ilha de Trindade and Ilha de Martim Vaz.
14   Fernando de Noronha.

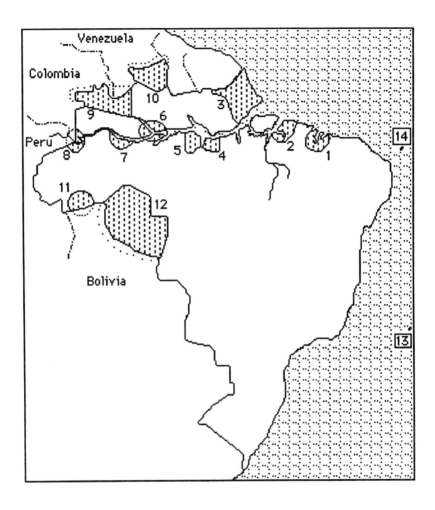

Extra sites worth visiting include the Ilha de Marajó (Pará), the islands and north bank of the Amazon between Obidos and Prianha (Pará), the Serra dos Carajás (Pará), Alta Floresta (Mato Grosso), the Jaú National Park (Amazonas) and Cruzeiro do Sul (Acre). Species previously recorded at the first and last of these locations have been included within parentheses for the Amapá and Rio Branco sites respectively.

São Luís

This site, in northern Maranhão, encompasses the Baía de São Marcos and includes Rosário, Arari, Pindaré-Mirim and São Bento. The habitat is mangroves, Amazonian forest and agricultural land. Away from the Rio Pindaré delta, the habitat quickly changes to drier forest and open areas.

The nearest good mangroves to São Luís are at the tiny village of Iguaiba. Although the villagers are not used to taking tourists out and use their boats purely for fishing, it may be possible to enlist someone's help. We were lucky to find Pitou who took us to the nearby Scarlet Ibis roost downriver. Little Wood-Rail was common here and Mangrove Cuckoo also encountered. The São Luís area also has Rufous Crab-Hawk on mangrove islands in the Baía de São Marcos. If you have time to spare, take the ferry from the Estação Marítima in São Luís to Alcântara, across the Baía de São Marcos. Stay in the historical town of Alcântara itself and hire a canoe to nearby mangroves. Anyone with more time to explore would find the Rio Pindaré area very productive.

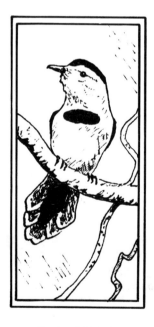

Pectoral Antwren

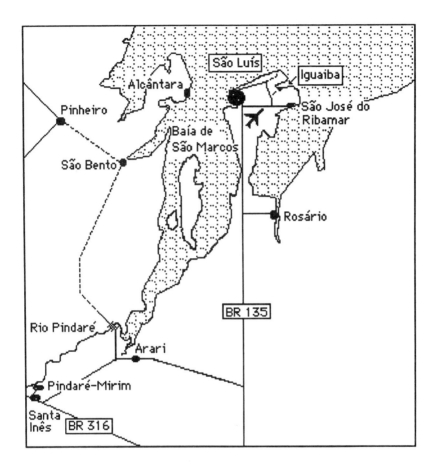

Other good birds of the area are Jandaya Parakeet, Glossy Antshrike, Pectoral Antwren, Hooded Gnateater, Ash-throated Casiornis, Para Oropendola and White-bellied Seedeater. São Luis is on the eastern edge of the Amazon and the northern edge of the caatinga region, and so has many species not encountered elsewhere in one or the other region.

Ref: C7, C8, C10, F8, F12, G6, H1, H4, H6, M5, M11, N3, P12, P19, P28, P37, S6, S16, S35, V5.

Belém

The Belém area of northeastern Pará includes here the area east to Bragança and south to Igarapé-Miri, though most of the birds on the accompanying list are found around the city of Belém itself. Beware the months of January to March, which have exceedingly heavy rains, though any time of the year is fairly wet here. The temperature in this area remains constant throughout the year - hot.

The Guamá Ecological Research Area (EMBRAPA) is worth seeing, if you are confining yourself to Belém and its immediate surroundings. It is located on the southeast edge of the city. Seek permission through Dr. Pinheiro, Instituto Evandro Chagas, or alternatively from Dr.Skaff, Museu Paraense Emilio Goeldi. (The addresses of these can be found at rear of guide.)

Once permission has been obtained and you arrive at the EMBRAPA administration, ask to see Henri Bento Baptista who speaks good English. Ideally, you want to visit the forest near the tower. Unfortunately, this tower, which once gave excellent views over the tree tops, is no longer safe. The forest gradually descends into varzea and wellingtons become very useful. To reach the tower, drive towards the CEASA (market) and, at the entrance, just before a gasoline station, turn off left, and pass through an EMBRAPA gate onto a dirt road. Follow this until encountering forest on your right. A footpath, once a dirt road, leads down into the forest passing the tower and continuing towards the Rio Guamá. We saw Red-necked Aracari, in this good forest, in 1992.

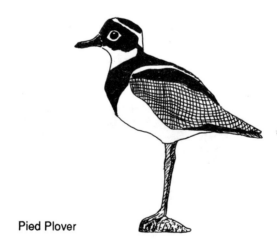

Pied Plover

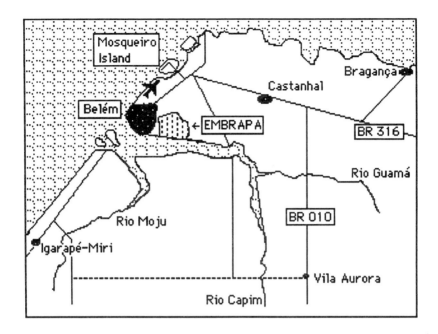

Mosqueiro Island, 65 kms north of Belém, may be worth a day`s visit. Seven kms after crossing the bridge over to the island look for footpaths leading into the forest on your left. The forest is better farther south on the island but it is difficult to obtain access to this region. Buses run every hour during the day between Belém and Mosqueiro, which is a tourist resort.

It is sad to note that the State of Pará is being extensively deforested. I suspect that the farther you travel from Belém the better the birds, so it is probably a case of driving until a good patch of forest is found. A glance at the accompanying Belém list will indicate that the area has a very impressive number of species, so perseverance in searching for mature forest should pay off. Interesting species to look for include Brazilian Tinamou, White-browed Hawk, Slaty-backed Forest-Falcon, Yellow-breasted Crake, Pearly Parakeet, Scarlet-shouldered Parrotlet, Vulturine Parrot, Dot-eared Coquette, Crimson Topaz, Striolated Puffbird, Rufous-necked Puffbird, Hooded Gnateater, White-tailed Cotinga, Opal-crowned Manakin, Black-chested Tyrant and Para Oropendola.

Ref: A3, A7, F7, F11, F12, G5, G6, G10, H1, H5, I5, I6, L5, M5, M6, M7, M11, N6, N12, N14, N15, N17, O3, O6, P19, P28, P29, P31, P33, P37, R9, R11, R19, R20, S13, S16, S20, S24, S25, T5, T14, W7.

Amapá

Amapá is almost an unknown quantity. It is Brazil's most northeastern state, sandwiched between the Amazon and French Guiana. Amapá would appear to have three basic habitats that run vertically up the State: From Macapá north to the town of Amapá (308 kms) and presumably farther north to Oiapoque (601 kms) is a dry cerrado type of habitat with many species normally associated with central Brazil. In recent years, it has been blemished with the introduction of commercial plantations. From Porto Grande (108 kms) west towards Serra do Navio (198 kms) is rainforest. The eastern edge of the state is aquatic, with severely indented tidal mud-flats which are prolific with northern waders on migration. Riverine edges hold Band-tailed Antshrike, Black-chinned Antbird and Silvered Antbird. Species in parentheses on this site list refer to the Ilha de Marajó and Ilha Mexiana. Though both are in the State of Pará, their habitat is similar to that of coastal Amapá, and their fauna is thus similar. The record of White-bellied Spinetail, also in parenthesis, refers to the French Guianan side opposite Oiapoque.

One of the few inland accessible areas is the Serra do Navio, from where many of Amapá's unusual hummingbirds have been claimed. These include "Bronze-tailed" Barbthroat, probably a subspecies of Pale-tailed Barbthroat; *Bronze Barbthroat* and *Sooty Barbthroat* , both probably variants of Bronze-tailed Barbthroat. Serra do Navio is a private mining town where public accommodation is not available, though I expect permission could be granted if pressed. Instead, we stayed at the very cheap and pleasant Recanto Ecologica Sonho Meu, six kilometres west of Porto Grande. The road from here to Serra do Navio was superb; highlights being Todd`s Antwren, Spot-tailed Antwren, Crimson Topaz and Crimson Fruitcrow. As well as the surrounding area, it would well be worth exploring the newly constructed road to the west, being part of Brazil`s *Perimetral Norte* which if ever completed will link up the states on Brazil`s northern border. As new roads are constructed, the pioneering birdwatcher has a chance to penetrate virgin land and even near the end of the twentieth century, it is possible to go to areas never previously investigated ornithologically.

Amapá has several reserves but, unfortunately, all are difficult to reach. The Cabo Orange National Park in the north corner is the best known. This reserve encompasses tropical forest, grassland and mangroves with Greater Flamingo and Scarlet Ibis. The Oiapoque Biological Reserve is also in the north, on the border with French Guiana, and west of the town of Lourenço. The Lago Piratuba Biological Reserve is on the marshy coast of Amapá, as is the Maraca-Jipioca Ecological Station.

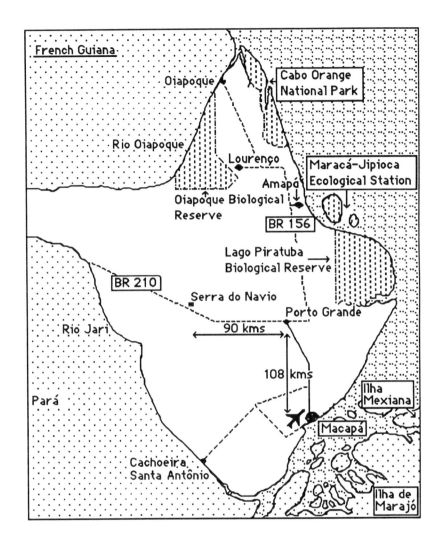

Anyone visiting Amapá would fly into Macapá, from which a plane can be chartered to reach the remoter sites or a car hired to explore the state's few existing roads. The San Marino Hotel in Macapá was very comfortable and excellent value.

Ref: C8, F11, F12, G6, G10, H1, H7, I5, I7, M5, M11, N2, N12, N16, O1, P19, P28, P31, P37, R11, R19, S5, S8, S16, S24, S25, T3, T14, V1, W29.

Santarém

Santarém is on the south bank of the Amazon in the state of Pará. Although it will probably be used as a stepping-off place en route to the Amazônia National Park, Santarém has its own variety of birds which can complement a visit to the National Park.

Alter do Chão, 38 kms west of Santarém, has very dry habitat, almost cerrado, which one does not normally associate with the Amazon. Here are found certain species more characteristic of central Brazil, such as Narrow-billed Wood-creeper, Rusty-backed Antwren and Campo Suiriri. To reach the best habitat, drive to the village of Alter do Chão and commandeer a boat to take you across to the peninsula to the north. We were fortunate enough to meet up with Ana who was studying monkeys on the peninsula and was able to ferry us across for the day.

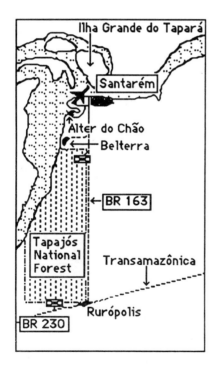

To the south lies Belterra, a mystical town about 50 kms from Santarém. It was set up by Henry Ford in the 1930s as a rubber plantation but was never very successful. The town is a sprawled out community with a village atmosphere. The official hotel was in disrepair and closed but at the rear was a pleasant, colonial-style, clean and spacious accommodation where we were asked for a token payment. There was no food and no official restaurant in town. However, one hospitable family agreed to act as our hosts and provided a most superb meal. Belterra was idyllic, as if we were in a time capsule, cut off from the outside world. From Belterra south runs the Tapajós National Forest, run by IBAMA. The neighbouring habitat, being rainforest with a network of footpaths for the rubber tappers, proved ideal for birdwatching. Kawall's Parrot, should be looked for in this area.

There are several islands in the Amazon off Santarém, including the Ilha Grande do Tapará which contain varzea species. Regrettably, we found neither Varzea Piculet nor Scaled Spinetail but we did find Lower Amazonian Piculet, a dark race of White-barred Piculet, to be both numerous and visually striking.

Other species of particular note in the area are White-browed Hawk, Dark-winged Trumpeter, Vulturine Parrot, Scaled Ground-Cuckoo, Silky-tailed Nightjar, Concolor Woodcreeper, Glossy Antshrike, Sclater's Antwren, Klages's Antwren, White-eyed Antwren, Bare-eyed Antbird, Black-bellied Gnateater, White-tailed Cotinga, Guianan Red-Cotinga, Black-necked Red-Cotinga, Opal-crowned Manakin, Pale-bellied Mourner, Zimmer's Tody-Tyrant and Tooth-billed Wren.

Ref: A7, C8, D2, F8, F11, G6, G7, G9, G10, G12, G14, H1, I6, L3, M5, M6, M7, N2, N3, N8, P19, P26, P28, P29, P31, P37, R9, R11, R14, R19, S8, S16, S24, S25, S36, T14, V1, W7, Z1.

Amazônia National Park

The Amazônia or Tapajós National Park is on the west bank of the Rio Tapajós 53 kms southwest of Itaituba, in western Pará. Itaituba can be reached from Santarém: overland via 370 kms of very rough dirt road; by boat, leaving Santarém every other evening and arriving at Itaituba the following morning; or by plane, either the scheduled Taba Airways flight which operates three times a week (Mon, Wed & Fri) or by hiring an air-taxi which holds up to five people. The travel agencies Curua-Una Turismo in both Santarém and Itaituba were both extremely helpful in arranging the two flights and in arranging our meeting with the park director, Dr. Egydio Castro, in Itaituba.

Despite being well known and often advertised, this national park has had few visitors since it opened in 1974. The director welcomes visitors. On both our visits, we were given transport from Itaituba to the reserve, accommodation, a cook (we supplied the food) and a guide while we were there. In return, we were charged a nominal sum. On our first visit, we paid the exact amount it cost the administrator to buy some much needed water pipes. Obviously, the park director works on a shoestring and the more who visit the reserve, the better the facilities will become.

It is gratifying to know that your presence is much appreciated. This isolated reserve will provide great satisfaction to anyone willing to make the effort required to get there. The reserve encompasses over one million hectares of tropical forest. In contrast, the journey from Itaituba to the reserve is very disheartening, since all the forest has been felled along the roadside. Even within the reserve, there is apparently an illegal gold mine causing habitat destruction which the few wardens are powerless to prevent. Such activities need to be brought under control.

Accommodation is available to researchers at Uruá, 12.5 kms past the entrance at Tracoá. This is a good place to see the spectacular Golden Parakeet, Brazil's national bird. The Ramal da Capelinha Trail is 16.6 kms farther along the recently built Transamazônica Highway, that runs through the reserve. This footpath is becoming overgrown in places, but leads for 30 kms through varying types of forest, to a religious shrine. Without doubt this trail is one of the best in the Amazon for seeing *Formicariidae*. Look out for Tassel-ear Marmoset, Silvery Marmoset and Tamandua Anteater.

Though the list for this site refers almost exclusively to the Amazônia National Park, it also includes species seen north to Parintins and east to the west side of the mouth of the Rio Tapajós. Historical records refer mostly to various localities along the west bank of the Rio Tapajós. Interesting species to look for include White-crested Guan, Red-throated Piping-Guan, Razor-billed Curassow, Dark-winged Trumpeter, Vulturine Parrot, Brown-chested Barbet, Hoffmann's Woodcreeper, Striped Woodhaunter, Saturnine Antshrike, Ornate Antwren, Harlequin Antbird, Dot-backed Antbird, Pale-faced Antbird, Snow-capped Manakin, Flame-crested Manakin, Black-chested Tyrant and the ultimate raptor, Harpy Eagle.

Ref: A7, C8, C10, F8, F11, G6, G10, H1, H3, I7, L2, M5, N5, N7, P4, P7, P19, P27, P28, P37, R9, R19, S2, S16, S24, S25, S36.

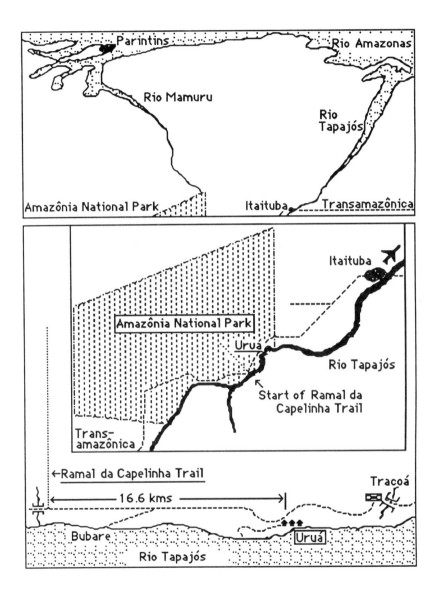

Parintins

Rio Amazonas

Rio Mamuru

Rio Tapajós

Amazônia National Park   Itaituba   Transamazônica

Itaituba

Amazônia National Park

Uruá

Rio Tapajós

Start of Ramal da Capelinha Trail

Trans-amazônica

←Ramal da Capelinha Trail

Tracoá

←——— 16.6 kms ———→

Bubare   Uruá

Rio Tapajós

Manaus

The city of Manaus lies near the confluence of the two largest waterways of western Amazonia: just east of the city, the Rio Negro runs into the Rio Solimões to form the most awesome river in the world, the Amazon. From the air it hardly seems a river, for one can see water from one horizon to the other as water glints within the green blanket below. This is várzea, or flooded forest. It then becomes clear that the Amazon shown on maps is not the whole river but merely the centre of its course, and in this instance, it has extended its width by flooding its banks on either side. The river's width and height fluctuate enormously according to season. At Manaus, the water level varies up to twelve metres. About one fifth of the world's freshwater flows down the mighty Amazon.

The area covered by the Manaus site includes the land surrounding the confluence of the Rios Negro and Solimões. The three sections of land within the site have evolved separately due to the nature of the two rivers. The Rio Negro is a blackwater river, transparent and dark brown, while the Rio Solimões is a whitewater river, muddy and saturated with inorganic matter. The fauna of each land area, has developed independently, and since each has an extensive variety of bird species, it is advantageous to visit all three. However, the emphasis here is on the section north of Manaus, since it is the easiest to visit. The habitat within this site is predominantly tropical forest, both terra firme and várzea, though the campina habitat north of Manaus is a very different, dry scrubby forest in contrast to the vast majority of the site.

Manaus is perhaps not the best Amazonian site but it offers easy access and good accommodation. Within Manaus itself, two sites of interest are the grounds of the Tropical Hotel on Ponta Negra west of the city and the "Japanese colony" north of the city, near Parque Dez. Areas farther from the city will be richer in birds, such as the Ducke Biological Reserve, on the Itacoatiara road and the W.W.F. site on the ZF-2 dirt track farther north, (the entrance to which is 50 kms from the police check-point, on the north side of Manaus). Each of these are excellent localities to visit for a few days. Both have towers, which is unusual in Brazil. These offer excellent opportunities to study canopy birdlife or soaring birds of prey and provide breathtaking views of early morning mist rising through the trees as in a Japanese painting. Of the two towers the second is the better since it can accommodate a large group of people. In recent years, Olive-green Tyrannulet has been seen on several occasions from the top of the tower. At the junction to the reserve, a sign reads *"Reserva Biologica dos Cuieiras Projeto Bacia Modelo, Km 14."* The footpath to the tower is located 15.5 kms down this very rough track. Accommodation is available at these sites, though prior permission is required. To visit Ducke contact I.N.P.A., and for the W.W.F. sites contact Roger Hutchison.

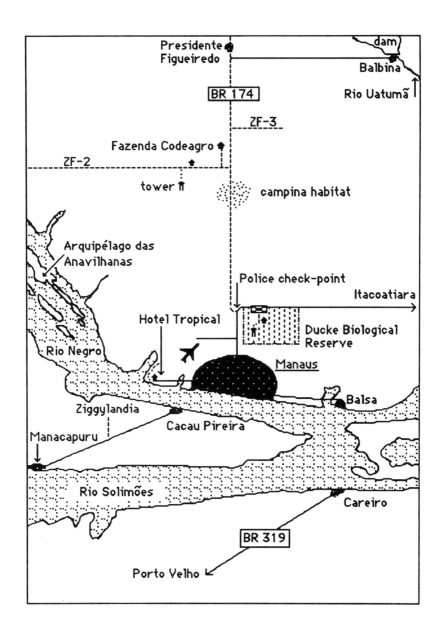

Incidentally, the W.W.F. sites are now run by an offshoot of the Smithsonian Institute, since the project *Biological Dynamics of Forest Fragments* has ceased to be run by W.W.F..

Other places of interest are Careiro, Manacapuru, Balbina, Itacoatiara and Anavilhanas. Careiro, on the south side of the Amazon, can be surveyed in a day if required. In 1992, we found a family of White-eyed Attilas here. Ferries leave from Balsa, at 7.00, 11.00 and 16.00, returning from the south side at 9.00, 14.00 and 18.00. A ferry also crosses from Manaus to the Manacapuru peninsula, between Porto São Raimundo and Cacau Pireira. Ziggylandia, 50 kms along the road to Manacapuru is worth visiting. A sign at the entrance reads *"Grande Reserva Ecologica"* which is on Fazenda São Francisco. It is run by IBAMA. Hiring a boat for the day, to travel the side waterways on the south side of the Rio Negro in the area known as Lagoa Janauary may prove rewarding, as it was for us. This area on the peninsula, near the *meeting of the waters* can attract many waterbirds depending upon the level of flooding. We were fortunate to see Agami Herons, Sungrebes, Azure Gallinules, Long-billed Woodcreepers and Amazonian Black-Tyrant. Balbina has some very good moist forest containing several species normally found farther east. Caverna Refugio do Mangoara, a unique series of caves, is 6.5 kms along the Balbina asphalt road just south of Presidente Figueiredo. A small but well made footpath leads through the forest to the caves which are a local tourist attraction. Here, we were fortunate to see, at the only known locality in this part of Brazil, Guianan Cock-of-the-Rock. The forest is also good in its own right for several quality species including Yellow-billed Jacamar and Brown-bellied Antwren. In 1992, Andrew Whittaker found Stygian Owl, Black Manakin, Pale-bellied Mourner and Rufous-crowned Elaenia, on the Itacoatiara road. Most of the historical records in the list are from either Manacapuru or Itacoatiara. The Arquipélago das Anavilhanas, run by IBAMA, offers yet another habitat, Igapo, or forest that is continuously underwater.

Interesting birds to look for include Maroon-tailed Parakeet, Red-lored Parrot, Red-fan Parrot, Crimson Topaz, Chestnut-headed Nunlet, Guianan Toucanet, Plain Softtail, Ruddy Foliage-Gleaner, Black-throated Antshrike, Band-tailed Antshrike, Leaden Antwren, Spot-backed Antwren, Black-headed Antbird, Pompadour Cotinga, Dusky Purpletuft, Glossy-backed Becard, Crimson Fruitcrow, Saffron-crested Tyrant-Manakin, Tiny Tyrant-Manakin, Dotted Tanager and Fulvous Shrike-Tanager. In recent years, much study has been devoted to the area`s nightbirds and Manaus might now be considered the "potoo capital" of the world. No fewer than five species are now known to occur here, including the rare and colourful Rufous Potoo. Perhaps more interesting still is the recent discovery in the area of White-winged Potoo, and now that its call has been identified, it has been found to be reasonably common.

My particular thanks to Mario Cohn-Haft for his invaluable help in supplying information for this site.

Ref: A7, D6, F6, F8, F11, F12, F13, G6, G14, H1, H4, I3, I4, K4, L3, M5, M10, N2, O2, P19, P27, P28, P37, R9, R11, R13, R19, S5, S14, S16, S28, S32, S36,T14, W7, W16, W17, W30, W31.

Tefé

Tefé is halfway between Tabatinga and Manaus on the south bank of the Rio Solimões. It can be reached by boat, or by air from either Tabatinga or Manaus. Despite its isolation, the town contains several hotels.

Within this site are included several study sites from outlying areas. These are: a study by Peres and Whittaker of the birds at the Petrobrás oil field on the Rio Urucu, directly south of Tefé Lake; birds observed by Dr. Andrew Johns at Ponta da Castanha, on the south western edge of Lake Tefé; and birds seen by various observers, including Paul Sterry at Lake Mamirúa which is on an island on the north side of the Rio Solimões. To visit any of these sites requires either a large boat or helicopter. If, however, you are confined to more modest transport, as we were, then birdwatching at Tefé is best done either on foot or motorised canoe, though there are a few short roads. There is a taxi rank for these canoes which we found to be very reasonably priced once negotiated.

During our stay at Tefé, it appeared to us that there was little birdlife near the town. It was unclear whether this was due to the spraying of insecticides or just general human disturbance. The two islands directly out from town were very disappointing, but after further exploration we did find two exceptionally good islands, rich in birdlife, farther down river. One was flooded forest (A on map), the other complementing this was grassy (B). This was in July; at other times of the year the water level will be quite different and may not prove so good. Species of interest seen on the flooded forest island were Short-tailed Parrot, Festive Parrot, Ash-colored Cuckoo, Ladder-tailed Nightjar, White-bearded Hermit, White-chinned Jacamar, Long-billed Woodcreeper, Red-and-white Spinetail, Orange-fronted Plushcrown, Black-and-white Antbird,

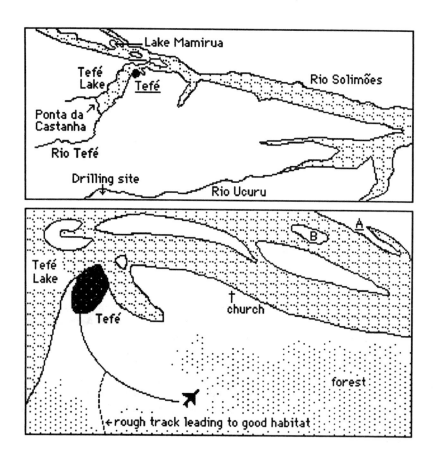

Cinnamon Attila, Brownish Elaenia, Pearly-breasted Conebill, Bicolored Conebill Scarlet-crowned Barbet and Oriole Blackbird. On the nearby grassy island, we saw White-bellied Spinetail, Dark-breasted Spinetail, Lesser Wagtail-Tyrant, Riverside Tyrant, River Tyrannulet and Yellow-hooded Blackbird. Tefé is obviously a good location for várzea specialties.

Some other species known for the Tefé site include Bartlett's Tinamou, Slaty-backed Forest-Falcon, Pale-winged Trumpeter, Red-winged Wood-Rail, Black-banded Crake, Needle-billed Hermit, Gould's Jewelfront, Fiery Topaz, Rufous Motmot, Brown-banded Puffbird, Rufous-necked Puffbird, Lanceolated Monklet, Lemon-throated Barbet, Bar-bellied Woodcreeper, Curve-billed Scythebill, Short-billed Antwren, Rufous-tailed Antwren, Ihering's Antwren, Slate-colored Antbird, Plumbeous Antbird, Hairy-crested Antbird, Plum-throated Cotinga, Citron-bellied Attila and Blue-backed Tanager. Peres and Whittaker encountered a previously undescribed puffbird of the genus *Nystalus* on the Rio Urucu. Clearly Tefé will reward the enterprising birdwatcher.

Ref: C9, F8, G14, H1, H3, H5, J1, M5, N2, N3, P17, P19, P28, P37, R19, S8, S13, S16, S25, S29, W7, Z2.

Tabatinga and Benjamin Constant

On the border with Colombia and Peru, this site covering the north and south sides of the Rio Solimões has understandably a very impressive list of species. If coming from within Brazil, you will fly into Tabatinga but may wish to cross over into nearby Colombia where there are more hotels available in Leticia, though the Solimões Hotel on the army camp at Tabatinga is good value. You can also reach Leticia by ferry from Manaus. Border crossing is unrestricted, so long as you do not intend to wander too far into Colombia. If this is the case, seek out the immigration authorities and have your passport stamped. While at Leticia, you may well wish to explore the Amaca Yacú National Park and other nearby sites in Colombia. Alternatively, you may wish to cross the river and base yourself at Benjamin Constant, as we did. The ferry goes three times a day from Tabatinga and in 1989 cost about $1.50.

At Benjamin Constant, there are a couple of hotels near the harbour. If you hire a boat for the day and go upriver, you will be birdwatching equally in Peru. Although we found such a boat trip very pleasant and good value, we found the terra firme birdwatching to be superior. Some birdwatchers in the past have gone for several days upriver, and this may be worthwhile. We were fortunate to find some easily accessible good quality forest which to reach took us about forty minutes of fast walking or an hour at a more leisurely pace. For ideal birding, this has to be done in the dark so that you are on site for dawn. I suggest a reconnaissance during daylight since the trail leads precariously over large areas of marsh where logs, forming bridges, have to be negotiated.

From the pier, walk west along the main road through town until it turns sharp left. At the next junction, turn right onto a track and follow it until you reach a small church. Behind the church, follow a very small footpath around a fenced enclosure and you will reach a second major path. Turn right here, and you are now on the edge of town. Continue west and go onto a small footpath which will lead past a very small pool, then a second larger one. The path now splits. Take the right fork and follow it for about five minutes until the path again splits. Again, take the right fork. This will now lead across a large marsh in which you have to cross a series of logs. After several hills interspersed with marshes, you will come to a small community of huts. Follow the path through them, bearing right, and ultimately you will come to what seems to be the local school hut. Here, turn sharp left and after going through a clearing you will reach mature forest. The walk from town is through open country and so is very hot during the day.

Although the site list is extensive, care should be taken to establish which species are to be found on the north as opposed to the south side of the Solimões. Ideally, you will spend equal time on both sides, though Brazil on the north side has few access points. The site list also includes historical records from São Paulo de Olivenca. As with nearby Colombian sites, records from the Peruvian side of the Rio Javari are in parentheses.

Birdwatching on the north side is particularly good in Colombia, though the birds there could doubtless be found in adjacent Brazil. Santena flies out of Leticia every Friday to Ipiranga on the Rio Iça in Brazil close to the Colombian border. From here a boat service meets the plane and takes passengers upriver to Tarapaca in Colombia, where the river now changes its name to the Río Putumayo. From here, a boat can be hired through INDERENA to take you up the Río Cotuhe to Lorena which is on the northern border of the Amaca Yacú National Park. One could interrupt this journey and make Ipiranga serve as a base for birdwatching in Brazil.

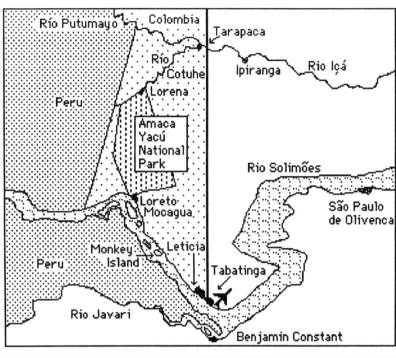

Río Putumayo
Colombia
Tarapaca
Río Cotuhe
Ipiranga
Rio Içá
Lorena
Peru
Amaca Yacú National Park
Rio Solimões
São Paulo de Olivenca
Loreto Mocagua
Leticia
Monkey Island
Peru
Tabatinga
Rio Javari
Benjamin Constant

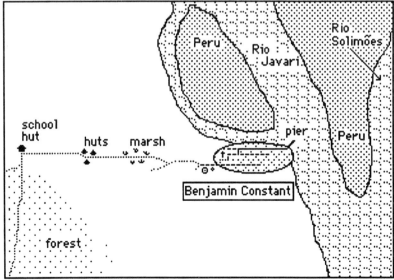

Peru
Rio Javari
Rio Solimões
school hut
huts
marsh
pier
Peru
Benjamin Constant
forest

Access to the southern section of the Amaca Yacú National Park is possible via the village of Loreto Mocagua about 50 kms northwest of Leticia. Arrangements for accommodation and transport can be made through the INDERENA office in Leticia (the director is Oscar Tamayo, INDERENA, AA006, Leticia, Amazonas, Colombia). Upriver from Leticia, Monkey Island has a comfortable lodge and is well worth visiting for several days. Arrangements for staying here, as well as for general natural history excursions around the Leticia area, can be made through Mike Tsalickis who runs Turamazonas, Hotel Parador Ticuna, Leticia. Mike will set up excursions to any of the three countries within the site and will supply equipment, accommodation, transport and a guide.

If you are staying around Leticia, it would be worthwhile following some dirt tracks out of town to the north. Although these are heavily disturbed, they still allow some interesting birdwatching. Point-tailed Palmcreeper is a recent colonist in Colombia but it is thriving in the *Mauritia* Palms around Leticia. During the austral winter, thousands of Southern Martins swarming to their evening roost in Leticia provide a memorable spectacle. When visiting Amazonian islands, bear in mind that the time of year can dramatically affect the habitat and consequently the birds. The Solimões is highest from March to June and lowest in August and September. The difference between the two is astonishing, as much as 25 feet or more. When the river is high, some islands disappear while others change from ones to walk on to ones where you have to canoe among the treetops!

Species known for the Brazilian part of this site include Black-banded Crake, Red-billed Ground-Cuckoo, Needle-billed Hermit, Golden-tailed Sapphire, Dark-breasted Spinetail, Bluish-slate Antshrike, Rufous-tailed Antwren, Chestnut-shouldered Antwren, Black Antbird, Slate-colored Antbird, Elusive Antpitta, Johannes' Tody-Tyrant, Gray Wren and Yellow-crested Tanager. Doubtless many of the species recorded in parentheses, on the site list, occur regularly in Brazil as well, and should be looked for.

Ref: C8, F8, G6, G14, H1, H5, I5, K1, L2, L4, M5, N2, P3, P10, P19, P28, P29, P37, R1, R19, S16, S25, T14, W11, W12, W15, W25, Z1.

Pico da Neblina

The site includes the northwestern corner of Brazil from the Rio Padauari west along the Rio Negro to the Rio Uaupés and north to the borders of Colombia and Venezuela. The habitat is predominantly Amazonian forest with the Serra Tapirapeco mountain spine running along the Venezuelan border eastward toward Roraima. Pico da Neblina (3,014m),"the mountain of mist", is the highest peak in Brazil and lies on the western edge of this ridge.

Very little recent ornithological work has been carried out in this region, despite its avian richness. Nearly all the records listed come from H. Friedmann's book on the bird collections by Ernest G. Holt during two expeditions between 1929 and 1931. Recent records from Neblina mostly come from the Venezuelan side (see Willard,*et al.*). Other birds listed in parentheses are known from the area north along the Caño Casiquiare or the Sierra de Unturán, both in southern Venezuela and Mitú in eastern Colombia.

To visit the area still involves almost the same expeditionary tactics as required of Holt. Today's adventurer will probably base himself at São Gabriel da Cachoeira which has good pensãos. Either sail up the Rio Negro to Barcelos (boats leave Manaus twice a week) or fly with Taba Airways

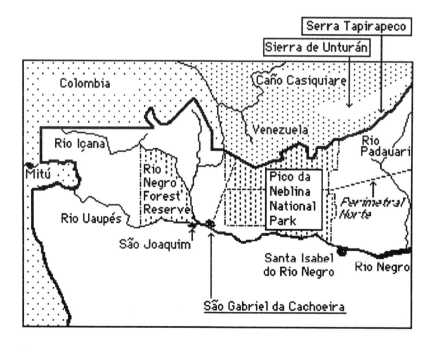

(flights are three times per week). At Barcelos stay at the Hotel Oasis, run by a German family offering jungle tours. Continue by boat to São Gabriel da Cachoeira which will take a further two days.

Pico da Neblina National Park encompasses 2,200,000 hectares and is contiguous with the Venezuelan Serranía de La Neblina National Park. There is a rough road leading from São Gabriel da Cachoeira north to the Venezuelan border and the *Perimetral Norte* branches east through the centre of the park, but transport would be difficult to obtain. With permission to visit the National Park from IBAMA, transport might be possible. The *Perimetral Norte* is still under construction but should eventually link up with Roraima. Regrettably, its course through this remote reserve will inevitably lead to the desecration of this unique site. Nearby is also the Rio Negro Forest Reserve. However, the area at present has a high level of military presence, due to its sensitive border location. Birdwatchers in camouflaged gear and with binoculars might raise suspicion. Add to this the fact that to the east lies the territory of the Yanomami Indians, who are being pushed out of their land by garimpeiros (gold diggers) in Roraima, and to the west live the Maku Indians. Then consider the additional factor of narcotics smuggling across the border from Colombia and you find yourself in a potentially hazardous area. Despite these facts, this site offers adventure and ornithological rewards for those willing to spend time and no doubt experience some discomfort in locating new species for the area and perhaps for Brazil.

Particularly interesting species known for this area include Gray-legged Tinamou, Barred Tinamou, White-tipped Swift, Fiery Topaz, Tawny-tufted Toucanet, Undulated Antshrike, Cherrie`s Antwren, Gray-bellied Antbird, Chestnut-crested Antbird, Pelzeln`s Tody-Tyrant, Azure-naped Jay and Scaled Flower-piercer. The *Neblina Foliage-gleaner* has proven to be merely a synonym of White-throated Foliage-gleaner.

Ref: B2, F12, F13, G6, G11, H5, I5, I7, M3, M5, M6, M7, N12, P19, P23, P27, P28, P37, R11, R19, R20, S16, S25, V1, W8.

Northern Roraima

The site includes the Rio Anaua Biological Reserve in central Roraima, north through the savanna region to the Tepuis, a series of disjunct mountain ranges that form Roraima's northern border with Venezuela and Guyana. Within this region lies the Ilha de Maracá and its Ecological Reserve, which has the longest bird list for any one area in Roraima. The *Perimetral Norte* section has been completed within Roraima and gives access to parts of the state not previously open to birdwatchers, and may well prove worth exploring. Like the *Transamazônica*, it will no doubt be hard going in wet weather.

Most people will reach Roraima by flying into Boa Vista from Manaus. The Hotel Praia was very comfortable, but bear in mind that hotels, flights and buses are often fully booked. The Rio Branco Antbird, a species of very limited distribution, is the main target bird for the immediate area. Although living in thickets along the Rio Branco north of Boa Vista, we only located it on the Ilha São José, about 10 kms up river from Boa Vista, where it was very easily seen. Substantial wetlands across the river from town are worth exploring. Here we saw Boat-billed Heron, Black-crested Antshrike, Cinnamon Manakin, Slate-headed Tody-Flycatcher, Pale-tipped Tyrannulet and Gray Seedeater.

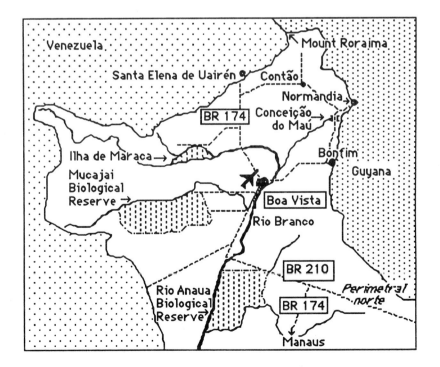

Having encountered the *cercomacra,* the next target should be the Hoary-throated Spinetail. Historically known for the Rio Cotingo, we endeavoured to follow its scent in 1992. Our search took us into some very remote parts of Brazil, where there was no public gasoline stations, nor hotels. Locals rallied round to provide sufficient fuel and basic sleeping quarters. In this part of Roraima everyone knows the problems and is willing to help. Ornithologically, this is an almost virgin land and we found many unusual birds, including the first records for Brazil of Pale-eyed Pygmy-Tyrant which we found to be reasonably common at Contão. After several days search, we eventually found one pair of the spinetails in seasonally flooded forest on the south side of the Rio Tacutu at Conceição do Maú which is only a couple kilometres from the Guyanan border. The spinetails preferred to remain close to the ground within the dense thickets and vines but would venture high when agitated. We were very fortunate to see this little known species, and were delighted that perseverance with both this and the *cercomacra* paid off. Just prior to going to press, Davis Finch informed me that he visited the Guyanan side, in January 1993, and located both the *poecilurus* and the *cercomacra* in suitable breeding habitat. Thus both species are no longer endemics to Brazil.

The Biological Reserves of Mucajai and Rio Anaua both have difficult access, though their edges can be reached by road. Likewise, the much publicised Ilha de Maracá Biological Reserve has access problems unless you have work permits for the site. However, its edge too can be reached by road. Most of the sites in Roraima can be approached tantalisingly close but defy easy access. Mount Roraima can be be climbed from the Venezuelan side but it takes several days to explore. Other Tepuis in the region all require full expeditions to reach the top, though they beckon intrepid birdwatchers. These unique rock formations, shrouded in mist, are one of the great mystical wonders of South America but remain unattainable to mere mortals.

Species of note, in northern Roraima, include Double-striped Thick-knee, Fiery-shouldered Parakeet, Tepui Parrotlet, Rufous-winged Ground-Cuckoo, Tepui Swift, Blue-fronted Lancebill, Brown Violetear, Sparkling Violet-ear, Peacock Coquette, Tepui Goldenthroat, White-chested Emerald, White-bellied Piculet, Tepui Spinetail, Roraiman Barbtail, Plain-winged Antwren, Roraima Antwren, Brown-breasted Antpitta, Scaled Antpitta, Rose-collared Piha, Scarlet-horned Manakin, Olive Manakin, Sierran Elaenia, Bicolored Wren, Tepui Wren, Tepui Greenlet, Golden-tufted Grackle, Tepui Redstart, Two-banded Warbler, Greater Flower-piercer, Finsch's Euphonia, Speckled Tanager, Olive-backed Tanager, Paramo Seedeater and Tepui Brush-Finch. Many of these species are Pantepui endemics.

Ref: C2, C8, D4, F4, F6, F11, F12, G6, G9, H1, I 5, I 7, M3, M5, M6, M7, N11, N12, O1, P19, P21, P22, P28, P37, R11, R19, S8, S16, S18, S19, S25, S32, S33,S36, V1, W6, W8.

Rio Branco

This site is the eastern corner of Acre with Rio Branco, the state capital, being the base from which excursions will be made. Areas covered include Plácido de Castro and Brasiléia on the border with Bolivia, Porto Acre to the north and Sena Madureira near the Rio Purús to the northwest. Two forms of parentheses are used on the site list. Those species shown with normal curved brackets refer to birds seen around Cruzeiro do Sul, in the far west of the state (this area has been visited by only a few, including myself in 1992, but offers great potential). The second form of bracketing, is as has been used for most sites, denoting species that have been observed outwith Brazil but close to the site border. Here, the area dealt with is the Bolivian border opposite Brasiléia. There is no doubt that most of these occur undetected on the Brazilian side.

Acre, like Rondônia, is now ignominious for having been *burnt to the ground* as part of the *taming of the Amazon policy* that was prevalent through the 1980s. Fortunately, the situation is not as bad as portrayed. Admittedly, when driving Acre roads, one is presented with a grim vista of cattle ranches with poor soil and scattered burnt trees standing like gravestones of a once lush flora, with forest now confined to the horizons. Roads deep in dust makes overtaking a nightmarish operation, only to be attempted after suffering the recurring choking dust raised by the timber lorry in front, and requiring a few prayers before pulling out into the path of unseen traffic coming head-on. Normally,

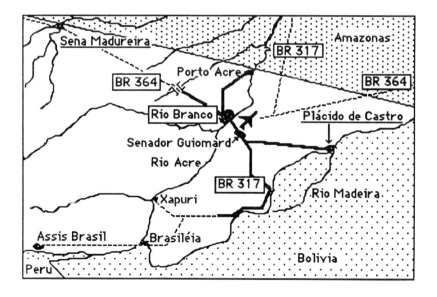

one has to drive perhaps fifty kilometres before finding even a fragment of forest close to the road. However, if one takes to the air, the true picture is quickly apparent. There are few roads in the Amazon. Where there are roads, the forest has been surgically removed but one can fly for hundreds of kilometres without seeing a road, everything beneath being a sea of virgin green. It may be difficult for birdwatchers to get into forest but the birds are still there. One can only hope that Brazil will see the futility of cutting down forest for agriculture and ranching when the soil is so obviously lacking in nutrients. Like the roads, such ground quickly turns to desert. Dirt roads in this part of Brazil are impassable for much of the year due to heavy rains. The road between Rio Branco and Cruzeiro do Sul is generally only open, for heavy vehicles, between June and August.

In 1989, good forest was found along the roadside about 20 kms before Plácido de Castro, which lies 100 kms east of Rio Branco. On our return in 1992, the same patch of forest remained intact. This area has a network of footpaths created for the practice of rubber tapping, but proves particularly useful to the birdwatcher. Stay at the Hotel Carioca in Plácido de Castro.

While at Plácido de Castro, I was impressed by the number of species colonizing this part of Brazil. In 1989, we encountered several Emerald Toucanets which I believe had not previously been recorded in the country. We also had apparently the second Brazilian record of Blue-headed Macaw. Rock Parakeets, which had only been recorded once previously in Brazil, were found in reasonable numbers. Regrettably, on our second visit we failed to find these species, but did turn up other interesting birds. These included Bluish-fronted Jacamar, Peruvian Recurvebill, Bluish-slate Antshrike, Sclater`s Antwren, Rio Suno Antwren, Cinnamon Manakin, Johanne`s Tody-Tyrant and Plain Tyrannulet. On both trips, we saw at this site, Pheasant Cuckoo and Lemon-throated Barbet. As at other Amazonian sites bordering neighbouring countries, it may be quite easy to discover new species for Brazil.

Other species of particular interest here include Pale-winged Trumpeter, Chestnut Jacamar, Striolated Puffbird, Yellow-billed Nunbird, Curl-crested Aracari, Rufous-breasted Piculet, Bamboo Antshrike, Black Antbird, Goeldi's Antbird, Sooty Antbird, Ash-throated Gnateater and Opal-crowned Tanager.

Ref: C8, F4, F6, F8, F11, F12, G6, G14, L4, M5, N10, P5, P15, P16, P19, P24, P28, P35, P37, R2, S13, W7.

Rondônia

Due to the lack of available information on specific localities, this site includes the whole State of Rondônia, though most records come from the centre of the state around Jaru or the northwestern corner along the border with Bolivia near Guajará-Mirim. Records in parentheses are from localities along the border in Bolivia. Most of the historical records are from either the southern part of the state near the border with Mato Grosso, or the northern corner at the mouth of the Rio Jiparaná. There are three large reserves within the state, but regrettably all have poor access. They are the Pacaás Novos National Park, a mountainous region with tropical forest and cerrado, the Jaru Biological Reserve, a tract of tropical forest, and the Guaporé Biological Reserve, an important wetland. Previously known as the *Territorio de Guaporé*, Rondônia only became a state in 1981.

Most people entering the state fly into Porto Velho, though there is now a reasonable road from Cuiabá and there is access from the north although the road to Manaus as with the Transamazônica, can only be travelled in good weather. Like the previous site in Acre, this area has been extensively deforested in recent years. Though there are still large untouched areas of Rondônia, areas near roads have been cleared. The BR 364 south to Jaru is typical of the poor accessibility of good untouched forest. Between Porto Velho and Ariquemes, a distance of 202 kms, there is little of interest for the birdwatcher. From here to Jaru, a further 93 kms, there are several trails leading off the main highway. The best that we visited led south of the road, 55 kms from Ariquemes. This track skirted the edge of good forest which had a fairly good footpath with such Rondônia specialties as Crimson-bellied Parakeet and Red-necked Aracari. Continuing farther on, the track led to a typical large clearing where we encountered that much coveted species, the awe-inspiring Harpy Eagle.

To reach Guajará-Mirim, take the BR 364 out of Porto Velho to Abuna. This road in 1990 was in perfect condition, possibly the best road in Brazil at the time. From Abuna south, the road is being upgraded from a very rough dirt track to asphalt. The old road follows the route of the original railway line, even using the railway bridges themselves. Amazingly, one bridge still had the railway track, which had not been removed before converting the bridge for cars. Steam-engine enthusiasts will enjoy the old railway station at Guajará-Mirim, now a museum, as well as the rusty engines from bygone days. Of equal interest are the old rotting river boats that line the banks and bring to mind the Mississippi of the 1800's. Surprisingly, this border town seems like a metropolis considering the wild terrain encountered on the way here. The Lima Palace Hotel was excellent value and can be strongly recommended. As in several other remote areas of Brazil, one has a feeling of stepping back in time when encountering *garimpeiros* and the like, and one is reminded of the pioneering days of the original American settlers who "opened up the west".

In 1990, we found excellent birdwatching southeast of town. Turn left at the roundabout on entering Guajará-Mirim. Follow the asphalt surface for 6 kms where it veers left. At this point, take the dirt track that continues straight ahead. After a farther 2 kms, take the right fork and in a farther kilometre cross over the bridge and you are now in unspoiled Amazonian forest with a dirt track that appears to continue for some considerable distance. We saw Dusky-headed Parakeet, Pale-rumped Swift, Pavonine Quetzal, Black-girdled Barbet, Curl-crested Aracari, White-eyed Antwren, Purple-throated Fruitcrow, Bare-necked Fruitcrow and White-winged Shrike-Tanager. Blue-tufted Starthroat was encountered along the asphalt section of the road.

Recent studies in Rondônia and northern Bolivia suggest that many species associated with central Pará have a much wider distribution. It now seems apparent that these three areas are linked together forming an important and individual biosphere. Although Brazil never did have many endemics within her Amazonian region, most of these have now been lost, having recently been found in northern Bolivia, and most of the remaining few will almost certainly go the same way.

Noteworthy species for Rondônia are Fiery-tailed Awlbill, White-throated Woodpecker, Concolor Woodcreeper, Hoffman's Woodcreeper, Dusky-billed Woodcreeper, Chestnut-throated Spinetail, Crested Foliage-gleaner, Gray-throated Leaftosser, Bamboo Antshrike, Glossy Antshrike, Rondônia Bushbird, Striated Antbird, Chestnut-shouldered Antwren, White-breasted Antbird, Black-bellied Gnateater, Snow-capped Manakin, Tricolored Tody-Tyrant, Plain Tyrannulet, Tooth-billed Wren, Buff-rumped Warbler and Yellow-shouldered Grosbeak.

Ref: B3, C8, F9, F12, G6, G13, H1, H3, J2, L2, M5, N2, P8, P9, P19, P27, P28, P37, R3, R11, R12, R15, R19, S5, S9, S16, S25, S32, S36, V1.

White-banded and Black-collared Swallows

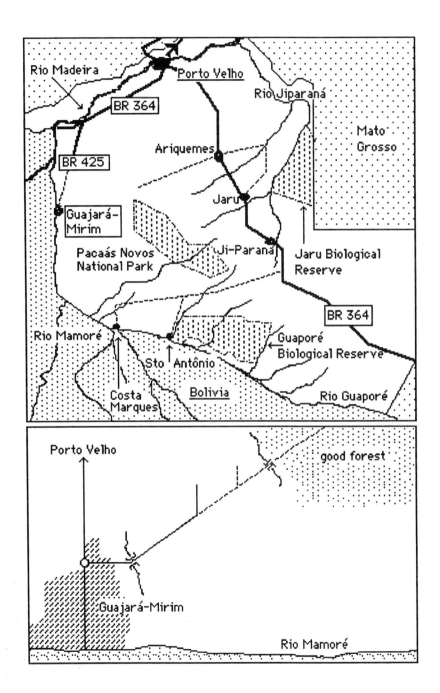

Ilha de Trindade and Ilha de Martim Vaz

These islands are about 1,200 kms off the coast of Espírito Santo, so it is unlikely that many non-Brazilians will make the effort to visit them. Of the two, Trindade is certainly the better because of its seabird colonies. Both this site and the following are included here only because they have recorded a few species not encountered elsewhere in Brazil.

Resident species on the Ilha de Trindade include Herald Petrel, Red-footed Booby, Great Frigatebird and Fairy Tern. Species which breed on the island, but depart for part of the year include Masked Booby, Lesser Frigatebird, Black Noddy, Brown Noddy and Sooty Tern.

Ref: N9, S16, T10, W9.

Red-billed Tropicbird

Fernando de Noronha

Fernando de Noronha lies 350 kms northeast of Natal (Rio Grande do Norte). Travel agencies in Brazil can organise tours to the island. Alternatively, fly direct with Nordeste Airlines which fly daily from Recife, departing 7.00, 12.00 and 18.15. It is surprising to learn that the population on this isolated island is about 1,300. As might be expected, the only local food is fish. Water and other foods are flown in. There is one hotel on the island, Hotel Esmeralda do Atlantico.

The entire archipelago has recently been made a National Park. Bird-watchers visiting here will be particularly interested in the seabirds and the Noronha Vireo, until recently considered a race of Red-eyed Vireo but now judged to be a separate species and a Brazilian endemic. Thankfully, it is found to be quite common. Another recent split is the Noronha Elaenia, previously thought to be a subspecies of Large Elaenia. The island also has a race of the Eared Dove.

The islands are of importance as a vagrant trap that would rank with the Isles of Scilly, Fair Isle or Sable Island (Nova Scotia), if a bird observatory were established. Little Shearwater, Purple Heron, Squacco Heron, Northern Pintail, Bar-tailed Godwit and Franklin's Gull have occurred here as vagrants, being "firsts" for Brazil. What else has turned up, in the past, can only be surmised, but for certain there would have been several more "firsts". The islands are also famous for Green Sea Turtles and Spinner Dolphins.

Ref: A9, F2, M5, N1, O4, O5, P19, S16, T3, T5, T8, T10, W9.

Key to Lists

The Brazilian Checklist, beginning on page194, forms the index to the three sets of sites. The order followed in this guide is based on Meyer de Schauensee`s *The species of birds of South America and their distribution* , though certain sections have been rearranged, following more recent publications by specialists of individual families. Where nomenclature is in doubt or where doubt has arisen in the past, alternatives are given. Likewise, English names have also changed over time and as with scientific names alternative English names are provided. Many of these have been influenced by Robert Ridgely, Nigel Collar (BirdLife International), and Sibley and Munroe.

Key to symbols used in the lists:-

E   - Endemic to Brazil (though includes species that have occurred as a non-breeding accidental in a neighbouring country).
" "   - Species within inverted commas are, either subspecies of the previous species, or in some cases, are colour morphs that have been considered as separate species in the past.
{ }   - Species not yet recorded in Brazil but close to border in a neighbouring country.
( )   - Two Amazonian sites have records within curved brackets, these refer to areas adjoining the site but still within Brazil.
[ ]   - Species that have only been recorded as dead at site, having been
      - washed ashore.

H   - Historical Record. Species having not been recorded since 1960.
A   - Accidental. Only up to two records since 1960 and hence involves predominantly migrant birds.
M   - Migrant. Presumed to occur regularly at site for only part of the year.
S   - Presumed resident but scarce.
X   - Presumed resident.
I   - Introduced species to that site.

The above six categories are all subjective. Whilst all species listed for individual sites have been recorded at their given location, their individual status often has to be guessed, bearing in mind its known status at a neighbouring site. Thus many "S"s should perhaps be "M"s and some "M"s perhaps "A"s. Although not one hundred per cent accurate, the status of individual species for any one particular site is a good guideline from which to build on. Regrettably, a few sites have seldom been visited, thus even common species may have been overlooked e.g. the lack of White-faced Ibis for the Barranquinho site is presumably due more to lack of coverage than to a lack of birds.

It should be said that, in a few cases, certain species have been regularly misidentified by many contributors. Closely allied pairs of species such as Sulphur-rumped Flycatcher and Black-tailed Flycatcher plus Wied's Tyrant-Manakin and Pale-bellied Tyrant-Manakin have been particularly troublesome. The status of these species at particular sites should be viewed with caution.

Agami Heron

Southern and Central Brazil

Tai - Taim Ecological Station, Rio Grande do Sul. Covers the area from the border with Uruguay, at Chuí, north up coast to São Lourenço.

Mos - Mostardas Peninsula. Coastal marshes on the peninsula north to the Santa Catarina border.

Apa - Aparados da Serra National Park south to São Francisco de Paula.

Uru - Uruguaiana south to Barra do Quaraí.

FdI - Foz do Iguaçu (Iguaçu National Park). Species in parentheses have been recorded on the Argentine side.

CtC - Campo Grande to Corumbá (Southern Pantanal).

Tra - Transpantaneira (Northern Pantanal). From Cuiabá to Porto Jofre.

Bar - Barranquinho. Covers the area from Cáceres to Descalvados.

CdG - Chapada dos Guimarães.

Ema - Emas National Park.

Can - Serra da Canastra National Park.

Cip - Serra do Cipó National Park south to Lagoa Santa.

Bra - Brasilia National Park.

Ara - Araguaia National Park and the Ilha do Bananal.

Serra Antwren

Species	Tai	Mos	Apa	Uru	FdI	CtC	Tra	Bar	CdG	Ema	Can	Cip	Bra	Ara
Rockhopper Penguin	A	[A]	-	-	-	-	-	-	-	-	-	-	-	-
Macaroni Penguin	A	-	-	-	-	-	-	-	-	-	-	-	-	-
Magellanic Penguin	M	M	-	-	-	-	-	-	-	-	-	-	-	-
Greater Rhea	S	X	-	X	-	X	X	X	S	X	X	H	X	S
Solitary Tinamou	-	-	S	-	S	-	-	-	-	-	-	-	-	-
Cinereous Tinamou	-	-	-	-	-	-	-	-	-	-	-	-	-	S
Little Tinamou	-	-	-	-	-	-	-	-	S	S	-	-	-	-
Brown Tinamou	-	-	S	X	-	X	-	-	-	-	-	S	H	-
Undulated Tinamou	-	-	-	-	-	X	X	X	X	X	-	-	-	X
E Yellow-legged Tinamou	-	-	H	S	-	-	-	-	-	-	-	S	-	-
Brazilian Tinamou	-	-	-	-	-	-	-	-	-	-	-	-	-	S
Small-billed Tinamou	-	-	-	-	S	S	X	X	X	-	-	H	X	X
Tataupa Tinamou	-	-	S	-	X	X	S	X	X	S	S	H	-	-
Red-winged Tinamou	H	S	X	-	-	X	X	X	X	X	X	-	X	S
E Lesser Nothura	-	-	-	-	-	H	-	-	H	S	S	S	S	-
Spotted Nothura	X	X	X	X	{S}	H	-	-	-	X	X	X	X	-
Dwarf Tinamou	-	-	-	-	-	-	-	-	-	-	-	H	S	-
Least Grebe	S	-	X	-	{S}	X	X	S	S	-	X	H	X	S
White-tufted Grebe	X	X	S	M	-	A	-	-	-	-	-	-	-	-
Great Grebe	X	X	-	X	-	A	-	-	-	-	-	-	-	-
Pied-billed Grebe	X	X	X	X	{X}	-	-	-	-	-	-	S	X	S
Wandering Albatross	M	A	-	-	-	-	-	-	-	-	-	-	-	-
Royal Albatross	[A]	A	-	-	-	-	-	-	-	-	-	-	-	-
Black-browed Albatross	M	M	-	-	-	-	-	-	-	-	-	-	-	-
Shy Albatross	-	[A]	-	-	-	-	-	-	-	-	-	-	-	-
Yellow-nosed Albatross	M	M	-	-	-	-	-	-	-	-	-	-	-	-
Light-mantled Sooty Albatr.	[A]	-	-	-	-	-	-	-	-	-	-	-	-	-
Southern Giant-Petrel	M	M	-	-	-	-	-	-	-	-	-	-	-	-
Southern Fulmar	M	[A]	-	-	-	-	-	-	-	-	-	-	-	-
Cape Petrel	M	-	-	-	-	-	-	-	-	-	-	-	-	-
Kerguelen Petrel	A	-	-	-	-	-	-	-	-	-	-	-	-	-
White-headed Petrel	A	-	-	-	-	-	-	-	-	-	-	-	-	-
Atlantic Petrel	M	[M]	-	-	-	-	-	-	-	-	-	-	-	-
Soft-plumaged Petrel	M	[H]	-	-	-	-	-	-	-	-	-	-	-	-
Broad-billed Prion	A	[A]	-	-	-	-	-	-	-	-	-	-	-	-
"Antarctic (Dove)" Prion	M	-	-	-	-	-	-	-	-	-	-	-	-	-
Thin-billed Prion	M	[M]	-	-	-	-	-	-	-	-	-	-	-	-
Gray Petrel	A	[H]	-	-	-	-	-	-	-	-	-	-	-	-
White-chinned Petrel	M	M	-	-	-	-	-	-	-	-	-	-	-	-
Cory's Shearwater	M	[M]	-	-	-	-	-	-	-	-	-	-	-	-
Great Shearwater	M	M	-	-	-	-	-	-	-	-	-	-	-	-
Sooty Shearwater	M	[M]	-	-	-	-	-	-	-	-	-	-	-	-

	Tai	Mos	Apa	Uru	FdI	CtC	Tra	Bar	CdG	Ema	Can	Cip	Bra	Ara
Manx Shearwater	M	M	-	-	-	-	-	-	-	-	-	-	-	-
Wilson's Storm-Petrel	M	M	-	-	-	-	-	-	-	-	-	-	-	-
Magellanic Diving-Petrel	A	-	-	-	-	-	-	-	-	-	-	-	-	-
Neotropic Cormorant	X	X	X	X	X	X	X	X	M	-	X	X	X	X
Anhinga	A	S	-	X	S	X	X	X	S	S	S	H	X	X
Magnificent Frigatebird	A	M	-	-	-	-	-	-	-	-	-	-	-	-
Cocoi (White-necked) Heron	X	X	-	X	S	X	X	X	-	-	-	H	X	X
Great (Common) Egret	X	X	S	X	X	X	X	X	X	X	X	X	X	X
Snowy Egret	X	X	S	X	X	X	M	X	H	X	X	H	X	X
Little Blue Heron	-	A	-	-	-	M	M	-	-	-	-	-	-	-
Tricolored Heron	-	-	-	-	-	-	-	-	-	-	-	-	A	-
Striated Heron	M	X	S	M	X	X	X	X	X	-	-	S	X	X
Agami Heron	-	-	-	-	-	A	-	A	-	-	-	-	-	X
Cattle Egret	X	X	X	X	X	X	X	X	X	X	X	-	X	X
Whistling Heron	X	X	X	X	S	X	X	X	-	X	X	-	X	-
Capped Heron	-	-	-	-	-	X	X	X	-	-	-	H	X	X
Black-crowned Night-Heron	X	X	-	X	X	X	X	X	-	-	-	H	X	X
Yellow-crowned Night-Heron	A	H	-	-	-	-	-	-	-	-	-	-	-	S
Boat-billed Heron	A	-	-	-	-	X	X	X	H	-	-	H	-	X
Rufescent Tiger-Heron	-	S	-	M	{S}	X	X	X	-	X	-	H	-	X
Fasciated Tiger-Heron	-	-	-	-	{H}	-	-	-	H	-	-	-	-	-
Stripe-backed Bittern	S	S	-	-	-	-	-	-	-	-	-	-	-	-
Least Bittern	-	-	-	-	{S}	S	S	-	-	-	-	-	-	S
Pinnated Bittern	S	S	-	M	-	-	-	-	-	-	-	-	-	-
Wood Stork	M	M	-	M	{M}	M	M	M	M	-	-	-	A	M
Maguari Stork	X	X	S	X	{A}	X	X	X	-	-	-	H	-	S
Jabiru	-	-	-	-	-	X	X	X	-	-	-	H	A	X
Plumbeous Ibis	X	S	-	-	{S}	X	X	X	-	-	-	-	-	-
Buff-necked Ibis	-	S	X	-	-	X	X	X	H	X	X	-	X	X
Green Ibis	A	-	-	-	X	M	M	M	-	X	-	-	X	X
Bare-faced Ibis	X	X	-	X	-	X	X	X	-	-	-	H	A	X
White-faced Ibis	X	X	-	X	-	M	M	-	-	-	-	H	-	-
Roseate Spoonbill	X	X	-	X	{A}	X	X	X	-	-	-	H	-	X
Chilean Flamingo	M	X	-	-	-	-	-	-	-	-	-	-	-	-
Andean Flamingo	-	A	-	-	-	-	-	-	-	-	-	-	-	-
Horned Screamer	-	-	-	-	-	S	H	H	-	-	-	-	-	X
Southern Screamer	X	X	-	X	-	X	X	X	-	-	-	-	-	-
Fulvous Whistling-Duck	X	X	-	X	-	M	M	M	-	-	-	-	M	M
White-faced Whistling-Duck	X	X	-	X	-	X	X	X	-	-	M	-	M	M
Black-bellied Whistling-Duck	-	-	-	-	-	X	X	X	-	-	-	-	M	M
Coscoroba Swan	X	X	-	-	-	-	-	-	-	-	-	-	-	-
Black-necked Swan	X	X	-	-	-	-	-	-	-	-	-	-	-	-

	Tai	Mos	Apa	Uru	FdI	CtC	Tra	Bar	CdG	Ema	Can	Cip	Bra	Ara
Orinoco Goose	-	-	-	.	-	-	-	..	-	-	-	-	-	S
Speckled Teal	X	X	X	X	-	-	-	-	-	-	-	-	-	-
Chiloe Wigeon	A	A	-	-	-	-	-	-	-	-	-	-	-	-
White-cheeked Pintail	A	A	-	-	-	-	-	-	-	-	-	-	-	M
Yellow-billed Pintail	X	X	X	X	{A}	-	-	-	-	-	-	-	-	-
Silver Teal	X	X	-	X	-	-	-	-	-	-	-	-	-	-
Blue-winged Teal	-	A	-	-	-	-	-	-	-	-	-	-	-	-
Cinnamon Teal	H	-	-	-	-	-	-	-	-	-	-	-	-	-
Red Shoveler	M	M	-	-	-	-	-	-	-	-	-	-	-	-
Ringed Teal	S	A	-	A	{A}	M	M	-	-	-	-	-	-	-
Rosy-billed Pochard	X	X	-	X	{A}	-	-	-	-	-	-	-	-	-
Southern Pochard	-	-	-	-	-	-	-	-	-	-	-	-	M	-
Brazilian Duck (Teal)	X	X	X	X	X	X	X	X	-	-	S	X	M	X
Comb Duck	-	-	-	-	{A}	M	M	-	-	-	-	-	M	M
Muscovy Duck	S	X	-	-	X	X	X	Y	H	X	-	-	X	X
Brazilian Merganser	-	-	-	-	A	-	-	-	-	A	S	-	-	-
Lake Duck	M	A	-	-	-	-	-	-	-	-	-	-	-	-
Masked Duck	S	S	-	S	{X}	H	-	-	-	-	-	H	A	-
Black-headed Duck	S	-	-	-	-	-	-	-	-	-	-	-	-	-
Andean Condor	-	-	-	-	-	-	-	-	A	-	-	-	-	-
King Vulture	-	-	A	-	S	S	S	S	S	S	S	S	S	S
Black Vulture	S	X	X	-	X	X	X	X	X	X	X	X	X	X
Turkey Vulture	S	S	X	X	X	X	X	X	X	X	X	X	X	X
Lesser Yellow-headed Vulture	S	S	-	X	{X}	X	X	X	-	X	-	-	-	X
Greater Yellow-headed Vulture	-	-	-	-	-	-	-	-	-	-	-	-	-	S
White-tailed Kite	X	X	X	X	X	X	X	X	X	X	X	X	M	-
Pearl Kite	-	-	-	-	-	X	X	X	X	X	-	-	X	-
Swallow-tailed Kite	-	M	M	-	M	M	-	M	M	M	-	-	-	M
Gray-headed Kite	H	-	-	-	{S}	S	S	S	S	-	S	S	S	S
Hook-billed Kite	-	-	-	-	-	S	S	H	H	S	-	-	-	S
Rufous-thighed Kite	-	-	-	A	-	X	S	S	-	S	-	-	-	-
Plumbeous Kite	-	M	M	-	M	M	M	M	M	M	-	H	M	M
Snail Kite	X	X	-	X	M	X	X	X	-	-	-	H	A	X
Slender-billed Kite	-	-	-	-	-	-	-	-	-	-	-	-	-	S
Bicolored Hawk	-	-	S	-	S	H	S	H	H	-	-	-	-	-
Tiny Hawk	-	-	-	-	{A}	A	A	-	A	-	-	-	H	-
Gray-bellied Hawk	-	-	-	H	{H}	-	-	-	-	-	-	-	-	-
Sharp-shinned Hawk	-	-	-	S	A	{S}	-	S	-	S	-	S	H	-
Black-chested Buzzard-Eagle	H	-	A	S	-	S	-	-	-	-	X	X	-	-
White-tailed Hawk	X	-	X	X	-	X	X	-	X	X	X	X	X	-
Zone-tailed Hawk	-	-	-	-	-	-	S	S	-	-	-	-	S	-
Swainson's Hawk	A	-	-	-	A	-	-	-	-	-	-	-	-	-

	Tai	Mos	Apa	Uru	FdI	CtC	Tra	Bar	CdG	Ema	Can	Cip	Bra	Ara
Broad-winged Hawk	-	-	-	-	-	H	-	-	-	-	-	-	-	-
Roadside Hawk	X	X	X	X	X	X	X	X	X	X	X	X	X	X
White-rumped Hawk	-	-	S	-	{S}	-	-	-	-	-	-	-	-	-
Short-tailed Hawk	-	-	A	A	-	S	H	S	-	S	-	S	-	S
Gray-lined Hawk	-	-	-	-	-	S	S	H	S	S	-	-	-	X
Harris's Hawk	-	-	A	-	A	-	M	M	-	H	-	A	-	-
White Hawk	-	-	-	-	-	-	-	-	H	-	-	-	-	-
Mantled Hawk	-	-	A	-	-	S	-	-	-	-	-	-	-	-
E White-necked Hawk	-	-	-	-	-	-	-	-	-	-	-	A	-	-
Black-collared Hawk	-	-	-	-	A	{S}	X	X	X	-	-	-	-	X
Savanna Hawk	X	X	X	X	S	X	X	X	X	X	X	X	X	X
Great Black-Hawk	X	-	S	A	X	X	X	X	S	-	S	H	-	X
Crowned Eagle	H	H	A	-	-	S	-	H	A	X	-	-	S	-
Crested Eagle	-	-	-	-	A	-	-	-	-	-	-	-	-	-
Harpy Eagle	-	-	-	-	-	-	-	-	A	-	-	H	-	S
Black-and-white Hawk-Eagle	H	-	A	-	-	S	S	S	S	-	-	-	-	-
Ornate Hawk-Eagle	-	-	-	-	-	S	-	S	S	S	-	H	-	S
Black Hawk-Eagle	A	-	A	-	-	S	-	-	S	-	-	-	-	S
Cinereous Harrier	M	A	-	A	-	-	-	-	-	-	-	-	-	-
Long-winged Harrier	X	A	-	X	-	H	M	-	-	X	A	-	-	-
Crane Hawk	H	A	-	-	{S}	X	X	X	S	X	-	H	S	X
Osprey	-	-	A	-	-	M	M	M	-	-	-	-	M	M
Laughing Falcon	-	-	-	-	A	X	X	X	X	X	X	-	-	-
Collared Forest-Falcon	-	-	S	-	S	-	S	H	-	-	S	H	S	S
Barred Forest-Falcon	H	H	S	-	X	S	S	S	S	-	-	H	-	-
Red-throated Caracara	-	-	-	-	-	-	-	-	H	-	-	-	-	-
Yellow-headed Caracara	S	X	X	-	X	X	X	X	X	X	X	X	X	S
Chimango Caracara	X	X	X	X	A	S	-	-	-	-	A	-	-	-
Crested Caracara	X	X	X	X	X	X	X	X	X	X	X	X	X	X
Peregrine Falcon	A	-	A	-	M	M	-	H	-	-	-	-	-	-
Orange-breasted Falcon	-	-	-	-	-	-	H	-	-	-	-	A	-	-
Bat Falcon	-	-	-	-	S	X	X	X	X	-	X	H	S	X
Aplomado Falcon	M	A	X	X	{S}	X	X	X	X	X	X	X	M	X
American Kestrel	X	X	X	X	X	X	X	X	X	X	X	X	X	-
Speckled Chachalaca	S	S	S	-	-	-	-	-	-	-	-	-	-	-
Chaco Chachalaca	-	-	-	-	-	X	X	X	-	-	-	-	-	-
Rusty-margined Guan	H	-	-	-	S	-	S	H	X	-	X	H	X	X
Dusky-legged Guan	H	-	X	X	X	-	-	-	-	-	-	-	-	-
E Chestnut-bellied Guan	-	-	-	-	-	-	S	H	-	-	-	-	-	S
Blue-throated Piping-Guan	-	-	-	-	X	X	X	-	-	-	-	-	-	-
"Red-throated" Piping-Guan	-	-	-	-	-	S	-	-	-	-	-	-	-	-
Black-fronted Piping-Guan	-	-	A	-	S	-	-	-	-	-	-	-	-	-

	Taj	Mos	Apa	Uru	FdI	CtC	Tra	Bar	CdG	Ema	Can	Cip	Bra	Ara
Bare-faced Curassow-	-	-	-	-	-	X	X	X	-	X	-	-	X	X
Marbled Wood-Quail -	-	-	-	-	-	-	-	-	-	-	-	-	-	S
Spot-winged Wood-Quail -	-	-	S	-	S	-	-	-	-	-	-	H	-	-
Limpkin-	X	X	-	X	S	X	X	X	-	-	-	-	A	X
Plumbeous Rail -	X	X	X	-	{S}	-	-	-	-	-	-	-	-	-
Blackish Rail -	-	S	X	-	X	M	M	-	-	-	-	X	M	-
Spotted Rail -	S.	S	-	-	-	-	-	-	-	-	-	-	-	-
Uniform Crake -	-	-	-	-	-	-	-	-	A	-	-	-	-	-
Gray-necked Wood-Rail -	S	X	S	S	{A}	X	X	X	X	-	X	H	X	X
Giant Wood-Rail-	X	-	-	X	{H}	-	-	-	-	-	-	-	-	X
Slaty-breasted Wood-Rail	-	X	X	-	X	-	-	-	-	-	-	H	-	-
Ash-throated Crake -	-	-	S	-	{X}	-	S	S	S	S	-	H	S	-
Yellow-breasted Crake -	-	-	-	-	-	-	-	-	-	-	-	H	-	-
Gray-breasted Crake-	-	-	-	-	-	-	S	-	-	-	-	-	-	-
Rufous-faced Crake -	-	-	-	-	-	-	-	-	-	-	-	-	S	-
Rufous-sided Crake -	-	X	S	-	X	-	S	-	-	-	-	-	-	S
Red-and-white Crake-	-	-	-	S	-	-	-	-	-	-	-	-	-	-
Russet-crowned Crake -	-	-	-	-	-	-	H	-	A	-	-	-	A	A
Ocellated Crake-	-	-	-	-	-	-	-	-	-	A	-	-	S	-
Paint-billed Crake -	-	-	-	-	-	-	M	-	-	-	-	-	-	-
Spot-flanked Gallinule -	X	X	X	X	{S}	-	-	-	-	-	-	-	-	-
Common Gallinule (Moorhen) -	X	X	S	X	S	S	S	-	-	-	S	H	X	S
Purple Gallinule -	M	M	M	-	{M}	X	X	X	-	-	-	H	A	X
Azure Gallinule-	-	-	-	-	{H}	M	M	H	-	-	-	H	-	-
Red-gartered Coot -	S	X	-	-	-	-	-	-	-	-	-	-	-	-
White-winged Coot -	X	X	-	X	-	-	-	-	-	-	-	-	-	-
Red-fronted Coot -	S	-	-	-	-	-	-	-	-	-	-	-	-	-
Sungrebe -	-	-	-	-	-	S	S	S	S	-	-	-	-	-
Sunbittern -	-	-	-	-	-	-	X	X	-	-	-	-	-	X
Red-legged Seriema -	S	-	X	-	-	X	X	X	X	X	X	X	X	X
Wattled Jacana -	X	X	S	X	X	X	X	X	X	-	X	X	X	X
South American Painted-Snipe	A	A	-	-	-	-	-	-	-	-	-	-	-	-
American Oystercatcher -	X	X	-	-	-	-	-	-	-	-	-	-	-	-
Southern Lapwing -	X	X	X	X	X	X	X	X	X	X	X	X	X	X
Pied Plover-	-	-	-	-	-	-	X	X	X	-	-	H	X	X
Black-bellied Plover	M	M	-	-	-	M	-	-	-	-	-	-	-	-
American Golden Plover -	M	M	-	-	{M}	M	M	H	H	-	-	H	M	M
Semipalmated Plover-	M	M	-	-	-	-	-	-	-	-	-	-	-	-
Two-banded Plover -	M	M	-	-	-	-	-	-	-	-	-	-	-	-
Collared Plover-	X	X	-	X	X	X	X	X	H	-	-	H	A	X
Rufous-chested Plover -	M	M	-	M	-	-	-	-	-	-	-	-	-	-
Tawny-throated Dotterel-	M	M	-	-	-	-	-	-	-	-	-	-	-	-

	Tai	Mos	Apa	Uru	FdI	CtC	Tra	Bar	CdG	Ema	Can	Cip	Bra	Ara
Ruddy Turnstone	M	M	-	-	-	-	-	-	-	-	-	-	-	-
Solitary Sandpiper	M	M	-	M	M	M	M	M	H	-	-	H	M	M
Lesser Yellowlegs	M	M	-	M	M	M	M	M	-	-	-	H	M	M
Greater Yellowlegs	X	X	-	M	{M}	M	M	-	-	-	-	-	M	M
Spotted Sandpiper	-	A	-	-	M	M	M	-	-	-	-	-	M	M
Willet	A	A	-	-	-	-	-	-	-	-	-	-	-	-
Red Knot	X	X	-	-	-	-	A	-	-	-	-	-	-	-
Least Sandpiper	-	A	-	-	-	-	A	-	-	-	-	-	-	-
Baird's Sandpiper	M	A	-	-	-	A	-	-	-	-	-	-	-	A
White-rumped Sandpiper	M	M	-	M	{M}	M	M	M	-	-	-	-	M	M
Pectoral Sandpiper	M	M	-	M	{M}	M	M	-	-	-	-	H	M	M
Semipalmated Sandpiper	-	A	-	-	-	-	-	-	-	-	-	-	-	-
Sanderling	M	M	-	-	-	-	-	-	-	-	-	-	-	A
Stilt Sandpiper	M	M	-	-	-	-	M	-	-	-	-	-	-	-
Buff-breasted Sandpiper	M	M	-	-	-	-	M	M	-	-	-	-	-	-
Ruff	A	-	-	-	-	-	-	-	-	-	-	-	-	-
Upland Sandpiper	M	M	-	-	{M}	M	M	M	H	M	-	-	A	-
Whimbrel	M	M	-	-	-	-	-	-	-	-	-	-	-	-
Eskimo Curlew	-	-	-	-	-	-	-	H	-	-	-	-	-	-
Hudsonian Godwit	M	M	-	-	-	-	M	M	-	-	-	-	-	-
Short-billed Dowitcher	A	A	-	-	-	-	-	-	-	-	-	-	-	-
South American Snipe	X	X	X	X	{S}	X	X	H	X	-	X	X	X	X
Giant Snipe	-	-	S	-	-	-	-	-	H	-	S	S	X	-
White-backed Stilt	X	X	-	X	{M}	X	X	X	-	-	-	H	M	-
Wilson's Phalarope	M	M	-	-	-	-	M	-	-	-	-	-	-	-
Least Seedsnipe	-	A	-	-	-	-	-	-	-	-	-	-	-	-
Snowy Sheathbill	M	-	-	-	-	-	-	-	-	-	-	-	-	-
Antarctic Skua	M	M	-	-	-	-	-	-	-	-	-	-	-	-
Chilean Skua	[A]	-	-	-	-	-	-	-	-	-	-	-	-	-
South Polar Skua	[A]	-	-	-	-	-	-	-	-	-	-	-	-	-
Pomarine Skua (Jaeger)	A	A	-	-	-	-	-	-	-	-	-	-	-	-
Arctic Skua (Parasitic Jaeger)	M	M	-	-	-	-	-	-	-	-	-	-	-	-
Long-tailed Skua (Jaeger)	M	-	-	-	-	-	-	-	-	-	-	-	-	-
Olrog's Gull	A	A	-	-	-	-	-	-	-	-	-	-	-	-
Kelp Gull	X	X	-	-	-	-	-	-	-	-	-	-	-	-
Gray-hooded Gull	M	M	-	M	-	-	-	-	-	-	-	-	-	-
Brown-hooded Gull	X	X	-	A	-	-	-	-	-	-	-	-	-	-
Black Tern	-	A	-	-	-	-	-	-	-	-	-	-	-	-
Large-billed Tern	X	X	-	X	-	X	X	X	-	X	-	-	A	X
Gull-billed Tern	M	M	-	-	-	M	-	-	-	-	-	-	-	-
South American Tern	M	M	-	-	-	-	-	-	-	-	-	-	-	-
Common Tern	M	M	-	-	-	M	M	-	-	-	-	-	-	M

	Tai	Mos	Apa	Uru	Fdl	CtC	Tra	Bar	CdG	Ema	Can	Cip	Bra	Ara
Arctic Tern	-	A	-	-	-	-	-	-	-	-	-	-	-	-
Trudeau's Tern	X	X	-	-	-	-	-	-	-	-	-	-	-	-
Yellow-billed Tern	X	X	-	X	X	X	X	X	-	-	-	H	S	X
Royal Tern	X	X	-	-	-	-	-	-	-	-	-	-	-	-
Cayenne Tern	M	M	-	-	-	-	-	-	-	-	-	-	-	-
Black Skimmer	M	M	-	M	M	M	M	M	-	-	-	-	A	M
Scaled Pigeon	-	-	-	-	(M)	-	M	M	M	-	M	-	M	X
Picazuro Pigeon	M	M	M	M	M	M	M	M	M	M	X	X	X	X
Spot-winged Pigeon	-	-	-	-	X	-	-	-	-	-	-	-	-	-
Pale-vented Pigeon	-	M	M	-	X	X	X	X	X	X	X	S	M	X
Plumbeous Pigeon	H	-	S	-	-	-	-	-	-	-	X	H	X	S
Eared Dove	X	X	S	X	X	M	M	M	H	X	X	H	A	X
Scaled Dove	-	-	-	-	-	(H)	X	X	X	X	X	X	X	X
E Blue-eyed Ground-Dove	-	-	-	-	-	-	-	A	-	-	-	-	-	-
Common Ground-Dove	-	-	-	-	-	-	-	-	-	-	-	-	X	-
Plain-breasted Ground-Dove	-	-	-	-	-	X	X	X	S	-	S	S	X	X
Ruddy Ground-Dove	X	X	X	S	X	X	X	X	X	X	X	X	X	X
Picui Ground-Dove	X	X	X	X	X	X	X	X	-	-	S	-	-	-
Blue Ground-Dove	-	-	-	-	X	X	X	X	H	S	-	H	-	X
Purple-winged Ground-Dove	-	-	-	-	-	(A)	-	-	-	-	-	H	-	-
Long-tailed Ground-Dove	-	-	-	-	-	X	X	X	X	-	-	-	-	X
White-tipped Dove	X	X	X	X	X	X	X	X	S	X	X	H	X	X
Gray-fronted Dove	-	X	X	-	X	-	S	-	S	S	S	H	X	S
Ruddy Quail-Dove	-	-	S	S	-	S	-	S	-	H	-	H	-	S
Violaceous Quail-Dove	-	-	-	-	-	S	-	-	-	-	-	-	-	-
Hyacinth Macaw	-	-	-	-	-	X	X	X	-	-	-	-	-	X
Blue-and-yellow Macaw	-	-	-	-	-	S	S	-	H	X	-	-	A	S
Scarlet Macaw	-	-	-	-	-	-	-	-	-	-	-	-	-	S
Red-and-green Macaw	-	-	-	-	-	X	S	X	X	-	-	-	-	X
Chestnut-fronted Macaw	-	-	-	-	-	-	-	-	-	-	-	-	-	H
Blue-winged Macaw	-	-	-	H	-	H	H	-	S	-	S	H	S	S
Golden-collared Macaw	-	-	-	-	-	X	X	X	S	-	-	-	-	X
Red-bellied Macaw	-	-	-	-	-	-	S	-	S	X	-	-	-	X
Red-shouldered Macaw	-	-	-	-	-	H	X	X	X	X	-	-	X	X
Blue-crowned Parakeet	-	-	-	-	-	X	S	X	X	-	-	-	-	-
White-eyed Parakeet	-	-	-	-	X	X	X	X	X	S	X	H	X	S
E Golden-capped Parakeet	-	-	-	-	-	-	-	-	-	-	S	S	-	-
Peach-fronted Parakeet	-	-	-	-	-	A	X	X	X	X	X	X	X	X
Black-hooded Parakeet	-	-	-	-	-	X	X	X	-	-	-	-	-	-
Reddish-bellied Parakeet	-	X	X	-	X	H	-	-	-	-	-	X	H	-
Blaze-winged Parakeet	-	-	-	-	-	-	S	S	-	-	-	-	-	-
Green-cheeked Parakeet	-	-	-	-	-	-	X	S	-	-	-	-	-	-

	Tai	Mos	Apa	Uru	FdI	CtC	Tra	Bar	CdG	Ema	Can	Cip	Bra	Ara
Monk Parakeet	X	X	-	X	{S}	X	X	X	-	-	-	-	-	-
Blue-winged Parrotlet	-	-	-	-	X	S	S	-	-	S	X	X	X	-
E Plain Parakeet	-	-	-	-	-	-	-	-	-	-	-	X	-	-
Yellow-chevroned Parakeet	-	-	-	-	{X}	X	X	X	X	X	X	X	X	X
Red-capped Parrot	H	-	X	-	X	-	-	-	-	-	-	-	-	-
Blue-headed Parrot	-	-	-	-	-	S	S	H	X	X	-	-	-	X
Scaly-headed Parrot	-	-	X	-	X	X	X	X	-	S	X	X	X	-
Red-spectacled Parrot	H	-	S	-	{A}	-	-	-	-	-	-	-	-	-
Yellow-faced Parrot	-	-	-	-	-	S	S	-	-	X	-	-	X	S
Turquoise-fronted Parrot	-	-	-	-	X	X	X	X	H	X	-	H	X	-
Orange-winged Parrot	-	-	-	-	-	H	S	X	S	-	-	-	S	X
Vinaceous-breasted Parrot	-	-	H	X	-	H	-	-	-	-	-	-	-	-
E Blue-bellied Parrot	-	-	S	S	{A}	-	-	-	-	-	-	H	-	-
Ash-colored Cuckoo	A	-	-	A	-	A	-	A	A	-	-	-	-	-
Yellow-billed Cuckoo	-	A	A	A	A	H	-	-	-	-	-	-	M	-
Pearly-breasted Cuckoo	-	-	-	-	M	-	-	-	H	-	-	-	-	-
Dark-billed Cuckoo	M	A	-	M	M	H	M	M	H	M	-	-	A	M
Squirrel Cuckoo	S	X	X	X	X	X	X	X	X	X	X	X	X	X
Little Cuckoo	-	-	-	-	-	H	X	X	S	-	-	-	-	S
Greater Ani	-	-	-	-	-	X	X	X	X	X	-	-	-	X
Smooth-billed Ani	X	X	X	X	X	X	X	X	X	X	X	X	X	X
Hoatzin	-	-	-	-	-	-	-	-	-	-	-	-	-	X
Guira Cuckoo	X	X	X	X	X	X	X	X	X	X	X	X	X	X
Striped Cuckoo	X	X	S	X	X	X	X	X	X	-	X	X	X	X
Pheasant Cuckoo	-	-	-	-	S	H	-	H	S	-	-	H	-	S
Pavonine Cuckoo	-	-	-	-	S	-	-	-	S	-	-	-	-	-
Barn Owl	S	S	-	-	{S}	-	S	H	A	S	-	S	S	S
Tropical Screech-Owl	X	X	X	X	X	H	S	-	S	-	-	H	S	-
Variable(Black-c.)Screech-Owl	-	-	-	-	{S}	-	-	-	-	-	-	S	-	-
"Long-tufted" Screech-Owl	-	X	X	-	-	-	-	-	-	-	-	-	-	-
Great Horned Owl	A	A	-	A	-	S	X	H	S	-	S	-	-	-
Spectacled Owl	-	-	A	-	A	-	S	-	H	-	-	H	-	-
Tawny-browed Owl	-	-	-	-	S	-	-	-	-	-	-	-	-	-
Ferruginous Pygmy-Owl	-	-	-	-	X	X	S	H	X	-	X	X	X	X
Burrowing Owl	X	X	X	X	X	X	X	-	X	X	X	X	X	-
Black-banded Owl	-	-	-	-	-	-	-	-	H	S	-	H	-	-
Mottled Owl	-	-	-	-	S	-	-	-	-	-	-	H	-	-
Rusty-barred Owl	H	-	X	-	S	-	-	-	-	-	-	-	-	-
Striped Owl	X	X	-	-	-	-	-	-	H	-	-	H	A	-
Stygian Owl	-	-	-	-	-	H	-	-	-	-	-	H	-	-
Short-eared Owl	A	-	-	-	-	-	-	-	-	X	X	-	-	-
Buff-fronted Owl	H	-	-	-	{S}	-	-	-	-	-	-	-	A	-

	Tai	Mos	Apa	Uru	FdI	CtC	Tra	Bar	CdG	Ema	Can	Cip	Bra	Ara
Great Potoo	-	-	-	-	-	-	S	-	-	-	-	H	-	-
Large-tailed Potoo	-	-	-	-	-	-	-	-	-	-	-	H	-	-
Common Potoo	H	-	-	-	X	S	S	-	H	S	-	-	X	X
Short-tailed Nighthawk	-	-	-	-	M	-	X	-	A	S	H	-	-	S
Least Nighthawk	-	-	-	-	-	-	M	-	M	-	X	X	M	-
Sand-colored Nighthawk	-	-	-	-	-	-	-	-	-	-	-	-	-	S
Lesser Nighthawk	-	-	-	⌐	-	-	M	-	-	-	-	-	M	M
Common Nighthawk	A	A	-	-	-	(M)	M	-	M	M	-	-	-	M
Band-tailed Nighthawk	-	-	-	-	-	-	-	H	M	M	-	-	-	X
Nacunda Nighthawk	-	-	M	-	-	M	(M)	M	M	M	M	H	M	M
Pauraque	-	S	-	S	-	X	M	M	M	X	X	X	X	X
Ocellated Poorwill	-	-	-	-	-	-	-	-	-	-	-	H	-	-
Rufous Nightjar	H	-	-	-	-	-	M	-	-	-	-	H	A	S
Band-winged Nightjar	-	-	-	A	-	-	-*	-	-	-	-	M	A	-
White-winged Nightjar	-	-	-	-	-	-	-	A	-	-	S	-	-	-
Spot-tailed Nightjar	-	-	-	-	-	-	M	-	-	-	M	M	-	M
Little Nightjar	-	-	-	-	M	(M)	M	M	M	H	M	-	M	M
Ladder-tailed Nightjar	-	-	-	-	-	-	-	-	-	-	-	-	-	S
Scissor-tailed Nightjar	-	S	-	M	M	(M)	M	M	M	M	X	X	X	S
Long-trained Nightjar	-	-	-	H	-	-	-	-	-	-	-	S	-	-
Sickle-winged Nightjar	-	-	-	-	-	(A)	-	-	-	-	-	H	A	-
White-collared Swift	-	X	-	X	-	X	-	-	-	M	-	X	X	X
Biscutate Swift	-	-	-	X	-	M	-	-	-	M	-	H	-	-
Great Dusky Swift	-	-	-	-	-	X	-	-	-	M	X	S	-	-
Sooty Swift	-	-	-	M	-	M	-	-	-	-	-	M	-	-
Gray-rumped Swift	-	-	-	S	-	M	-	-	-	-	-	S	A	-
Ashy-tailed Swift	-	M	-	M	-	M	M	-	-	M	M	M	M	M
Short-tailed Swift	-	-	-	-	-	-	A	-	-	-	-	-	-	-
Lesser Swallow-tailed Swift	-	-	-	-	-	-	-	-	-	-	-	-	-	X
Fork-tailed Palm-Swift	-	-	-	-	-	-	S	S	-	S	X	-	X	S
Rufous-breasted Hermit	-	-	-	-	-	-	M	M	H	-	-	-	A	-
Red-billed Long-tailed Hermit	-	-	-	-	-	-	-	-	-	-	-	-	-	S
Scale-throated Hermit	-	-	-	S	-	X	-	-	-	-	-	-	-	-
White-bearded Hermit	-	-	-	-	-	-	-	-	H	-	-	-	-	-
Planalto Hermit	-	-	-	-	-	(A)	S	X	S	X	-	X	X	X
Buff-bellied Hermit	-	-	-	-	-	-	-	X	S	-	-	-	-	-
Cinnamon-throated Hermit	-	-	-	-	-	-	-	-	X	-	-	-	-	S
Reddish Hermit	-	-	-	-	-	-	-	A	-	-	-	-	S	S
Swallow-tailed Hummingbird	-	-	-	-	-	X	S	X	X	X	X	X	M	X
Black Jacobin	-	S	-	S	-	-	-	-	-	-	-	H	A	-
White-vented Violetear	-	-	-	H	-	A	-	-	H	X	X	X	X	M
Green-throated Mango	-	-	-	-	-	-	-	-	-	-	-	-	-	S

| | Tai | | Apa | | FdI | | Tra | | CdG | | Can | | Bra | |
		Mos		Uru		CtC		Bar		Ema		Cip		Ara
Black-throated Mango	-	M	M	-	M	M	M	M	M	-	-	H	M	M
Ruby-topaz Hummingbird	-	-	-	-	-	-	M	-	H	M	-	H	-	M
Violet-crested Plovercrest	S	S	X	-	S	-	-	-	-	-	-	-	-	-
Dot-eared Coquette	-	-	-	-	-	-	-	-	S	S	-	-	-	S
E Frilled Coquette	-	-	-	-	-	-	-	-	-	M	-	M	M	M
Blue-chinned Sapphire	-	-	-	-	-	-	-	-	-	-	-	-	-	S
Blue-tailed Emerald	-	-	-	-	-	-	-	-	-	-	-	-	-	S
Glittering-bellied Emerald	M	M	M	M	M	X	X	X	X	X	X	X	X	-
Fork-tailed Woodnymph	-	-	-	-	-	M	H	M	M	X	X	X	X	S
Violet-capped Woodnymph	-	-	-	X	-	X	-	-	-	-	X	H	-	-
Rufous-throated Sapphire	-	-	-	-	-	{M}	-	M	H	M	-	-	-	-
White-chinned Sapphire	-	-	-	-	-	-	-	-	-	-	-	-	-	M
Gilded Sapphire	X	X	-	X	M	M	M	M	M	M	-	-	A	-
White-throated Hummingbird	-	S	X	-	{S}	-	-	-	-	-	-	-	-	-
White-tailed Goldenthroat	-	-	-	-	{A}	M	M	H	H	X	X	X	A	X
Many-spotted Hummingbird	-	-	-	-	-	-	-	H	-	-	-	-	-	-
White-bellied Hummingbird	-	-	-	-	-	-	-	H	A	-	-	-	-	-
Versicolored Emerald	-	S	S	-	X	-	S	X	X	-	-	-	A	X
Glittering-throated Emerald	-	-	-	-	-	-	-	M	M	M	-	-	S	X
"Big" Glittering-thr. Emerald	-	A	-	-	-	-	-	-	-	-	-	-	-	-
Sapphire-spangled Emerald	-	-	-	-	-	-	-	-	-	A	-	M	-	M
Plain-bellied Emerald	-	-	-	-	-	-	-	-	-	-	-	-	-	X
E Sombre Hummingbird	-	A	A	-	-	-	-	-	-	-	-	H	-	-
E Hyacinth Visorbearer	-	-	-	-	-	-	-	-	-	-	-	X	-	-
Horned Sungem	-	-	-	-	-	-	-	M	M	-	-	M	M	-
Long-billed Starthroat	-	-	-	-	-	-	-	-	-	-	-	-	M	M
E Stripe-breasted Starthroat	-	-	-	-	-	-	-	-	-	-	S	H	-	-
Blue-tufted Starthroat	H	-	-	-	M	{M}	M	M	M	M	-	-	-	-
Amethyst Woodstar	H	-	H	-	-	-	M	M	H	M	-	M	M	M
Black-tailed Trogon	-	-	-	-	-	-	-	H	-	-	-	-	-	S
White-tailed Trogon	-	-	-	-	-	-	-	-	-	A	-	-	-	X
Black-throated Trogon	-	-	-	-	-	X	-	-	-	-	-	-	-	-
Surucua Trogon	-	-	H	X	-	X	-	-	-	-	-	S	-	-
"Brazilian" Trogon	-	-	-	-	-	-	-	-	-	-	-	S	-	-
Blue-crowned Trogon	-	-	-	-	-	X	X	X	X	-	-	-	-	X
Ringed Kingfisher	X	X	X	X	X	X	X	X	X	X	X	X	X	X
Amazon Kingfisher	X	X	S	X	X	X	X	X	H	X	X	X	X	X
Green Kingfisher	X	X	X	X	X	X	X	X	H	-	X	X	X	X
Green-and-rufous Kingfisher	-	-	-	-	-	S	S	S	-	-	-	-	-	S
American Pygmy Kingfisher	-	-	-	-	-	S	S	S	S	-	-	-	-	S
Rufous-capped Motmot	-	-	H	-	-	X	-	-	-	-	S	H	S	-
Blue-crowned Motmot	-	-	-	-	-	S	S	S	X	-	-	-	-	S

	Tai	Mos	Apa	Uru	FdI	CtC	Tra	Bar	CdG	Ema	Can	Cip	Bra	Ara
Brown Jacamar	-	-	-	-	-	S	-	S	S	-	-	-	-	-
E Three-toed Jacamar	-	-	-	-	-	-	-	-	-	-	-	H	-	-
Rufous-tailed Jacamar	-	-	-	-	-	X	X	X	X	X	X	H	X	X
Buff-bellied Puffbird	-	-	-	-	X	-	-	-	-	-	-	-	-	-
Pied Puffbird	-	-	-	-	-	-	-	-	-	-	-	-	-	S
White-eared Puffbird	-	-	-	-	X	H	S	S	X	X	X	X	X	X
Spot-backed Puffbird	-	-	-	-	-	X	X	-	X	-	-	-	-	X
E Crescent-chested Puffbird	-	-	-	-	-	-	-	-	-	-	-	H	-	-
Rusty-breasted Nunlet	-	-	-	-	-	S	-	-	-	-	-	H	-	-
Rufous-capped Nunlet	-	-	-	-	-	-	-	H	S	-	-	-	-	-
Black-fronted Nunbird	-	-	-	-	-	-	X	X	X	-	-	-	-	X
Swallow-wing	-	-	-	-	-	-	A	X	S	-	-	H	-	X
Black-necked Aracari	-	-	-	-	-	-	-	-	-	-	-	H	-	S
Chestnut-eared Aracari	-	-	-	-	X	X	X	X	X	-	-	H	-	S
Lettered Aracari	-	-	-	-	-	-	-	-	S	-	-	-	-	S
Spot-billed Toucanet	-	-	-	H	-	X	-	-	-	-	-	H	-	-
Saffron Toucanet	-	-	-	-	-	X	-	-	-	-	-	-	-	-
Red-breasted Toucan	-	-	-	S	-	X	-	-	-	-	X	H	X	-
Channel-billed Toucan	-	-	-	-	-	-	-	-	-	-	-	-	-	S
Yellow-ridged Toucan	-	-	-	-	-	-	-	-	X	-	-	-	-	-
Cuvier's Toucan	-	-	-	-	-	-	-	-	-	-	-	-	-	S
Toco Toucan	H	H	S	-	X	X	X	X	X	X	X	X	X	X
Mottled Piculet	-	-	-	A	S	-	-	-	-	-	-	-	-	-
White-wedged Piculet	-	-	-	-	-	X	X	X	X	X	-	H	X	X
White-barred Piculet	-	-	-	-	-	-	-	-	-	X	X	X	-	-
"Pilcomayo" Piculet	-	-	-	-	-	-	H	-	-	-	-	-	-	-
Ochre-collared Piculet	-	-	-	S	-	X	-	-	-	-	-	-	-	-
Campo Flicker	-	-	-	-	-	X	X	X	X	X	X	X	X	X
"Field" Flicker	X	X	X	X	X	-	-	-	-	-	-	-	-	-
Green-barred Woodpecker	-	X	X	X	X	X	X	S	X	X	X	X	X	X
Yellow-throated Woodpecker	-	-	-	-	-	-	-	-	-	-	-	-	-	X
White-throated Woodpecker	-	-	-	-	-	-	-	-	-	-	-	-	-	S
Golden-green Woodpecker	-	-	-	-	-	X	X	X	-	X	-	-	-	X
White-browed Woodpecker	-	-	-	S	-	S	-	-	-	-	-	-	-	-
Blond-crested Woodpecker	-	-	S	X	-	X	-	-	-	X	X	H	X	X
Pale-crested Woodpecker	-	-	-	-	-	X	X	X	X	-	-	-	-	-
Cream-colored Woodpecker	-	-	-	-	-	-	A	A	-	-	-	-	-	X
Ringed Woodpecker	-	-	-	-	-	-	-	S	-	-	-	-	-	-
Lineated Woodpecker	H	-	S	-	X	X	X	X	X	X	X	X	X	X
Helmeted Woodpecker	-	-	-	H	-	S	-	-	-	-	-	-	-	-
Yellow-tufted Woodpecker	-	-	-	-	-	-	-	-	X	X	-	-	-	X
Yellow-fronted Woodpecker	-	-	-	-	-	X	-	-	-	-	-	H	X	-

	Tai	Mos	Apa	Uru	FdI	CtC	Tra	Bar	CdG	Ema	Can	Cip	Bra	Ara
White Woodpecker	-	-	-	-	{S}	X	X	X	X	X	X	X	X	X
White-fronted Woodpecker	-	-	-	-	-	A	-	-	-	-	-	-	-	-
White-spotted Woodpecker	X	S	X	X	X	-	-	-	-	-	-	S	-	-
Little Woodpecker	-	-	-	-	-	X	X	X	X	-	X	X	X	X
Red-stained Woodpecker	-	-	-	-	-	A	-	-	-	-	-	-	-	-
E Yellow-eared Woodpecker	-	-	-	-	-	-	-	-	-	-	-	H	-	-
Checkered Woodpecker	-	-	-	-	X	-	H	S	-	S	-	S	S	-
Crimson-crested Woodpecker	-	-	-	-	-	X	X	X	X	-	-	H	S	X
Cream-backed Woodpecker	-	-	-	H	{A}	X	-	-	-	-	-	-	-	-
Red-necked Woodpecker	-	-	-	-	-	-	-	-	S	S	-	-	-	-
Robust Woodpecker	-	-	-	H	-	X	-	A	-	-	-	H	-	X
Plain-brown Woodcreeper	-	-	-	-	-	-	-	S	-	-	-	-	-	S
Thrush-like Woodcreeper	-	-	S	-	-	X	S	-	-	-	-	-	-	-
Olivaceous Woodcreeper	-	-	X	X	-	X	X	X	X	X	X	X	-	X
Wedge-billed Woodcreeper	-	-	-	-	-	-	-	-	A	-	-	-	-	-
Long-billed Woodcreeper	-	-	-	-	-	-	-	-	-	-	-	-	-	X
White-throated Woodcreeper	-	-	X	-	X	-	-	-	-	-	X	-	X	-
Great Rufous Woodcreeper	-	-	-	-	-	-	X	X	X	-	-	-	-	-
Pale-billed Woodcreeper	-	-	-	-	-	S	-	X	-	-	-	-	-	-
Planalto Woodcreeper	-	X	X	-	X	S	S	-	X	S	S	H	-	-
Straight-billed Woodcreeper	-	-	-	-	-	-	X	X	S	-	-	-	-	X
Buff-throated Woodcreeper	-	-	-	-	-	H	X	X	X	-	-	-	-	S
Narrow-billed Woodcreeper	-	-	-	X	-	X	X	X	X	X	X	X	X	X
Scaled Woodcreeper	-	-	X	-	{S}	-	-	-	S	-	S	-	-	-
Lesser Woodcreeper	-	X	X	-	X	-	A	-	-	-	-	-	-	-
Scimitar-billed Woodcreeper	-	-	-	X	-	-	-	-	-	-	-	-	-	-
Red-billed Scythebill	-	-	-	-	-	-	X	X	X	-	-	-	-	-
Black-billed Scythebill	-	-	S	-	A	-	-	-	-	-	-	-	-	-
Campo Miner	-	-	-	-	-	-	-	-	H	X	X	H	X	-
Common Miner	X	X	-	X	-	-	-	-	-	-	-	-	-	-
Bar-winged Cinclodes	M	M	-	M	-	-	-	-	-	-	-	-	-	-
E Long-tailed Cinclodes	-	-	X	-	-	-	-	-	-	-	-	-	-	-
Rufous Hornero	X	X	X	X	X	X	X	X	X	X	X	X	X	X
Pale-legged Hornero	-	-	-	-	-	X	X	X	S	-	-	-	A	X
E Wing-banded Hornero	-	-	-	-	-	-	-	-	-	-	X	-	-	X
Curve-billed Reedhaunter	X	X	-	-	-	-	-	-	-	-	-	-	-	-
Straight-billed Reedhaunter	-	-	X	-	-	-	-	-	-	-	-	-	-	-
Wren-like Rushbird	X	X	-	-	{A}	-	-	-	-	-	-	-	-	-
E Striolated Tit-Spinetail	-	A	S	-	-	-	-	-	-	-	-	-	-	-
Tufted Tit-Spinetail	-	-	-	-	X	-	-	-	-	-	-	-	-	-
Araucaria Tit-Spinetail	-	-	X	-	-	-	-	-	-	-	-	-	-	-
Chotoy Spinetail	S	S	-	X	-	X	X	X	-	-	-	-	-	-

	Tai	Mos	Apa	Uru	FdI	CtC	Tra	Bar	CdG	Ema	Can	Cip	Bra	Ara
Rufous-capped Spinetail	-	-	-	X	-	X	-	-	-	-	-	-	-	-
Sooty-fronted Spinetail	-	M	-	X	-	X	S	S	X	X	X	X	X	-
Chicli Spinetail	X	X	X	-	X	-	-	-	-	-	X	X	-	-
Cinereous-breasted Spinetail	-	-	-	-	-	S	S	-	-	-	-	-	-	S
Pale-breasted Spinetail	-	-	-	-	A	-	H	S	-	S	-	X	-	X
White-lored Spinetail	-	-	-	-	-	-	X	X	X	-	X	-	-	-
Gray-bellied Spinetail	-	H	-	S	-	X	-	-	-	-	-	-	-	-
Ochre-cheeked Spinetail	-	-	-	-	-	-	H	-	S	S	-	-	S	-
Sulphur-bearded Spinetail	-	S	-	-	-	-	-	-	-	-	-	-	-	-
E Gray-headed Spinetail	-	-	-	-	-	-	-	-	-	-	-	-	-	S
Stripe-crowned Spinetail	-	X	-	-	X	-	-	-	-	-	-	-	-	-
Olive Spinetail	-	X	X	X	-	X	-	-	-	-	-	-	-	-
Rusty-backed Spinetail	-	-	-	-	-	-	X	X	X	-	X	-	-	X
E Pallid Spinetail	-	-	-	-	-	-	-	-	-	-	-	H	X	-
Yellow-chinned Spinetail	-	X	X	X	X	X	X	X	X	-	-	X	X	X
Bay-capped Wren-Spinetail	-	S	-	-	-	-	-	-	-	-	-	-	-	-
Hudson's Canastero	-	A	-	-	-	-	-	-	-	-	-	-	-	-
Short-billed Canastero	-	-	-	-	X	-	-	-	-	-	-	-	-	-
E Cipo Canastero	-	-	-	-	-	-	-	-	-	-	-	X	-	-
Canebrake Groundcreeper	-	-	-	-	-	{S}	-	-	-	-	-	-	-	-
Rufous-fronted Thornbird	-	-	-	-	-	X	X	X	X	X	X	X	X	-
E Rufous-breasted Thornbird	-	S	X	-	-	-	-	-	-	-	-	-	-	-
Greater Thornbird	-	-	-	X	-	X	X	X	S	S	-	-	X	X
Freckle-breasted Thornbird	-	X	S	-	-	-	-	-	-	-	-	-	-	-
Lark-like Brushrunner	-	-	-	-	X	-	-	-	-	-	-	-	-	-
Firewood-gatherer	-	X	X	X	X	-	-	-	-	-	X	X	X	-
Point-tailed Palmcreeper	-	-	-	-	-	-	-	-	-	A	-	-	-	-
Rufous Cacholote	-	-	-	-	-	-	X	X	X	-	-	-	-	-
Brown Cacholote	-	-	-	-	X	-	-	-	-	-	-	-	-	-
Buff-browed Foliage-gleaner	X	X	X	S	S	-	-	-	-	-	-	-	-	-
White-browed Foliage-gleaner	-	-	-	-	{H}	-	-	-	-	-	-	-	-	-
Black-capped Foliage-gleaner	-	S	S	-	X	-	-	-	-	-	-	-	-	-
Rufous-rumped Foliage-gleaner	-	-	-	-	-	-	-	-	-	-	-	-	-	S
Russet-mantled Foliage-gleaner	-	-	-	-	-	-	-	-	-	-	X	S	-	X
Ochre-breasted Foliage-gleaner	-	-	-	-	X	S	-	-	-	-	-	-	-	-
Buff-fronted Foliage-gleaner	-	-	-	-	S	-	S	-	S	-	X	H	X	-
White-eyed Foliage-gleaner	-	A	-	-	X	-	-	-	-	-	-	H	X	-
Chestnut-capped Foliage-glean.	-	-	-	-	-	S	S	-	-	A	-	H	S	-
Sharp-billed Treehunter	-	-	-	X	-	S	-	-	-	-	-	-	-	-
Streaked Xenops	-	S	X	-	X	S	S	S	X	-	X	-	X	S
Plain Xenops	-	-	-	-	-	S	-	-	-	S	-	-	-	X
Rufous-breasted Leaftosser	-	S	X	-	S	-	-	-	-	H	-	H	-	-

-131-

	Tai	Mos	Apa	Uru	FdI	CtC	Tra	Bar	CdG	Ema	Can	Cip	Bra	Ara
Sharp-tailed Streamcreeper	-	-	X	-	X	-	-	-	X	-	X	X	X	-
Spot-backed Antshrike	-	-	-	-	X	-	-	-	-	-	-	H	-	-
Giant Antshrike	-	-	X	-	S	-	-	-	-	-	-	-	-	-
Large-tailed Antshrike	-	-	X	-	X	-	-	-	-	-	-	-	-	-
Tufted Antshrike	-	-	-	-	X	-	-	-	-	-	-	-	-	-
Great Antshrike	-	-	-	-	-	X	X	X	X	-	-	S	-	X
E Glossy Antshrike	-	-	-	-	-	-	-	-	-	-	-	-	-	X
White-bearded Antshrike	-	-	-	-	S	-	-	-	-	-	-	-	-	-
Barred Antshrike	-	-	-	-	{A}	X	X	X	X	X	-	-	-	X
Eastern Slaty-Antshrike	-	-	-	-	-	X	S	X	X	-	X	X	-	-
Amazonian Antshrike	-	-	-	-	-	-	-	H	-	-	-	-	-	S
Variable Antshrike	X	X	X	S	X	X	-	-	S	-	X	X	X	-
Rufous-winged Antshrike	-	-	-	-	-	H	H	-	X	X	X	X	X	-
Rufous-capped Antshrike	-	-	X	X	X	X	{H}	-	-	-	-	-	-	-
Spot-breasted Antvireo	-	-	-	-	-	{H}	-	-	-	-	-	-	-	-
Plain Antvireo	-	-	X	X	-	X	-	X	H	X	X	H	-	X
Cinereous Antshrike	-	-	-	-	-	-	-	-	-	-	-	-	-	S
Streaked Antwren	-	-	-	-	-	-	-	-	-	-	-	-	-	S
White-flanked Antwren	-	-	-	-	-	-	-	A	-	-	-	-	-	X
Long-winged Antwren	-	-	-	-	-	-	-	-	-	-	-	-	-	S
E Unicolored Antwren	-	S	-	-	-	-	-	-	-	-	-	-	-	-
Stripe-backed Antbird	-	-	-	-	-	X	S	-	A	-	-	-	-	-
Black-capped Antwren	-	-	-	-	-	X	-	-	-	-	-	X	X	X
Large-billed Antwren	-	-	-	-	-	H	S	X	X	X	-	-	S	X
Rufous-winged Antwren	-	-	-	-	X	-	-	-	-	-	-	-	-	-
White-fringed Antwren	-	-	-	-	-	-	-	S	-	-	-	-	-	X
Black-bellied Antwren	-	-	-	-	-	H	-	X	-	-	-	-	-	S
E Serra Antwren	-	-	-	-	-	-	-	-	-	-	-	H	-	-
Rusty-backed Antwren	-	-	-	-	-	X	X	X	X	-	-	H	-	X
Bertoni's Antbird	-	-	-	S	X	-	-	-	-	-	-	-	-	-
Striated Antbird	-	-	-	-	-	-	-	S	-	-	-	-	-	-
Dusky-tailed Antbird	-	-	-	S	S	-	-	-	-	-	-	-	-	-
Streak-capped Antwren	-	-	-	-	X	-	-	-	-	-	-	-	-	-
E Bananal Antbird	-	-	-	-	-	-	-	-	-	-	-	-	-	X
Mato Grosso Antbird	-	-	-	-	-	X	X	X	H	-	-	-	-	-
White-backed Fire-eye	-	-	-	-	-	S	-	X	X	-	-	-	-	-
White-shouldered Fire-eye	-	-	-	-	X	-	-	-	-	-	X	H	-	-
Black-faced Antbird	-	-	-	-	-	-	-	A	-	-	-	-	-	-
Warbling Antbird	-	-	-	-	-	-	-	A	-	-	-	-	-	-
Band-tailed Antbird	-	-	-	-	-	X	X	-	-	-	-	-	-	X
Black-throated Antbird	-	-	-	-	-	-	-	H	H	S	-	-	-	-
E Squamate Antbird	-	-	X	X	-	-	-	-	-	-	-	-	-	-

Species	Tai	Mos	Apa	Uru	FdI	CtC	Tra	Bar	CdG	Ema	Can	Cip	Bra	Ara
Plain-backed Antbird	-	-	-	-	-	-	-	-	H	-	-	-	-	-
Short-tailed Antthrush	-	-	X	-	X	-	-	-	-	-	-	H	-	-
E Brazilian Antthrush	-	-	S	-	-	-	-	-	-	-	-	-	-	-
Rufous-capped Antthrush	-	-	S	-	-	-	-	-	-	-	-	-	-	-
Variegated Antpitta	-	-	S	-	X	-	-	-	-	-	-	-	-	-
Speckle-breasted Antpitta	-	-	X	-	{S}	-	-	-	-	-	-	-	-	-
Rufous Gnateater	-	-	X	X	-	X	-	-	-	-	X	-	X	-
Collared Crescentchest	-	-	-	-	-	-	-	-	S	S	-	S	X	S
Spotted Bamboowren	-	-	-	-	-	S	-	-	-	-	-	-	-	-
Mouse-colored Tapaculo	-	-	X	-	{X}	-	-	-	-	-	-	-	-	-
E Brasilia Tapaculo	-	-	-	-	-	-	-	-	-	S	S	S	-	-
E White-breasted Tapaculo	-	-	A	-	-	-	-	-	-	-	-	-	-	-
Swallow-tailed Cotinga	-	H	-	M	-	S	-	-	-	-	-	-	-	-
E Hooded Berryeater	-	-	S	S	-	-	-	-	-	-	-	-	-	-
Screaming Piha	-	-	-	-	-	-	-	-	-	-	-	-	-	X
E Cinnamon-vented Piha	-	-	-	-	-	-	-	-	-	-	-	A	-	-
White-naped Xenopsaris	-	-	-	-	-	S	S	-	-	-	-	-	-	S
Green-backed Becard	-	-	S	X	S	X	S	X	X	-	-	H	S	S
Chestnut-crowned Becard	-	-	S	-	X	-	-	-	-	-	X	-	-	-
White-winged Becard	M	M	M	M	X	X	X	H	X	X	X	-	X	S
Black-capped Becard	-	-	-	-	-	-	-	-	H	-	-	H	-	-
Crested Becard	-	-	M	-	M	X	X	X	X	-	-	X	A	X
Black-tailed Tityra	-	-	M	-	S	X	X	X	X	-	-	H	X	X
Masked Tityra	-	-	-	-	-	-	-	X	X	-	-	-	-	-
Black-crowned Tityra	-	-	-	A	-	X	X	S	X	X	-	-	X	X
Purple-throated Fruitcrow	-	-	-	-	-	-	-	-	-	-	-	-	-	S
Red-ruffed Fruitcrow	-	-	-	H	-	X	-	-	-	-	S	H	S	-
Amazonian Umbrellabird	-	-	-	-	-	-	H	-	-	-	-	-	-	-
Bare-necked Fruitcrow	-	-	-	-	-	-	-	-	S	-	-	-	-	S
Bare-throated Bellbird	-	-	-	M	-	S	-	-	-	-	-	H	-	-
Sharpbill	-	-	-	-	-	S	-	-	-	-	-	-	-	-
White-tipped Plantcutter	-	-	-	A	-	-	-	-	-	-	-	-	-	-
Red-headed Manakin	-	-	-	-	-	-	-	-	-	-	-	-	-	X
Band-tailed Manakin	-	-	-	-	S	H	S	X	X	-	-	H	-	X
Helmeted Manakin	-	-	-	-	-	-	X	H	X	X	X	H	X	-
Blue-backed Manakin	-	-	-	-	-	-	-	-	-	-	-	-	-	X
Swallow-tailed Manakin	-	-	S	X	X	-	X	-	-	-	-	H	-	-
E Pin-tailed Manakin	-	-	-	-	-	-	-	-	-	-	S	H	-	-
White-bearded Manakin	-	-	-	-	X	-	-	-	-	-	-	H	-	-
Fiery-capped Manakin	-	-	-	-	-	-	-	-	M	-	-	-	-	S
Flame-crested Manakin	-	-	-	-	-	-	-	-	-	-	-	-	-	S
Pale-bellied Tyrant-Manakin	-	-	-	-	-	S	-	-	S	S	-	H	-	X

	Tai	Mos	Apa	Uru	FdI	CtC	Tra	Bar	CdG	Ema	Can	Cip	Bra	Ara
E Black-capped Manakin	-	-	-	S	-	-	-	-	-	-	-	-	-	-
Wing-barred Manakin	-	-	-	-	X	-	-	-	-	-	-	-	-	-
Greenish Manakin	-	-	S	X	-	X	-	-	-	-	-	-	X	-
Thrush-like Manakin	-	-	-	-	-	-	-	-	-	-	-	-	-	S
Chocolate-vented Tyrant	A	-	-	-	-	-	-	-	-	-	-	-	-	-
Black-and-white Monjita	X	X	X	-	{S}	-	-	-	-	-	-	-	-	-
Gray Monjita	X	X	X	X	-	H	X	X	X	X	X	X	X	S
White-rumped Monjita	-	-	-	-	-	X	X	X	S	X	X	X	X	S
Black-crowned Monjita	-	-	-	M	-	-	-	-	-	-	-	-	-	-
White Monjita	X	X	S	X	-	X	-	-	-	-	-	-	-	-
Patagonain Negrito	M	M	-	M	-	-	-	-	-	-	-	-	-	-
Long-tailed Tyrant	-	-	-	-	X	-	S	S	S	-	X	H	X	S
Streamer-tailed Tyrant	-	-	-	-	{S}	H	-	-	S	X	X	H	X	-
Cock-tailed Tyrant	-	-	-	-	-	H	-	-	-	X	X	-	X	-
Strange-tailed Tyrant	-	H	-	-	{A}	-	-	-	-	-	-	-	-	-
Crested Black-Tyrant	-	S	-	X	-	-	-	-	X	A	X	X	X	-
E Velvety Black-Tyrant	-	-	-	S	-	-	-	-	-	-	X	X	-	-
White-winged Black-Tyrant	-	-	-	-	-	-	-	-	-	-	A	-	-	-
Riverside Tyrant	-	-	-	-	-	-	-	-	-	-	-	-	-	X
Blue-billed Black-Tyrant	X	X	X	X	{S}	H	A	-	-	-	S	H	-	-
Amazonian Black-Tyrant	-	-	-	-	-	-	-	-	-	-	-	-	-	S
Hudson's Black-Tyrant	-	-	-	-	-	-	-	-	H	-	-	-	-	-
Cinereous Tyrant	-	-	-	-	-	H	-	-	-	-	-	-	-	-
Spectacled Tyrant	X	X	-	X	{A}	-	-	-	-	-	-	-	-	-
Shear-tailed Gray Tyrant	-	-	-	M	{A}	-	-	-	-	-	-	H	-	-
Black-backed Water-Tyrant	-	-	-	-	-	X	X	X	-	-	-	-	A	X
Masked Water-Tyrant	-	-	-	-	-	-	-	-	-	-	X	X	-	-
White-headed Marsh-Tyrant	X	X	-	-	{S}	X	X	X	-	-	-	X	X	X
Vermilion Flycatcher	M	M	-	M	M	M	M	M	M	M	M	H	M	M
Yellow-browed Tyrant	X	X	X	X	{S}	M	M	M	-	-	X	H	M	M
Cattle Tyrant	X	X	X	X	X	X	X	X	X	X	X	-	M	X
Sirystes	-	-	M	-	X	-	-	X	X	-	X	H	A	S
Fork-tailed Flycatcher	M	M	M	M	M	M	M	M	M	M	M	M	M	M
Eastern Kingbird	-	-	-	-	M	-	-	-	-	-	-	-	-	-
Tropical Kingbird	M	M	M	M	M	X	X	X	X	X	X	X	M	X
White-throated Kingbird	-	-	-	-	-	-	M	M	H	M	M	H	M	M
Variegated Flycatcher	-	-	M	M	-	M	M	M	M	M	-	-	M	M
Crowned Slaty Flycatcher	-	-	-	M	-	M	M	M	H	M	-	H	M	M
Piratic Flycatcher	-	-	-	M	-	M	M	M	M	M	M	-	A	M
Three-striped Flycatcher	-	-	-	-	X	-	-	-	-	-	-	-	-	-
Boat-billed Flycatcher	-	-	M	-	-	M	X	X	X	X	X	X	X	X
Streaked Flycatcher	-	-	M	M	M	M	M	M	M	M	M	-	M	M

	Tai		Apa		FdI		Tra		CdG		Can		Bra	
		Mos		Uru		CtC		Bar		Ema		Cip		Ara
Rusty-margined Flycatcher	-	-	-	-	-	-	X	X	X	X	-	-	-	X
Social Flycatcher	-	M	-	-	M	M	M	M	M	M	X	X	X	S
Dusky-chested Flycatcher	-	-	-	-	-	-	-	-	-	-	-	-	-	S
Sulphury Flycatcher	-	-	-	-	-	-	-	-	A	-	-	-	-	S
Great Kiskadee	X	X	X	X	X	X	X	X	X	X	X	X	X	X
Lesser Kiskadee	-	-	-	-	-	-	X	X	X	-	-	-	-	X
White-eyed Attila	-	-	-	-	-	-	-	S	H	H	-	-	-	-
E Gray-hooded Attila	-	-	S	-	-	-	-	-	-	-	-	-	-	-
Cinnamon Attila	-	-	-	-	-	-	-	-	-	-	-	-	-	A
Rufous-tailed Attila	-	-	-	M	-	-	H	-	-	-	-	-	-	-
Rufous Casiornis	-	-	-	-	-	M	M	M	M	X	S	H	A	X
E Ash-throated Casiornis	-	-	-	-	-	-	-	-	-	-	-	-	-	H
Grayish Mourner	-	-	-	-	-	-	-	-	S	-	-	-	-	S
Short-crested Flycatcher	-	-	-	-	-	S	X	X	X	X	X	X	X	X
Brown-crested Flycatcher	-	-	-	-	H	{M}	M	M	M	X	X	X	S	X
Swainson's Flycatcher	A	-	M	M	M	M	M	M	M	M	M	M	M	M
Dusky-capped Flycatcher	-	-	-	-	-	-	S	X	S	-	S	-	-	-
Tropical Pewee	-	-	-	-	-	X	-	S	S	-	X	-	H	S
Euler's Flycatcher	M	M	M	M	M	M	M	-	X	-	X	-	X	X
Fuscous Flycatcher	-	M	-	-	-	-	-	-	-	-	-	-	-	-
"Audible" Fuscous-Flycatcher	-	-	-	-	M	X	X	X	X	-	-	X	X	S
Sulphur-rumped Flycatcher	-	-	-	-	-	-	-	-	-	-	-	S	-	-
Black-tailed Flycatcher	-	-	-	-	-	-	-	-	-	-	-	-	A	-
Bran-colored Flycatcher	M	M	M	M	M	X	X	X	X	-	X	X	X	X
Swallow Flycatcher	-	-	-	M	-	-	H	-	X	-	X	X	X	S
Atlantic Royal-Flycatcher	-	-	-	-	A	-	-	-	-	-	-	-	-	-
Russet-winged Spadebill	-	-	S	-	-	S	-	-	-	-	-	-	-	-
White-throated Spadebill	-	X	X	-	X	H	-	-	X	X	X	H	X	-
Yellow-olive Flycatcher	-	X	X	-	X	S	X	X	X	X	X	H	X	X
Yellow-margined Flycatcher	-	-	-	-	-	-	-	S	S	-	-	-	-	-
Yellow-breasted Flycatcher	-	-	-	-	-	-	-	-	-	-	-	-	S	X
Rufous-tailed Flatbill	-	-	-	-	-	-	-	-	H	-	-	-	-	S
Large-headed Flatbill	-	-	-	-	S	-	-	-	-	-	-	-	-	-
E Yellow-lored Tody-Flycatcher	-	-	-	-	-	-	-	-	-	-	-	S	-	-
Common Tody-Flycatcher	-	-	-	-	-	S	S	S	S	X	X	-	X	X
Spotted Tody-Flycatcher	-	-	-	-	-	-	-	-	-	-	-	-	-	X
Smoky-fronted Tody-Flycatcher	-	-	-	-	-	-	-	-	-	-	-	-	-	S
Ochre-faced Tody-Flycatcher	X	X	X	-	S	-	-	-	-	-	X	-	-	-
Rusty-fronted Tody-Flycatcher	-	-	-	-	-	X	X	X	S	-	-	-	-	S
Stripe-necked Tody-Tyrant	-	-	-	-	-	-	S	H	H	-	-	-	-	S
Pearly-vented Tody-Tyrant	-	-	-	-	{S}	X	X	X	X	-	X	-	X	S
E Eye-ringed Tody-Tyrant	-	-	S	-	-	-	-	-	-	-	-	-	-	-

Species	Tai	Mos	Apa	Uru	FdI	CtC	Tra	Bar	CdG	Ema	Can	Cip	Bra	Ara
Drab-breasted Bamboo-Tyrant-	-	-	-	-	S	-	-	-	-	-	-	-	-	-
E Brown-breasted Bamboo-Tyrant	-	-	X	-	-	-	-	-	-	-	-	-	-	-
Eared Pygmy-Tyrant -	-	-	-	-	-	X	-	-	-	-	-	-	-	-
Southern Bristle-Tyrant-	-	-	-	-	-	X	-	-	-	-	-	-	-	-
Bay-ringed Tyrannulet -	-	-	-	-	-	X	-	-	-	-	-	-	-	-
Mottle-cheeked Tyrannulet -	-	X	X	X	-	S	-	-	-	-	-	S	-	-
E Serra do Mar Tyrannulet-	-	-	S	-	-	-	-	-	-	-	-	-	-	-
Sao Paulo Tyrannulet	-	-	-	-	-	X	-	-	-	-	-	-	-	-
Yellow Tyrannulet -	-	-	-	-	-	S	-	-	S	-	-	X	H	-
Tawny-crowned Pygmy-Tyrant -	-	-	-	S	{X}	X	X	X	S	-	-	S	-	-
Rufous-sided Pygmy-Tyrant -	-	-	-	-	-	H	-	-	-	-	-	-	A	-
Crested Doradito -	X	-	-	-	{A}	-	-	-	-	-	-	-	-	-
Warbling Doradito -	X	X	-	-	{M}	-	-	-	-	-	-	-	-	-
Bearded Tachuri-	-	-	-	-	-	-	H	A	-	H	A	-	-	-
E Gray-backed Tachuri-	-	-	-	-	-	-	-	-	-	-	S	X	-	-
Sharp-tailed Tyrant-	-	-	-	-	-	-	-	-	H	X	X	-	X	-
Many-colored Rush-Tyrant	-	X	X	-	-	A	-	-	-	-	-	-	-	-
White-crested Tyrannulet-	-	X	X	X	X	X	X	X	X	X	-	X	X	M
White-bellied Tyrannulet-	-	A	-	-	-	-	H	A	-	-	-	-	-	-
Sooty Tyrannulet -	X	X	X	X	X	-	-	-	-	X	X	X	-	-
Pale-tipped Tyrannulet -	-	-	-	-	-	-	-	-	-	-	-	-	-	X
Plain Tyrannulet -	-	-	-	-	-	-	H	X	X	-	-	-	-	-
Yellow-bellied Elaenia -	-	S	-	-	X	H	X	X	X	X	X	X	X	S
Large Elaenia -	-	-	-	M	{M}	M	M	-	-	-	-	H	-	M
White-crested Elaenia -	-	-	-	-	-	H	-	-	M	-	M	-	-	-
Small-billed Elaenia -	M	M	M	M	{M}	M	M	M	M	M	-	M	M	-
Olivaceous Elaenia -	M	M	M	-	M	-	H	-	-	M	M	-	M	-
Plain-crested Elaenia -	-	-	-	-	-	-	S	-	X	X	X	X	X	X
Lesser Elaenia -	-	-	-	-	{M}	M	M	-	-	M	-	M	M	M
Highland Elaenia -	X	X	X	-	-	-	-	-	-	-	X	X	X	-
Forest Elaenia -	-	-	-	-	-	-	X	X	X	-	S	-	-	X
Gray Elaenia -	-	-	-	-	S	-	S	S	S	-	X	X	-	S
Greenish Elaenia -	-	-	-	-	-	M	M	M	M	M	-	-	-	M
Chaco Suiriri -	-	-	-	X	-	S	-	-	-	-	-	-	-	-
Campo Suiriri -	-	-	-	-	-	S	S	-	X	X	X	H	X	X
Southern Scrub-Flycatcher -	-	-	-	M	-	H	M	M	M	-	-	H	-	-
Mouse-colored Tyrannulet -	-	-	-	-	-	S	H	S	H	-	-	S	A	S
Southern Beardless-Tyrannulet	M	M	M	M	M	X	X	X	X	-	X	X	X	X
Greenish Tyrannulet-	-	-	-	S	-	S	-	-	-	-	-	H	-	-
Reiser's Tyrannulet-	-	-	-	-	-	-	-	-	-	-	-	-	A	A
Planalto Tyrannulet-	-	-	-	M	-	M	-	-	S	-	-	H	-	S
Rough-legged Tyrannulet-	-	-	-	-	S	-	-	-	-	-	-	-	-	-

	Tai	Mos	Apa	Uru	FdI	CtC	Tra	Bar	CdG	Ema	Can	Cip	Bra	Ara
White-lored Tyrannulet	-	-	-	-	-	-	-	-	-	-	-	-	-	S
Sepia-capped Flycatcher	-	X	X	-	X	X	X	-	X	X	X	X	X	-
Ochre-bellied Flycatcher	-	-	-	-	-	-	-	-	-	A	-	-	-	-
Gray-hooded Flycatcher	-	S	X	-	X	-	-	-	-	-	X	-	X	-
Ringed Antpipit	-	-	-	-	-	-	-	-	-	-	-	-	-	S
Southern Antpipit	-	-	-	-	X	-	H	-	S	S	-	H	X	-
White-winged Swallow	-	-	-	-	X	X	X	X	X	-	-	X	-	X
White-rumped Swallow	X	X	X	X	X	M	M	M	-	M	M	M	M	M
Chilean Swallow	M	M	-	-	{M}	-	-	-	-	-	-	-	-	-
Brown-chested Martin	M	M	M	M	M	M	M	M	M	M	M	H	M	X
Purple Martin	-	-	-	-	-	M	M	-	-	-	-	-	-	M
Gray-breasted Martin	M	M	M	M	M	M	M	M	M	M	M	M	X	M
Blue-and-white Swallow	-	X	X	X	-	X	M	M	M	M	X	X	M	X
Black-collared Swallow	-	-	-	-	X	-	-	-	-	-	-	-	-	-
Tawny-headed Swallow	-	X	X	-	X	-	-	-	-	H	M	M	M	M
Southern Rough-winged Swallow	M	M	M	-	M	M	M	M	M	M	M	M	M	M
Bank Swallow (Sand Martin)	-	M	-	-	M	-	M	-	-	A	-	-	-	M
Barn Swallow	M	M	M	-	M	M	M	M	-	M	-	-	M	M
Cliff Swallow	M	M	M	-	M	-	M	-	-	M	-	-	M	M
Azure Jay	S	-	X	-	{S}	-	-	-	-	-	-	-	-	-
Purplish Jay	-	-	-	-	-	X	X	X	X	-	-	-	-	-
Curl-crested Jay	-	-	-	-	-	H	X	-	X	X	X	X	X	S
Plush-crested Jay	-	A	-	X	X	X	S	S	-	-	-	-	-	-
E White-naped Jay	-	-	-	-	-	-	-	-	A	-	S	H	X	S
Plain-breasted Wren	-	-	-	-	-	X	X	X	X	-	-	-	-	-
Grass Wren	H	-	-	-	-	-	-	-	-	X	X	H	X	-
Moustached Wren	-	-	-	-	-	S	X	X	X	-	-	-	S	-
Buff-breasted Wren	-	-	-	-	-	S	X	X	X	-	-	-	X	X
Fawn-breasted Wren	-	-	-	-	-	X	S	H	-	S	-	-	-	-
House Wren	X	X	X	X	X	X	X	X	X	X	X	X	X	X
Black-capped Donacobius	-	-	-	-	{X}	X	X	X	-	-	X	-	A	X
Chalk-browed Mockingbird	X	X	X	X	S	X	X	X	X	X	X	X	X	X
White-banded Mockingbird	M	M	-	M	-	M	M	H	M	-	-	-	-	-
Veery	-	-	-	-	-	-	A	-	A	-	-	-	-	-
Yellow-legged Thrush	-	M	M	-	-	-	-	-	-	-	-	H	-	-
Eastern Slaty-Thrush	M	M	M	-	M	-	-	-	M	M	-	M	-	-
Rufous-bellied Thrush	X	X	X	X	X	X	X	X	X	X	X	X	X	X
Pale-breasted Thrush	-	-	S	-	X	X	S	X	X	X	X	X	X	X
Creamy-bellied Thrush	X	X	X	X	X	X	X	X	X	X	X	X	M	S
Cocoa Thrush	-	-	-	-	-	-	-	-	-	S	-	-	-	-
White-necked Thrush	S	X	X	-	X	-	S	-	X	-	-	H	M	-
Long-billed Gnatwren	-	-	-	-	-	S	-	-	S	-	-	-	-	S

Species	Tai	Mos	Apa	Uru	FdI	CtC	Tra	Bar	CdG	Ema	Can	Cip	Bra	Ara
Cream-bellied Gnatcatcher	-	-	-	-	X	-	-	-	-	-	-	-	-	-
Masked Gnatcatcher	X	X	-	X	-	X	X	X	X	X	-	-	X	X
Short-billed Pipit	X	X	-	X	-	-	-	-	-	-	-	-	-	-
Hellmayr's Pipit	S	-	X	-	-	-	-	-	-	-	X	X	-	-
Yellowish Pipit	X	X	-	X	{X}	X	X	X	-	-	X	X	X	X
Correndera Pipit	X	X	-	M	-	-	-	-	-	-	-	-	-	-
Ochre-breasted Pipit	H	H	A	-	-	-	-	-	-	-	-	-	-	-
Rufous-browed Peppershrike	X	X	X	A	X	X	X	X	X	X	X	X	X	X
Red-eyed Vireo	M	M	M	M	M	M	M	M	M	M	-	H	X	X
Rufous-crowned Greenlet	-	X	X	-	X	H	-	-	-	-	-	-	-	-
E Gray-eyed Greenlet	-	-	-	-	-	-	-	-	-	-	X	X	-	-
Ashy-headed Greenlet	-	-	-	-	-	-	X	H	S	-	-	-	-	X
Shiny Cowbird	X	X	X	X	X	X	X	X	X	X	X	X	X	X
Screaming Cowbird	X	-	-	X	S	S	-	-	A	S	-	-	-	-
Bay-winged Cowbird	X	X	S	X	S	X	X	X	-	-	-	-	-	-
Giant Cowbird	-	-	-	-	-	S	X	X	X	S	-	-	-	S
Crested Oropendola	-	-	-	-	-	S	X	X	X	X	X	X	X	-
Yellow-rumped Cacique	-	-	-	-	-	S	S	S	S	-	-	-	S	X
Red-rumped Cacique	-	-	-	-	-	X	H	-	-	-	-	H	-	S
Golden-winged Cacique	S	S	X	X	{S}	X	S	-	-	-	-	-	-	-
Solitary Cacique	-	-	-	-	-	S	X	X	X	S	-	-	-	X
Chopi Blackbird	S	S	X	-	S	X	X	X	X	X	X	X	X	X
Yellow-winged Blackbird	X	X	-	X	-	-	-	-	-	-	-	-	-	-
Chestnut-capped Blackbird	X	X	S	X	{S}	-	S	-	-	-	-	-	-	X
Unicolored Blackbird	-	-	-	-	{S}	X	X	X	-	-	-	-	-	X
Saffron-cowled Blackbird	H	-	X	-	-	-	-	-	-	-	-	-	-	-
Epaulet Oriole	-	S	X	X	X	X	X	X	X	X	-	-	-	X
Troupial (Campo Oriole)	-	-	-	-	-	-	-	-	-	-	-	H	-	X
"Orange-backed" Troupial	-	-	-	-	-	X	X	X	-	-	-	-	-	-
Scarlet-headed Blackbird	X	X	-	-	-	X	S	S	-	-	-	-	-	-
Yellow-rumped Marshbird	S	S	X	X	-	-	S	-	-	X	X	X	X	S
Brown-and-yellow Marshbird	X	X	S	X	-	-	A	-	-	-	-	-	-	-
White-browed Blackbird	X	X	X	X	X	X	X	H	-	X	-	-	-	S
Pampas Meadowlark	H	-	-	-	-	-	-	-	-	-	-	-	-	-
Bobolink	-	-	-	-	-	H	-	-	-	-	-	-	-	-
Prothonotary Warbler	-	-	-	-	-	-	A	-	-	-	-	-	-	-
Tropical Parula	X	X	X	X	X	X	X	X	X	X	X	-	X	X
Masked Yellowthroat	X	X	X	X	X	X	X	X	X	X	X	X	X	-
Rose-breasted Chat	-	-	-	-	-	-	-	-	-	-	-	-	-	S
Flavescent Warbler	-	-	-	-	-	X	X	X	X	X	-	X	X	-
Golden-crowned Warbler	X	X	X	-	X	H	-	-	-	-	-	X	X	-
White-bellied Warbler	-	-	-	-	-	X	S	-	X	X	X	X	X	-

	Tai	Mos	Apa	Uru	FdI	CtC	Tra	Bar	CdG	Ema	Can	Cip	Bra	Ara
E White-striped Warbler	-	-	-	-	-	-	-	-	H	X	X	-	X	-
White-rimmed Warbler	-	S	X	X	X	X	-	-	-	-	-	X	-	-
Neotropical River Warbler	-	-	-	-	X	-	-	-	S	-	-	-	-	-
Bananaquit	-	X	X	X	-	X	X	X	X	X	-	X	X	X
Chestnut-vented Conebill	-	-	-	-	-	X	X	X	X	X	X	X	-	X
Red-legged Honeycreeper	-	-	-	-	-	-	-	M	-	M	-	-	A	X
Blue Dacnis	-	X	X	-	X	H	S	X	X	X	X	X	X	X
Swallow-Tanager	-	-	-	-	X	-	S	-	X	-	X	X	X	S
Blue-naped Chlorophonia	-	-	S	S	-	X	-	-	-	-	-	-	-	-
Golden-rumped Euphonia	-	-	S	S	-	{S}	H	-	-	-	-	-	-	-
Orange-bellied Euphonia	-	-	-	-	-	-	-	-	S	-	-	-	-	-
Purple-throated Euphonia	-	S	X	X	-	X	X	X	X	X	X	-	X	X
Thick-billed Euphonia	-	-	-	-	-	-	-	S	-	X	-	-	-	-
Violaceous Euphonia	-	-	S	-	-	X	S	A	-	A	-	X	S	S
Chestnut-bellied Euphonia	-	-	X	X	-	X	-	-	-	-	-	-	H	-
Green-chinned Euphonia	-	-	-	S	-	S	-	-	-	-	-	-	-	-
Fawn-breasted Tanager	-	S	S	X	-	X	H	-	-	S	-	S	-	-
Green-headed Tanager	-	-	H	A	-	X	-	-	-	-	-	-	-	-
Red-necked Tanager	-	-	H	-	-	-	-	-	-	-	-	-	-	-
E Gilt-edged Tanager	-	-	-	-	-	-	-	-	-	-	-	X	-	-
Turquoise Tanager	-	-	-	-	-	-	-	-	S	-	-	-	-	S
Chestnut-backed Tanager	-	X	X	X	-	{S}	-	-	-	-	-	-	-	-
Burnished-buff Tanager	-	-	-	-	-	-	S	-	X	X	X	X	X	S
Diademed Tanager	-	X	X	X	S	{S}	-	-	-	-	-	-	-	-
Sayaca Tanager	X	X	X	X	X	X	X	X	X	X	X	X	X	X
E Azure-shouldered Tanager	-	-	M	A	-	-	-	-	-	-	-	-	-	-
E Golden-chevroned Tanager	-	-	-	A	-	-	-	-	-	-	-	-	H	-
Palm Tanager	-	-	S	-	-	{A}	X	X	X	X	X	X	X	X
Blue-and-yellow Tanager	-	X	X	X	X	-	-	-	-	-	-	-	-	-
Silver-beaked Tanager	-	-	-	-	-	X	X	X	X	X	-	-	X	X
Lowland Hepatic-Tanager	-	-	-	-	M	-	S	S	-	X	-	X	X	S
Red-crowned Ant-Tanager	-	-	S	-	-	X	-	-	S	-	-	-	-	-
White-lined Tanager	-	-	-	-	-	S	H	X	X	X	X	-	X	X
Ruby-crowned Tanager	-	-	X	X	-	X	H	-	-	-	-	X	H	-
Flame-crested Tanager	-	-	-	-	-	-	-	-	X	-	-	-	-	X
"Natterer's" Tanager	-	-	-	-	-	-	-	-	H	-	-	-	-	-
Red-shouldered Tanager	-	-	-	-	-	-	A	A	-	A	-	-	-	-
White-shouldered Tanager	-	-	-	-	-	-	-	-	-	-	-	-	-	S
Gray-headed Tanager	-	-	-	-	-	H	X	H	X	X	-	-	S	S
Black-goggled Tanager	-	-	X	X	-	X	-	-	S	X	-	H	X	-
White-rumped Tanager	-	-	-	-	-	-	S	-	X	X	X	X	X	X
Chestnut-headed Tanager	-	-	S	X	-	X	-	-	-	-	-	-	-	-

	Tai	Mos	Apa	Uru	FdI	CtC	Tra	Bar	CdG	Ema	Can	Cip	Bra	Ara
Hooded Tanager	-	-	-	-	-	X	X	X	X	-	X	X	X	X
E Rufous-headed Tanager	-	-	-	-	-	-	-	-	-	-	X	X	-	-
Guira Tanager	-	-	S	-	-	X	X	X	X	X	X	-	X	X
Orange-headed Tanager	-	-	-	-	S	S	S	S	H	S	X	H	X	X
E Scarlet-throated Tanager	-	-	-	-	-	-	-	-	-	-	-	-	-	S
White-banded Tanager	-	-	-	-	-	H	H	-	X	X	X	X	X	-
Magpie Tanager	-	-	-	-	X	-	S	S	-	-	X	H	-	-
E Cinnamon Tanager	-	-	-	-	-	-	-	-	-	-	X	X	-	-
Black-faced Tanager	-	-	-	-	-	S	H	X	X	-	-	-	X	X
Buff-throated Saltator	-	-	-	-	-	-	S	S	X	X	-	-	-	S
Grayish Saltator	-	-	-	-	A	X	X	X	S	-	-	-	-	X
Green-winged Saltator	-	-	S	S	M	X	X	S	S	S	X	X	X	S
Golden-billed Saltator	-	-	-	-	X	-	H	A	-	A	-	-	-	-
Thick-billed Saltator	-	-	-	X	-	-	-	-	-	-	-	-	-	-
Black-throated Saltator	-	-	-	-	-	X	X	X	X	X	X	X	X	-
Black-throated Grosbeak	-	-	-	-	S	-	-	-	-	-	-	-	-	-
Yellow Cardinal	-	-	H	-	A	-	-	-	-	-	-	-	-	-
Red-crested Cardinal	-	X	X	-	X	-	X	X	X	-	-	-	-	-
E Crimson-fronted Cardinal	-	-	-	-	-	-	-	-	-	-	-	-	-	S
Yellow-billed Cardinal	-	-	-	-	A	{S}	X	X	X	-	-	-	-	-
Black-backed Grosbeak	-	-	-	-	-	-	H	S	S	-	-	-	-	-
Blue-black Grosbeak	-	-	-	-	-	-	-	-	H	-	-	-	-	-
Ultramarine Grosbeak	-	X	X	X	X	X	X	S	S	-	X	H	S	S
Indigo Grosbeak	-	M	S	S	X	{S}	H	-	-	-	-	-	-	-
Blue Finch	-	-	-	-	-	-	H	H	-	X	-	X	X	M
Blue-black Grassquit	-	S	X	X	X	X	X	M	X	X	X	X	X	X
Sooty Grassquit	-	-	-	-	-	-	-	-	-	S	-	-	-	-
Buffy-fronted Seedeater	-	-	-	-	{A}	-	-	-	-	-	-	-	-	-
Temminck's Seedeater	-	-	-	-	M	-	-	-	-	-	-	-	-	-
Plumbeous Seedeater	-	-	-	-	{S}	-	M	-	X	X	X	H	X	S
Rusty-collared Seedeater	-	S	S	-	-	X	X	X	-	-	-	H	X	S
Lined Seedeater	-	-	-	-	-	-	M	M	M	-	-	-	M	-
Yellow-bellied Seedeater	-	-	-	-	-	H	H	S	H	-	X	X	X	S
E Dubois's Seedeater	-	-	-	-	-	-	-	-	-	-	X	X	-	-
Double-collared Seedeater	-	X	X	X	X	X	X	X	X	X	X	X	X	S
White-bellied Seedeater	-	-	-	-	-	X	M	X	X	-	X	H	-	X
Black-and-tawny Seedeater	-	-	-	-	-	A	M	M	-	-	-	-	-	-
Capped Seedeater	-	-	-	-	{M}	M	M	-	-	X	X	H	X	X
Tawny-bellied Seedeater	-	-	-	-	-	X	X	H	-	S	-	-	-	-
Dark-throated Seedeater	-	-	-	-	A	{S}	X	-	-	-	-	-	-	-
Marsh Seedeater	-	A	-	-	{A}	A	A	-	-	-	M	-	-	-
Rufous-rumped Seedeater	-	-	-	-	-	-	M	M	M	-	M	-	-	-

	Tai	Mos	Apa	Uru	FdI	CtC	Tra	Bar	CdG	Ema	Can	Cip	Bra	Ara
Chestnut Seedeater	-	-	-	-	-	-	M	-	-	-	M	-	-	-
E Black-bellied Seedeater	-	-	-	M	-	-	-	-	-	-	-	-	M	-
Great-billed Seed-Finch	-	-	-	-	-	-	H	-	H	S	S	-	S	-
Lesser Seed-Finch	-	-	-	-	X	H	X	X	X	-	X	H	S	X
Blackish-blue Seedeater	-	-	-	-	S	-	-	-	-	-	-	-	-	-
Stripe-tailed Yellow-Finch	-	-	-	-	S	-	X	-	X	X	X	X	X	-
Orange-fronted Yellow-Finch	-	-	-	-	-	-	-	-	-	-	-	-	-	X
Saffron Finch	X	X	X	X	X	X	X	X	X	X	X	H	X	-
Misto Yellow-Finch	X	X	M	X	{M}	M	M	-	M	-	M	H	-	-
Common Duica-Finch	-	-	-	H	-	-	-	-	-	-	-	-	-	-
Uniform Finch	-	S	X	-	X	-	-	-	-	-	-	-	-	-
Coal-crested Finch	-	-	-	-	-	-	H	-	X	X	-	H	X	-
Red Pileated-Finch	X	X	-	X	X	X	X	X	X	X	X	-	X	-
Gray Pileated-Finch	-	-	-	-	-	-	A	-	-	-	X	X	X	X
Pectoral Sparrow	-	-	-	-	-	-	-	-	-	X	-	-	-	X
Saffron-billed Sparrow	-	-	-	-	-	-	-	-	-	-	X	X	X	-
"Gray-backed" Sparrow	-	-	-	-	-	S	X	X	X	X	X	-	-	-
Grassland Sparrow	X	X	X	X	{X}	X	X	X	X	X	X	X	X	X
Yellow-browed Sparrow	-	-	-	-	-	-	-	-	-	-	-	-	-	X
Rufous-collared Sparrow	X	X	X	X	X	X	X	X	X	X	X	X	X	X
Wedge-tailed Grass-Finch	-	S	-	-	-	X	X	X	X	X	X	X	X	X
Lesser Grass-Finch	S	S	X	-	-	-	-	-	-	-	-	-	-	-
Black-masked Finch	-	-	-	-	-	-	-	-	-	H	X	X	-	X
Long-tailed Reed-Finch	X	X	X	-	-	-	-	-	-	-	-	-	-	-
E Bay-chested Warbling-Finch	-	-	S	-	-	-	-	-	-	-	-	-	-	-
Cinereous Warbling-Finch	-	-	-	-	-	-	H	-	H	A	-	S	A	-
Black-capped Warbling-Finch	-	-	-	X	-	A	-	A	-	-	-	-	-	-
Black-and-rufous Warbling-Finch	X	X	X	X	-	-	-	-	-	-	-	-	-	-
Gray-throated Warbling-Finch	S	S	X	S	-	-	-	-	-	-	-	-	-	-
Great Pampa-Finch	X	X	X	X	{S}	-	-	-	-	-	X	-	-	-
E Buff-throated Pampa-Finch	-	-	-	-	-	-	-	-	-	-	-	X	-	-
Hooded Siskin	X	X	X	-	{S}	M	M	-	-	X	X	X	-	-
House Sparrow	X	X	X	X	X	X	X	X	X	X	X	X	X	X
Common Waxbill	-	-	-	-	-	I	-	-	-	-	-	-	I	-

Eastern Brazil

SdM - Serra do Mar. Includes the Serra da Bocaina National Park,
Angra dos Reis, the Boraceia Forest Reserve and Ubatuba.

Ita - Itatiaia National Park.

SdO - Serra dos Orgãos. Includes the Serra dos Orgãos National Park,
Nova Friburgo, Cantagalo,Tinguá and Santo Aleixo.

RdJ - Rio de Janeiro. Tijuca National Park east to Lagoa de Saquarema.

PNC - Parque Natural do Caraça. Includes Estação Ambiental de Peti.

N.L - Nova Lombardia Biological Reserve including Santa Teresa.

FMC - Fazenda Montes Claros (Estação Biologica de Caratinga).

RDS - Rio Doce State Park.

Soo - Sooretama Biological Reserve and the Linhares Forest Reserve.

S.B - Southern Bahia. From Rio Jequitinhonha south to the Espirito
Santo border, incuding Porto Seguro and Monte Pascoal National
Park.

Sal - Salvador. From Santo Amaro south to Canavieiras, including
the Una Biological Reserve, Ilhéus and Valença.

B.N - Boa Nova. Including the area east to Gongoji.

N.B - Northeastern Bahia. From Juazeiro and Paulo Afonso south to
Itaberaba. Within the site lie the Raso da Catarina and Morro do
Chapéu.

Mac - Maceió. Includes the Parque Estadual da Pedra Talhada, Pedra
Branca and São Miguel dos Campos.

Black-bellied Antwren

	SdM	Ita	SdO	RdJ	PNC	N.L	FMC	RDS	Soo	S.B	Sal	B.N	N.B	Mac
Magellanic Penguin	-	-	-	-	M	-	-	-	-	-	-	A	-	H
Greater Rhea	-	-	-	-	-	-	-	-	-	-	-	-	S	-
Solitary Tinamou	X	S	S	H	-	S	X	X	X	X	-	H	-	S
Little Tinamou	-	-	S	-	-	S	X	X	X	X	X	X	-	H
Brown Tinamou	X	X	X	S	S	X	X	X	-	-	-	-	-	-
Undulated Tinamou	-	-	-	-	-	-	-	-	S	-	-	-	-	-
Variegated Tinamou	-	-	-	-	-	-	-	X	X	X	H	-	-	-
E Yellow-legged Tinamou	-	-	-	H	-	-	S	H	X	H	S	H	H	-
Brazilian Tinamou	-	-	-	-	-	-	-	-	-	-	-	-	-	H
Small-billed Tinamou	-	-	-	-	-	S	S	S	S	-	S	X	X	X
Tataupa Tinamou	S	H	S	S	S	-	-	X	X	S	-	S	S	-
Red-winged Tinamou	-	-	S	-	-	-	-	-	S	-	S	X	X	X
White-bellied Nothura	-	-	-	-	-	-	-	-	-	-	-	-	S	H
Spotted Nothura	-	H	-	-	-	-	-	-	-	-	S	-	-	S
Least Grebe	-	-	S	S	S	S	-	X	X	-	X	S	X	-
Pied-billed Grebe	M	-	-	-	S	S	-	X	X	-	X	S	S	-
Black-browed Albatross	-	-	-	M	-	-	-	-	-	-	-	-	-	-
Yellow-nosed Albatross	-	-	-	M	-	-	-	-	-	-	-	-	-	-
Gray-headed Albatross	-	-	-	[A]	-	-	-	-	-	-	-	-	-	-
Southern Giant-Petrel	-	-	-	M	-	-	-	-	-	-	-	-	-	-
Cape Petrel	-	-	-	M	-	-	-	-	-	-	-	-	-	-
Kerguelen Petrel	-	-	-	-	-	-	-	-	-	-	-	A	-	-
Atlantic Petrel	-	-	-	M	-	-	-	-	-	-	-	-	-	-
Antarctic (Dove) Prion	-	-	-	[A]	-	-	-	-	-	-	H	A	-	-
Thin-billed Prion	-	-	-	M	-	-	-	-	-	-	-	-	-	-
White-chinned Petrel	-	-	-	M	-	-	-	-	-	-	-	-	-	-
Cory's Shearwater	-	-	-	M	-	-	-	-	-	-	-	-	-	-
Great Shearwater	-	-	-	M	-	-	-	-	-	-	-	-	-	-
Sooty Shearwater	-	-	-	[A]	-	-	-	-	-	-	-	-	-	-
Manx Shearwater	M	-	-	M	-	-	-	-	-	-	-	-	-	-
Wilson's Storm-Petrel	-	-	-	M	-	-	-	-	-	-	-	-	-	-
Leach's Storm-Petrel	H	-	-	-	-	-	-	-	-	-	-	-	-	-
Brown Pelican	-	-	-	H	-	-	-	-	-	-	-	-	-	-
Masked Booby	-	-	-	-	-	-	-	-	-	-	M	-	-	-
Red-footed Booby	-	-	-	A	-	-	-	-	-	-	A	-	-	-
Brown Booby	X	-	-	X	-	-	-	-	-	-	-	-	-	-
Neotropic Cormorant	X	-	-	X	X	-	-	X	X	-	X	X	X	-
Anhinga	-	-	-	M	S	-	-	S	-	-	-	H	-	-
Magnificent Frigatebird	X	-	-	X	-	-	-	-	-	X	-	-	-	-
Cocoi (White-necked) Heron	S	-	-	X	-	-	-	X	-	-	-	-	-	-
Great (Common) Egret	X	M	-	X	-	-	S	X	X	-	X	S	X	X
Snowy Egret	X	M	-	X	S	-	S	X	X	-	X	S	S	-

Species	SdM	Ita	SdO	RdJ	PNC	N.L	FMC	RDS	Soo	S.B	Sal	B.N	N.B	Mac
Little Blue Heron	-	-	-	X	-	-	-	-	S	X	X	-	-	-
Striated Heron	S	S	H	X	-	S	X	X	X	-	X	X	X	X
Cattle Egret	X	X	-	X	-	-	X	X	X	-	-	X	X	-
Whistling Heron	-	-	-	H	-	-	-	-	S	-	-	-	-	-
Capped Heron	S	A	-	S	-	-	S	X	X	X	-	-	S	-
Black-crowned Night-Heron	-	-	-	X	-	-	-	X	X	-	-	-	-	-
Yellow-crowned Night-Heron	-	-	-	S	-	-	-	-	..	-	-	X	-	-
Boat-billed Heron	-	-	-	S	-	-	-	-	-	-	-	-	-	-
Rufescent Tiger-Heron	-	-	-	H	-	S	-	X	X	-	-	-	-	H
Stripe-backed Bittern	-	-	-	S	-	-	-	-	A	-	-	-	-	-
Least Bittern	H	-	-	H	-	-	-	-	S	-	S	-	-	H
Pinnated Bittern	-	-	-	S	-	-	-	-	-	-	-	-	-	-
Maguari Stork	-	-	-	H	-	-	-	-	M	-	-	-	-	-
Jabiru	-	-	-	H	-	-	-	-	-	-	-	-	-	-
Bare-faced Ibis	-	-	-	A	-	-	-	-	-	-	-	-	-	-
Scarlet Ibis	-	-	-	H	-	-	-	-	-	-	-	-	-	-
White-faced Ibis	-	-	-	A	-	-	-	-	S	-	-	-	-	-
Roseate Spoonbill	-	-	-	S	-	-	-	-	-	-	-	-	-	-
Horned Screamer	-	-	-	-	-	-	-	-	S	-	-	-	-	-
Fulvous Whistling-Duck	-	-	-	A	-	-	-	-	-	-	-	-	-	-
White-faced Whistling-Duck	-	M	-	M	-	-	M	M	M	-	-	-	M	H
Black-bellied Whistling-Duck	-	-	-	-	-	-	M	M	M	-	-	-	-	-
White-cheeked Pintail	-	-	-	M	-	-	-	-	-	-	-	-	M	H
Red Shoveler	-	-	-	A	-	-	-	-	-	-	-	-	-	-
Southern Pochard	-	-	-	X	-	-	-	-	-	-	M	-	M	M
Brazilian Duck (Teal)	-	-	S	H	X	X	-	X	X	X	S	-	X	X
Comb Duck	-	-	-	H	-	-	-	-	-	-	-	M	-	-
Muscovy Duck	-	-	-	H	S	-	-	X	X	-	X	-	-	-
Masked Duck	-	-	-	S	-	-	-	S	-	-	-	-	-	-
King Vulture	-	S	A	-	S	-	S	S	S	S	-	-	S	-
Black Vulture	X	X	X	X	X	X	X	X	X	X	X	X	X	X
Turkey Vulture	X	X	X	X	-	X	X	X	X	X	X	X	X	X
Lesser Yellow-headed Vulture	H	-	-	M	-	S	X	X	X	X	X	X	X	X
White-tailed Kite	S	X	-	S	-	S	-	X	X	X	X	X	X	X
Pearl Kite	-	-	-	-	-	-	-	-	-	-	H	H	X	-
Swallow-tailed Kite	M	M	H	M	-	M	M	M	M	M	M	H	-	-
E White-collared Kite	-	-	-	-	-	-	-	-	-	-	-	-	-	S
Gray-headed Kite	S	S	S	-	-	S	S	S	S	S	-	S	-	-
Hook-billed Kite	-	-	S	H	-	A	-	-	S	-	-	-	-	-
Rufous-thighed Kite	X	X	X	-	-	X	S	-	X	-	-	-	-	-
Double-toothed Kite	-	-	-	H	H	-	-	-	S	-	H	H	H	-
Plumbeous Kite	-	-	A	H	-	-	M	M	M	M	M	-	-	-

	SdM	Ita	SdO	RdJ	PNC	N.L	FMC	RDS	Soo	S.B	Sal	B.N	N.B	Mac
Snail Kite	-	-	-	-	-	-	-	-	S	-	-	-	-	A
Bicolored Hawk	-	S	H	H	-	-	-	-	-	H	-	-	H	S
Tiny Hawk	-	-	-	S	-	-	-	-	A	-	-	-	-	-
Gray-bellied Hawk	-	-	-	-	-	-	-	-	S	-	-	-	-	-
Sharp-shinned Hawk	X	X	X	S	S	S	-	-	-	-	-	-	H	-
Black-chested Buzzard-Eagle	-	-	A	-	S	-	-	-	-	-	-	-	S	S
White-tailed Hawk	H	X	X	X	-	S	-	S	S	S	-	S	-	S
Zone-tailed Hawk	-	-	-	-	-	-	-	-	-	-	-	S	-	S
Broad-winged Hawk	-	-	-	A	-	-	-	-	-	-	-	-	-	-
Roadside Hawk	X	X	X	X	X	X	X	X	X	X	X	X	X	X
White-rumped Hawk	S	S	-	A	-	-	-	S	-	-	-	-	-	-
Short-tailed Hawk	X	X	S	S	X	X	X	-	S	-	-	-	X	-
Gray-lined Hawk	-	-	-	H	-	-	-	-	S	S	-	-	-	S
Harris's Hawk	-	A	S	S	-	-	-	-	S	S	H	-	-	-
Mantled Hawk	S	S	S	-	-	S	-	S	H	H	-	S	-	S
E White-necked Hawk	S	-	S	S	-	S	-	S	S	S	H	-	-	S
Black-collared Hawk	-	-	-	H	-	-	-	-	-	-	-	-	-	-
Savanna Hawk	X	X	S	S	X	X	-	-	S	-	X	-	X	X
Great Black-Hawk	-	-	-	S	-	-	-	S	A	S	-	-	-	-
Crowned Eagle	S	-	A	-	-	-	-	-	-	-	-	-	A	-
Crested Eagle	-	-	-	H	-	-	-	-	-	-	-	-	-	-
Harpy Eagle	A	A	H	-	-	A	A	A	A	A	-	-	-	-
Black-and-white Hawk-Eagle	-	A	A	-	-	-	-	A	A	-	-	-	-	-
Ornate Hawk-Eagle	A	S	A	-	-	-	-	A	S	-	-	-	-	-
Black Hawk-Eagle	X	S	X	S	-	X	-	S	S	S	-	-	-	S
Long-winged Harrier	-	-	-	H	-	-	-	-	M	-	-	-	-	-
Crane Hawk	A	-	H	-	-	A	-	X	S	S	S	-	S	-
Osprey	-	-	A	M	-	-	-	-	-	-	-	-	-	-
Laughing Falcon	S	S	S	S	X	S	X	X	X	X	X	X	X	X
Collared Forest-Falcon	-	-	H	H	-	S	S	-	X	S	-	-	-	-
Barred Forest-Falcon	X	X	X	S	X	X	X	-	S	S	-	X	-	S
Lined Forest-Falcon	-	-	-	-	-	-	-	A	H	H	-	-	-	-
Yellow-headed Caracara	X	X	X	X	X	X	X	X	X	X	X	X	X	X
Crested Caracara	X	X	X	S	X	X	X	X	X	X	X	X	X	X
Peregrine Falcon	-	A	-	M	-	-	-	-	-	-	-	-	-	M
Bat Falcon	-	-	S	S	X	S	-	S	X	S	-	-	X	-
Aplomado Falcon	-	M	S	S	X	S	-	S	S	-	X	X	X	X
American Kestrel	S	X	X	X	-	X	X	X	X	X	X	X	X	X
Speckled Chachalaca	-	-	-	-	-	-	S	X	X	X	X	X	X	X
Rusty-margined Guan	-	-	S	S	X	S	S	X	X	X	X	-	S	-
Dusky-legged Guan	X	X	X	-	X	-	X	X	-	-	-	-	-	-
E White-browed Guan	-	-	-	-	-	-	-	-	-	-	-	-	S	-

	SdM	Ita	SdO	RdJ	PNC	N.L	FMC	RDS	Soo	S.B	Sal	B.N	N.B	Mac
Black-fronted Piping-Guan	S	A	S	-	-	-	-	S	H	S	S	-	-	-
E Red-billed Curassow	-	-	-	H	-	-	-	S	X	S	S	-	-	-
E Alagoas Curassow	-	-	-	-	-	-	-	-	-	-	-	-	-	S
Spot-winged Wood-Quail	X	X	X	S	-	X	-	-	X	X	X	-	-	X
Limpkin	H	-	-	S	-	-	-	-	S	-	-	-	-	S
Clapper Rail	-	-	H	S	-	-	-	-	-	H	X	-	-	-
Plumbeous Rail	-	-	-	S	-	-	-	-	-	-	-	-	-	-
Blackish Rail	S	X	H	S	X	X	X	X	X	X	X	X	X	X
Uniform Crake	S	-	-	-	-	-	-	-	S	S	-	-	-	S
E Little Wood-Rail	-	H	-	S	-	-	-	-	-	-	X	-	H	X
Gray-necked Wood-Rail	H	A	-	X	X	S	X	X	X	H	X	H	-	X
Slaty-breasted Wood-Rail	X	X	X	X	X	X	X	X	-	-	-	-	-	-
Ash-throated Crake	X	X	S	X	X	H	X	X	X	X	H	X	-	X
Yellow-breasted Crake	-	-	-	S	-	-	-	-	-	-	-	-	-	-
Gray-breasted Crake	-	-	-	-	-	-	-	-	A	-	-	-	-	-
Rufous-sided Crake	X	S	H	X	-	S	-	X	X	X	X	X	-	X
Red-and-white Crake	-	-	H	-	-	-	-	-	-	-	-	-	-	-
Russet-crowned Crake	-	-	-	X	-	-	-	-	S	S	-	-	-	S
Paint-billed Crake	-	-	-	-	-	-	-	-	-	-	-	-	-	S
Spot-flanked Gallinule	-	-	-	X	-	-	-	-	-	-	-	-	S	-
Common Gallinule (Moorhen)	-	S	H	X	X	-	X	X	X	-	X	X	X	X
Purple Gallinule	-	-	-	H	S	-	-	-	X	X	-	H	X	X
Red-gartered Coot	-	-	-	A	-	-	-	-	-	-	-	-	-	-
Sungrebe	-	-	-	-	-	-	-	-	H	-	H	-	-	-
Red-legged Seriema	-	S	S	S	X	-	X	X	-	-	-	-	S	-
Wattled Jacana	H	X	H	X	X	S	-	X	X	-	X	X	X	X
South American Painted-Snipe	-	-	H	S	-	-	-	-	-	-	-	-	-	-
American Oystercatcher	-	-	-	S	-	-	-	-	-	-	-	-	-	A
Southern Lapwing	X	X	S	X	-	S	X	X	X	X	X	X	X	X
Pied Plover	-	-	-	-	-	-	-	S	-	-	-	-	-	-
Black-bellied Plover	-	-	-	M	-	-	-	-	-	-	M	-	-	-
Semipalmated Plover	-	-	M	M	-	-	-	-	-	-	H	-	-	-
Collared Plover	-	-	H	X	-	-	-	-	-	-	H	-	-	-
Ruddy Turnstone	-	-	-	A	M	-	-	-	-	-	H	-	-	-
Solitary Sandpiper	-	-	-	M	-	-	-	-	M	M	M	-	-	H
Lesser Yellowlegs	-	-	-	H	M	-	-	-	-	-	H	-	-	-
Greater Yellowlegs	-	-	-	M	-	-	-	-	-	-	H	-	-	-
Spotted Sandpiper	-	-	M	M	-	-	-	-	-	-	M	-	-	-
Willet	-	-	-	-	-	-	-	-	-	-	M	-	-	-
Red Knot	-	-	-	A	-	-	-	-	-	-	-	-	-	-
Least Sandpiper	-	-	-	-	-	-	-	-	-	-	H	-	-	-
White-rumped Sandpiper	-	-	-	M	-	-	-	-	-	-	H	-	-	-

	SdM	SdO	PNC	FMC	Soo	Sal	N.B
	Ita	RdJ	N.L	RDS	S.B	B.N	Mas

Species	SdM	Ita	SdO	RdJ	PNC	N.L	FMC	RDS	Soo	S.B	Sal	B.N	N.B	Mas
Pectoral Sandpiper	-	-	-	-	M	-	-	-	-	-	M	-	-	M
Semipalmated Sandpiper	-	-	-	-	M	-	-	-	-	-	-	-	-	-
Sanderling	-	-	-	-	M	-	-	-	-	-	M	-	-	-
Whimbrel	-	-	-	-	M	-	-	-	-	-	M	-	-	-
Short-billed Dowitcher	-	-	-	-	H	-	-	-	-	-	-	-	-	-
South American Snipe	-	-	S	H	X	-	S	-	-	S	-	H	X	X
Giant Snipe	-	-	-	-	S	-	-	-	-	-	-	-	-	-
White-backed Stilt	-	-	-	-	H	-	-	M	-	-	-	-	-	-
Snowy Sheathbill	-	-	-	-	M	-	-	-	-	-	-	-	-	-
Great Skua	-	-	-	-	-	-	-	-	-	-	-	-	-	A
"Antarctic" Skua	-	-	-	-	M	-	-	-	-	-	-	-	-	-
Chilean Skua	-	-	-	-	M	-	-	-	-	-	-	-	-	-
Arctic Skua (Parasitic Jaeger)	-	-	-	-	M	-	-	-	-	-	-	-	-	A
Long-tailed Skua (Jaeger)	-	-	-	-	A	-	-	-	-	-	-	-	-	-
Kelp Gull	-	X	-	-	X	-	-	-	-	-	-	-	-	-
Gray-hooded Gull	-	-	-	-	A	-	-	-	-	-	-	-	-	-
Brown-hooded Gull	-	-	-	-	H	-	-	-	-	-	-	-	-	-
Large-billed Tern	-	-	-	-	H	-	-	-	-	-	-	-	-	-
Gull-billed Tern	-	-	-	-	A	-	-	-	-	-	-	-	-	S
South American Tern	-	-	-	-	M	-	-	-	-	-	M	-	-	-
Common Tern	-	-	-	-	M	-	-	-	-	-	-	-	-	-
Trudeau's Tern	-	-	-	-	M	-	-	-	-	-	-	-	-	-
Roseate Tern	-	-	-	-	-	-	-	-	-	-	[A]	-	-	-
Sooty Tern	-	-	-	-	A	-	-	-	-	-	-	-	-	-
Yellow-billed Tern	-	-	-	-	S	-	-	S	-	-	-	-	-	-
Least Tern	-	A	-	-	-	-	-	-	-	-	-	-	-	-
Royal Tern	-	M	-	-	M	-	-	-	-	-	-	-	-	-
Sandwich Tern	-	-	-	-	M	-	-	-	-	-	-	-	-	-
"Cayenne" Tern	-	M	-	-	X	-	-	-	-	-	-	-	-	M
Brown Noddy	-	-	-	-	[A]	-	-	-	-	-	-	-	-	-
Black Skimmer	-	-	-	-	M	-	-	-	-	-	-	-	-	-
Scaled Pigeon	-	-	-	-	S	-	-	-	X	X	-	-	-	H
Picazuro Pigeon	-	-	S	S	A	-	-	S	X	X	-	-	X	-
Pale-vented Pigeon	-	-	-	S	S	X	-	X	X	X	X	X	-	-
Plumbeous Pigeon	-	X	X	X	H	X	X	-	X	-	H	-	X	-
Eared Dove	-	-	S	-	H	H	-	-	-	-	-	-	X	-
Scaled Dove	-	-	-	A	-	-	S	-	X	X	X	-	H	X	X
Common Ground-Dove	-	-	-	-	-	-	-	-	-	-	-	-	S	-
Plain-breasted Ground-Dove	-	-	S	S	S	-	-	-	S	S	H	X	X	X
Ruddy Ground-Dove	-	X	X	X	X	X	X	X	X	X	X	X	X	X
Picui Ground-Dove	-	-	-	-	-	-	-	-	X	X	-	X	X	X
Blue Ground-Dove	-	-	S	S	S	S	-	S	X	S	S	H	H	S

		SdM		SdO		PNC		FMC		Soo		Sal		N.B	
			Ita		RdJ		N.L		RDS		S.B		B.N		Mac
	Purple-winged Ground-Dove	S	S	S	-	-	A	-	-	-	H	-	-	-	-
	White-tipped Dove	S	S	S	S	X	-	X	X	X	X	H	H	X	X
	Gray-fronted Dove	X	X	X	H	S	X	X	-	X	X	X	X	S	-
	Ruddy Quail-Dove	S	A	S	S	-	S	-	-	S	S	-	-	-	S
	Violaceous Quail-Dove	-	A	H	-	-	-	-	-	S	-	-	H	-	-
E	Indigo Macaw	-	-	-	-	-	-	-	-	-	-	-	-	S	-
E	Little Blue (Spix's) Macaw	-	-	-	-	-	-	-	-	-	-	-	-	S	-
	Red-and-green Macaw	-	-	-	-	-	-	-	-	S	H	-	-	-	-
	Blue-winged Macaw	-	A	A	-	-	-	X	X	X	-	-	-	S	-
	Red-shouldered Macaw	-	-	-	-	-	-	X	-	X	-	-	-	-	-
	Blue-crowned Parakeet	-	-	-	-	-	-	-	-	-	-	-	-	X	X
	White-eyed Parakeet	-	-	S	H	-	X	-	X	-	S	S	S	X	-
E	Golden-capped Parakeet	-	-	-	-	-	-	X	X	-	S	-	H	-	-
E	Caatinga (Cactus) Parakeet	-	-	-	-	-	-	-	-	-	-	-	X	X	-
	Peach-fronted Parakeet	-	-	-	H	-	-	-	S	-	X	X	X	X	-
E	Ochre-marked Parakeet	-	-	-	H	I	-	-	X	X	X	X	-	H	-
	Reddish-bellied Parakeet	-	X	X	X	X	-	X	S	-	-	-	X	H	-
	Maroon-faced Parakeet	-	-	-	H	I	-	-	X	X	X	X	-	H	X
	Monk Parakeet	-	-	-	I	-	-	-	-	-	-	-	-	-	-
	Blue-winged Parrotlet	X	X	X	X	X	X	-	X	X	X	X	X	X	X
E	Plain Parakeet	X	X	X	X	-	S	-	X	X	X	H	X	-	X
	Yellow-chevroned Parakeet	-	-	H	M	-	-	-	S	-	-	S	-	-	-
E	Black-eared Parrotlet	S	S	S	S	-	-	-	-	-	H	-	-	-	-
E	Golden-tailed Parrotlet	S	A	S	-	-	S	-	-	S	S	H	H	-	S
	Red-capped Parrot	X	X	X	S	-	X	-	-	-	S	-	S	-	-
	Blue-headed Parrot	-	-	H	-	S	-	-	-	S	X	-	-	-	-
	Scaly-headed Parrot	X	X	X	X	X	X	X	X	X	S	-	H	-	-
E	Red-browed Parrot	-	-	-	H	-	M	-	X	X	X	S	H	-	S
	Turquoise-fronted Parrot	-	-	-	H	-	-	-	S	S	S	S	S	-	S
	Orange-winged Parrot	-	-	-	S	-	S	-	-	X	X	-	-	-	-
	Mealy Parrot	-	-	-	H	-	-	-	X	X	X	X	-	-	-
	Vinaceous-breasted Parrot	S	-	H	-	-	-	X	-	-	-	-	-	-	-
E	Blue-bellied Parrot	S	-	S	-	-	S	-	-	A	H	-	-	-	-
	Yellow-billed Cuckoo	-	-	M	H	-	-	-	-	M	-	-	-	-	-
	Pearly-breasted Cuckoo	-	M	H	M	-	-	-	-	M	-	-	H	-	M
	Dark-billed Cuckoo	H	-	M	A	-	-	-	-	M	-	-	M	M	M
	Squirrel Cuckoo	X	X	X	S	X	X	X	X	X	X	X	X	X	X
	Greater Ani	-	-	-	H	-	-	X	X	X	H	-	-	-	-
	Smooth-billed Ani	X	X	X	X	X	X	X	X	X	X	X	X	X	X
	Guira Cuckoo	X	X	S	X	X	X	X	X	X	X	X	X	X	X
	Striped Cuckoo	X	S	X	X	X	X	X	X	X	X	X	X	X	X
	Pheasant Cuckoo	-	-	H	-	-	A	-	-	-	-	-	-	A	-

	SdM	Ita	SdO	RdJ	PNC	N.L	FMC	RDS	Soo	S.B	Sal	B.N	N.B	Mac
Pavonine Cuckoo-	A	-	H	-	-	-	-	-	-	-	A	-	-	-
Rufous-vented Ground-Cuckoo-	-	-	H	-	-	-	-	S	S	-	-	H	-	-
Barn Owl	S	-	S	S	-	-	S	S	S	-	H	-	-	S
Tropical Screech-Owl	S	S	S	X	X	X	X	X	X	X	X	X	X	X
Variable(Black-c.)Screech-Owl	A	X	X	-	-	-	-	-	X	S	-	-	-	-
Great Horned Owl	-	-	S	-	-	S	-	-	-	-	-	-	S	S
Spectacled Owl	-	-	-	-	-	S	-	-	X	X	X	H	-	-
Tawny-browed Owl	X	X	X	S	-	S	X	X	X	X	-	-	-	-
Least Pygmy-Owl-	A	-	-	-	-	S	-	-	S	S	S	-	-	-
Ferruginous Pygmy-Owl	S	X	X	X	X	X	X	X	X	X	X	X	-	X
Burrowing Owl	-	X	-	X	X	X	X	X	X	-	X	H	X	-
Black-banded Owl	-	-	-	H	-	S	-	S	-	-	-	-	-	-
Mottled Owl-	-	S	S	-	-	-	-	-	S	-	-	-	-	-
Rusty-barred Owl	S	X	-	-	-	-	-	S	-	-	-	-	-	-
Striped Owl-	-	H	-	S	-	-	-	S	S	-	-	-	-	-
Great Potoo-	-	-	H	-	-	-	-	S	S	-	-	-	-	-
Large-tailed Potoo	S	A	H	-	-	-	-	-	A	H	-	-	-	-
Common Potoo	S	S	S	H	-	-	-	-	X	-	-	-	X	H
E White-winged Potoo	-	-	-	-	-	-	-	-	-	H	-	-	-	-
Short-tailed Nighthawk	X	X	X	S	S	X	-	-	X	X	X	-	-	X
Least Nighthawk-	-	-	-	-	-	-	-	S	-	-	-	-	X	X
Lesser Nighthawk	-	-	-	-	M	-	-	-	-	A	-	-	-	A
Common Nighthawk	-	-	-	-	A	-	-	-	-	-	-	-	-	-
Nacunda Nighthawk	-	-	-	-	M	-	-	-	-	-	-	-	-	-
Pauraque	X	X	S	S	X	X	X	X	X	X	X	X	X	X
Ocellated Poorwill	-	-	H	A	-	S	X	-	-	X	S	-	H	-
Rufous Nightjar-	-	-	-	H	-	-	-	-	-	X	-	-	S	-
Band-winged Nightjar	-	-	M	M	M	M	-	-	-	-	-	-	M	-
Spot-tailed Nightjar	-	-	-	-	M	-	-	-	-	A	-	-	-	-
Little Nightjar-	-	-	-	-	M	M	-	-	-	-	H	H	M	H
E Pygmy Nightjar	-	-	-	-	-	-	-	-	-	-	-	-	A	A
Scissor-tailed Nightjar	-	-	S	X	X	X	-	X	X	-	-	S	X	X
Long-trained Nightjar	S	S	S	-	S	-	-	-	-	-	-	-	-	-
White-collared Swift	X	X	X	X	X	X	X	X	S	X	-	-	-	-
Biscutate Swift-	-	S	X	S	X	-	-	-	M	M	-	-	M	-
Sooty Swift-	X	X	X	H	S	S	-	-	-	S	-	-	-	-
Gray-rumped Swift	X	X	X	X	-	X	X	-	X	X	X	-	-	-
Band-rumped Swift	-	-	-	-	-	-	-	-	-	-	-	-	-	S
Ashy-tailed Swift	X	X	X	X	X	S	X	X	X	X	-	-	-	-
Lesser Swallow-tailed Swift-	M	M	M	A	-	-	-	-	S	S	-	-	S	X
Fork-tailed Palm-Swift -	-	-	M	-	-	-	-	-	S	-	-	S	X	X
E Saw-billed Hermit	X	-	S	S	-	S	-	-	-	-	-	-	-	-

	SdM	Ita	SdO	RdJ	PNC	N.L	FMC	RDS	Soo	S.B	Sal	B.N	N.B	Mac
E Hook-billed Hermit	-	-	-	-	-	-	-	-	A	S	-	H	-	-
Rufous-breasted Hermit	S	S	S	S	-	X	X	X	X	X	-	H	-	-
Klabin Farm Long-tailed Hermit	-	-	-	-	-	-	-	-	A	A	S	-	-	S
Scale-throated Hermit	X	X	X	S	X	X	-	-	-	-	-	X	-	-
"Black-billed" Hermit	-	-	-	-	-	X	-	-	-	-	-	-	-	-
Dusky-throated Hermit	S	S	S	X	X	S	S	-	-	-	-	-	-	-
Planalto Hermit	S	S	X	X	X	S	X	X	-	X	X	X	X	X
E Broad-tipped Hermit	-	-	-	-	-	-	-	-	-	-	-	X	X	A
Reddish Hermit	X	-	X	S	-	-	X	X	-	X	X	X	-	X
E Minute Hermit	-	-	H	-	-	A	-	X	X	-	-	-	-	-
Gray-breasted Sabrewing	-	-	-	-	S	-	-	-	-	-	-	-	-	-
Swallow-tailed Hummingbird	X	X	X	X	X	X	X	X	X	X	X	X	X	X
Black Jacobin	X	X	X	S	X	X	-	X	S	X	X	H	-	X
White-vented Violetear	S	X	S	X	X	X	X	X	-	-	-	-	X	-
Black-throated Mango	M	M	M	H	M	M	-	M	M	M	M	M	-	M
Ruby-topaz Hummingbird	-	-	-	M	-	M	-	-	M	-	H	-	M	M
Black-breasted Plovercrest	X	X	X	-	-	-	-	-	-	-	-	-	-	-
E Frilled Coquette	M	M	M	A	-	M	-	-	-	-	-	M	-	M
Festive Coquette	M	M	-	-	-	M	-	-	-	-	-	-	-	-
Black-bellied Thorntail	A	A	H	S	-	S	-	-	-	S	-	-	-	-
Racket-tailed Coquette	-	-	S	-	-	-	-	-	S	S	-	-	-	-
Blue-chinned Sapphire	-	-	-	S	-	S	-	-	-	S	-	-	-	S
Glittering-bellied Emerald	X	X	X	S	X	X	X	X	-	-	X	X	X	X
Fork-tailed Woodnymph	-	A	S	-	X	-	-	-	-	-	-	-	-	-
E Long-tailed Woodnymph	-	-	-	-	-	-	-	-	-	S	-	-	-	X
Violet-capped Woodnymph	X	X	X	X	S	X	X	X	X	X	X	X	-	X
Rufous-throated Sapphire	-	-	M	M	-	M	-	M	M	M	-	-	-	M
White-chinned Sapphire	M	-	M	M	M	M	M	M	M	M	M	-	-	-
White-throated Hummingbird	X	X	X	S	X	X	X	-	-	-	-	-	-	-
White-tailed Goldenthroat	-	-	H	S	-	-	-	-	X	-	-	-	-	S
Versicolored Emerald	M	M	M	A	-	S	-	X	X	X	H	-	-	X
Glittering-throated Emerald	-	M	M	M	-	X	X	X	-	-	-	H	X	X
"Big" Glittering-thr. Emerald	S	-	H	-	-	-	-	-	-	-	-	-	-	-
Sapphire-spangled Emerald	-	M	M	M	X	X	X	-	-	-	-	-	X	A
Plain-bellied Emerald	-	-	-	-	-	-	-	-	-	S	X	-	-	-
E Sombre Hummingbird	S	M	X	S	X	X	-	-	M	-	M	M	-	M
E Brazilian Ruby	X	X	X	S	X	X	-	-	-	-	-	-	-	-
E Hooded Visorbearer	-	-	-	-	-	-	-	-	-	-	-	-	S	-
E Hyacinth Visorbearer	-	-	-	-	X	-	-	-	-	-	-	-	-	-
Black-eared Fairy	H	M	H	M	-	-	-	-	S	X	-	X	-	X
Long-billed Starthroat	-	-	-	-	-	-	-	-	-	-	-	M	M	-
E Stripe-breasted Starthroat	-	-	-	A	-	-	-	-	-	-	-	A	-	-

	SdM	Ita	SdO	RdJ	PNC	N.L	FMC	RDS	Soo	S.B	Sal	B.N	N.B	Mac
Amethyst Woodstar	-	M	M	M	M	M	-	-	M	-	-	M	-	-
White-tailed Trogon	X	S	X	S	X	X	X	X	X	X	-	H	-	H
Collared Trogon	-	-	-	-	-	-	-	-	X	H	H	H	-	S
Black-throated Trogon	X	X	X	-	X	X	-	X	X	X	-	X	-	S
Brazilian Trogon	X	X	X	X	X	X	-	-	S	H	-	-	-	-
Blue-crowned Trogon	-	-	-	-	-	-	-	-	-	-	-	-	X	X
Ringed Kingfisher	X	S	H	X	X	S	X	X	X	X	X	-	X	X
Amazon Kingfisher	S	-	H	S	S	S	X	X	X	X	-	X	X	H
Green Kingfisher	S	-	H	S	S	X	X	X	X	X	X	X	X	X
Green-and-rufous Kingfisher	S	-	H	-	-	-	-	S	-	-	-	-	-	-
American Pygmy Kingfisher	-	-	S	-	-	S	-	-	-	-	S	-	-	-
Rufous-capped Motmot	S	X	X	X	-	S	X	X	X	X	-	H	H	-
Blue-crowned Motmot	-	-	-	-	-	-	-	-	-	-	-	-	-	S
E Three-toed Jacamar	-	-	S	-	-	S	-	-	-	-	-	-	-	-
Rufous-tailed Jacamar	S	-	S	H	S	-	X	X	X	X	X	H	X	X
Buff-bellied Puffbird	X	H	H	-	-	S	-	-	S	S	-	-	-	-
White-eared Puffbird	H	X	S	S	-	-	-	S	-	-	-	-	-	-
Spot-backed Puffbird	-	-	-	-	-	-	-	-	-	-	X	X	X	X
E Crescent-chested Puffbird	S	S	S	S	X	X	X	X	X	X	-	H	-	-
Rusty-breasted Nunlet	-	-	H	-	-	-	-	-	-	-	-	-	-	-
White-fronted Nunbird	-	-	-	-	-	-	-	-	X	X	H	H	-	-
Swallow-wing	-	-	H	H	-	-	X	X	X	X	H	X	-	-
Black-necked Aracari	-	-	-	S	-	S	X	X	X	X	-	X	-	X
Spot-billed Toucanet	S	X	X	S	-	X	X	-	X	X	X	-	-	-
Saffron Toucanet	X	X	X	M	-	S	S	-	-	-	-	-	-	-
Red-breasted Toucan	X	X	S	-	X	S	-	-	-	-	-	-	-	-
Channel-billed Toucan	S	S	X	X	-	X	X	X	X	X	X	X	-	X
E Tawny Piculet	-	-	-	-	-	-	-	-	-	-	-	-	-	S
Golden-spangled Piculet	-	-	-	-	-	-	-	-	S	-	-	X	-	X
White-barred Piculet	X	X	X	X	X	X	X	X	X	X	X	-	-	-
E Spotted Piculet	-	-	-	-	-	-	-	-	-	-	H	X	S	-
Campo Flicker	X	X	X	X	X	X	X	X	X	X	-	-	-	-
Green-barred Woodpecker	X	X	X	-	X	X	X	X	-	-	X	-	X	-
Yellow-throated Woodpecker	X	-	X	S	-	X	-	-	X	X	H	X	-	H
Golden-green Woodpecker	-	-	-	-	-	-	S	S	X	X	-	-	H	-
White-browed Woodpecker	X	X	X	H	X	X	-	-	-	-	-	-	-	-
Blond-crested Woodpecker	X	-	S	H	-	-	X	X	X	X	H	H	H	-
Cream-colored Woodpecker	-	-	-	-	-	-	-	-	X	X	-	-	-	H
Ringed Woodpecker	-	-	-	-	-	-	-	-	X	S	S	-	-	-
Lineated Woodpecker	X	X	S	-	X	X	X	X	X	X	X	X	-	-
Helmeted Woodpecker	A	-	-	-	-	-	-	-	-	-	-	-	-	-
Yellow-fronted Woodpecker	X	S	X	S	-	-	X	X	X	X	-	-	-	-

Species	SdM	Ita	SdO	RdJ	PNC	N.L	FMC	RDS	Soo	S.B	Sal	B.N	N.B	Mac
White Woodpecker	-	X	H	S	X	-	X	X	X	-	-	-	-	-
White-spotted Woodpecker	X	X	-	-	-	-	-	-	-	-	-	-	-	-
Little Woodpecker	-	-	S	-	S	-	-	X	S	X	-	X	X	X
Red-stained Woodpecker	-	-	-	-	-	-	-	-	X	-	S	H	-	X
E Yellow-eared Woodpecker	-	S	X	X	-	X	X	X	X	X	X	X	S	-
Crimson-crested Woodpecker	-	-	-	-	S	-	-	-	-	-	-	-	-	-
Robust Woodpecker	S	X	-	S	X	S	S	S	X	S	S	H	-	-
Thrush-like Woodcreeper	X	X	X	X	-	X	X	X	X	X	X	H	-	S
Olivaceous Woodcreeper	X	X	X	X	X	X	X	X	X	X	H	X	X	X
Wedge-billed Woodcreeper	-	-	-	-	-	-	-	X	X	X	-	-	-	X
White-throated Woodcreeper	X	X	X	S	-	X	X	-	X	X	X	H	X	-
E Moustached Woodcreeper	-	-	-	-	-	-	-	-	-	-	-	-	(H)	-
Barred Woodcreeper	-	-	-	-	-	-	-	-	-	-	-	-	-	H
Planalto Woodcreeper	X	X	X	S	S	X	X	X	X	X	X	H	H	-
Straight-billed Woodcreeper	-	-	-	-	-	-	-	-	-	H	X	X	H	X
Buff-throated Woodcreeper	-	-	S	H	-	S	-	X	X	X	X	H	-	S
Narrow-billed Woodcreeper	-	A	H	S	X	-	-	-	-	-	S	-	X	H
Scaled Woodcreeper	X	X	X	S	S	X	X	-	S	S	-	X	-	-
Lesser Woodcreeper	X	X	X	X	-	X	X	X	X	X	X	X	H	X
Red-billed Scythebill	-	-	-	-	-	-	-	-	-	H	H	H	X	H
Black-billed Scythebill	S	X	X	H	-	X	-	-	X	-	S	X	-	-
Rufous Hornero	X	X	X	X	X	X	X	X	X	-	X	X	-	S
Pale-legged Hornero	-	-	-	-	-	-	S	S	-	-	H	X	X	X
E Wing-banded Hornero	-	-	S	-	X	X	X	X	X	-	H	X	X	X
Araucaria Tit-Spinetail	S	S	-	-	-	-	-	-	-	-	-	-	-	-
E Itatiaia Spinetail	-	X	X	-	-	-	-	-	-	-	-	-	-	-
Rufous-capped Spinetail	X	X	X	-	X	X	X	X	-	-	-	X	-	-
E Pinto's Spinetail	-	-	-	-	-	-	-	-	-	-	-	-	-	X
Sooty-fronted Spinetail	-	-	-	-	S	S	-	-	S	X	X	X	X	X
Chicli Spinetail	X	X	X	S	X	X	X	X	X	S	-	S	-	-
Cinereous-breasted Spinetail	-	-	-	-	-	-	-	-	-	-	X	-	-	X
Pale-breasted Spinetail	-	S	S	-	S	-	-	-	S	-	-	X	X	X
Gray-bellied Spinetail	S	S	S	-	-	-	-	-	S	-	-	-	-	-
E Red-shouldered Spinetail	-	-	-	-	-	-	-	-	-	-	-	-	X	-
Ochre-cheeked Spinetail	-	-	-	-	-	-	-	-	-	-	-	X	H	X
E Gray-headed Spinetail	-	-	-	-	-	-	-	-	-	-	-	X	H	X
E Pallid Spinetail	X	X	X	-	X	X	-	-	-	-	-	X	-	-
Yellow-chinned Spinetail	S	X	S	X	X	-	X	X	X	H	X	X	X	X
E Striated Softtail	-	-	-	-	-	-	-	-	X	-	H	X	-	-
Canebrake Groundcreeper	A	-	-	-	-	-	-	-	-	-	-	-	-	-
Rufous-fronted Thornbird	-	-	S	-	X	S	X	X	-	X	H	X	X	X
E Red-eyed Thornbird	X	-	S	-	S	-	-	-	-	-	H	-	-	-

	SdM	Ita	SdO	RdJ	PNC	N.L	FMC	RDS	Soo	S.B	Sal	B.N	N.B	Mac
"Rufous-breasted" Thornbird-	-	S	-	-	-	-	-	-	-	-	-	-	-	-
Firewood-gatherer	-	-	-	-	-	-	-	-	-	-	S	S	X	-
Rufous Cacholote	-	-	-	-	-	-	-	-	-	-	-	S	X	-
E Pale-browed Treehunter	X	S	X	H	X	X	-	X	-	-	-	-	-	-
E White-collared Foliage-gleaner	X	X	X	S	-	X	S	-	-	-	-	S	-	-
Buff-browed Foliage-gleaner	X	X	X	-	X	S	-	-	-	-	-	-	-	-
White-browed Foliage-gleaner	X	S	S	S	-	S	-	-	-	-	-	-	-	-
Black-capped Foliage-gleaner	X	X	X	X	-	X	-	X	X	X	X	-	-	-
E Alagoas Foliage-gleaner-	-	-	-	-	-	-	-	-	-	-	-	-	-	S
Ochre-breasted Foliage-gleaner	X	S	S	X	S	S	-	-	-	-	S	S	-	-
Buff-fronted Foliage-gleaner	X	X	X	X	X	X	X	-	-	-	-	S	-	-
White-eyed Foliage-gleaner	-	X	X	X	X	S	X	X	X	X	X	X	H	-	X
Chestnut-capped Foliage-glean.	-	-	H	-	-	-	-	-	-	-	-	-	-	-
Sharp-billed Treehunter-	X	X	X	H	-	S	-	-	-	-	-	-	-	-
Streaked Xenops-	X	X	X	X	X	X	X	X	X	X	X	X	H	X
Plain Xenops	X	-	X	X	-	S	-	-	X	X	X	H	-	X
E Great Xenops	-	-	-	-	-	-	-	-	-	-	-	-	A	-
Rufous-breasted Leaftosser	X	S	X	X	-	S	X	X	-	-	-	-	H	S
Tawny-throated Leaftosser	-	S	-	-	-	S	-	-	S	S	-	-	-	-
Black-tailed Leaftosser-	-	-	-	-	-	-	-	-	S	H	H	-	-	S
Sharp-tailed Streamcreeper	X	X	X	X	X	X	X	X	-	H	H	X	-	-
Spot-backed Antshrike	X	-	S	H	-	X	-	-	-	-	-	X	-	-
Giant Antshrike-	X	X	X	H	-	S	-	-	-	-	-	-	-	-
Large-tailed Antshrike	S	X	X	S	X	-	X	-	-	-	-	-	-	-
Tufted Antshrike	X	X	X	S	X	S	-	-	-	-	-	X	-	-
Great Antshrike-	-	-	-	A	-	-	-	-	X	H	X	X	X	X
E Silvery-cheeked Antshrike	-	-	-	-	-	-	-	-	-	-	-	X	S	-
White-bearded Antshrike-	-	S	S	-	-	-	-	-	-	-	-	-	-	-
Barred Antshrike	-	-	H	-	-	-	-	-	-	-	-	X	X	X
Lined Antshrike-	X	-	S	X	-	S	X	X	X	X	H	H	-	X
White-shouldered Antshrike	-	-	-	-	-	-	-	-	-	-	-	-	-	S
Eastern Slaty Antshrike-	S	S	S	X	S	-	X	X	X	X	X	X	X	X
Variable Antshrike	X	X	X	-	X	X	S	X	-	X	X	-	-	X
Rufous-winged Antshrike-	-	-	H	-	-	-	-	-	-	H	X	X	X	-
Rufous-capped Antshrike-	X	X	X	-	-	X	-	-	-	-	-	-	-	-
Spot-breasted Antvireo	X	X	X	X	-	X	X	-	-	-	X	-	-	-
Plain Antvireo	X	X	X	X	X	X	X	X	S	-	X	-	X	-
E Rufous-backed Antvireo	X	X	X	S	-	-	-	-	-	-	-	-	-	-
E Plumbeous Antvireo	-	-	-	-	A	S	S	X	-	-	H	-	-	-
Cinereous Antshrike-	-	-	S	-	S	X	X	X	X	X	H	-	X	-
E Star-throated Antwren	X	X	X	X	-	S	-	-	-	-	-	-	-	-
White-flanked Antwren	-	X	X	-	-	X	X	X	X	H	H	-	H

-153-

	SdM	Ita	SdO	RdJ	PNC	N.L	FMC	RDS	Soo	S.B	Sal	B.N	N.B	Mac
E Rio de Janeiro Antwren	-	-	-	A	-	-	-	-	-	-	-	-	-	-
E Salvadori's Antwren-	-	-	X	-	S	-	-	S	-	-	S	S	-	-
E Unicolored Antwren	-	-	X	-	S	S	-	A	-	-	-	-	-	-
E Bititinga Antwren	-	-	-	-	-	-	-	-	-	-	-	-	-	S
E Band-tailed Antwren	-	-	-	-	-	-	-	-	X	X	X	H	-	-
Stripe-backed Antbird	-	-	-	-	-	-	-	-	-	-	-	X	X	S
Black-capped Antwren	-	-	-	H	-	X	-	-	-	-	X	X	X	X
E Pileated Antwren	-	-	-	-	-	-	-	-	-	-	S	S	-	-
E Pectoral Antwren	-	-	-	-	-	-	-	-	-	-	H	-	S	-
Large-billed Antwren	-	-	-	-	-	-	A	-	-	-	-	-	-	-
Rufous-winged Antwren	-	-	X	S	H	X	-	X	X	X	X	X	-	X
E Narrow-billed Antwren	-	-	-	-	-	-	-	-	-	-	-	X	H	-
White-fringed Antwren	-	-	-	-	-	-	-	S	-	X	X	-	-	X
Black-bellied Antwren	-	-	-	-	-	-	-	-	-	-	-	X	X	H
E Serra Antwren	-	-	-	H	-	S	S	S	-	-	-	-	-	-
"Restinga" Antwren	-	-	-	S	-	-	-	-	-	-	-	-	-	-
E Black-hooded Antwren	-	-	S	-	H	-	-	-	-	-	-	-	-	-
Rusty-backed Antwren	-	-	-	S	-	-	-	-	S	-	-	-	-	-
E Ferruginous Antbird-	-	-	X	X	X	S	X	X	-	-	X	X	-	-
Bertoni's Antbird	-	-	S	X	S	-	-	-	-	-	-	-	-	-
E Rufous-tailed Antbird	-	-	X	X	X	-	-	-	-	-	-	-	-	-
E Ochre-rumped Antbird	-	-	X	X	X	S	X	-	-	-	S	-	-	-
Dusky-tailed Antbird	-	-	X	A	S	S	X	-	-	-	-	-	-	-
E Scaled Antbird	-	-	X	-	X	X	X	-	X	X	X	H	-	X
Streak-capped Antwren	-	-	X	S	X	X	-	X	-	-	X	-	S	-
E Alagoas Antwren-	-	-	-	-	-	-	-	-	-	-	-	-	-	X
E Rio de Janeiro Antbird	-	-	A	S	H	A	-	A	-	-	S	-	-	-
E Lower Amazonian Antbird-	-	-	-	-	-	-	-	-	-	-	-	-	-	S
White-backed Fire-eye	-	-	-	-	-	-	-	-	-	-	-	-	-	X
E Fringe-backed Fire-eye	-	-	-	-	-	-	-	-	-	-	S	-	-	-
White-shouldered Fire-eye	X	X	X	X	X	X	X	-	X	X	X	X	H	-
E Slender Antbird-	-	-	-	-	-	-	-	-	-	-	S	-	-	-
E Scalloped Antbird	-	-	-	-	-	-	-	-	S	H	-	H	-	X
E Squamate Antbird	-	-	X	-	-	-	-	-	-	-	-	-	-	-
E White-bibbed Antbird	-	-	-	X	X	S	S	X	-	X	-	-	X	-
Short-tailed Antthrush	-	-	-	-	-	X	X	-	-	-	-	-	-	X
E Such's Antthrush	-	-	X	X	X	S	-	S	-	-	-	-	-	-
E Brazilian Antthrush-	-	-	X	X	X	-	S	-	-	S	-	-	-	-
Rufous-capped Antthrush-	-	-	X	-	S	-	-	-	X	X	X	X	H	X
Variegated Antpitta-	-	-	X	X	X	S	-	X	-	S	S	S	-	-
E White-browed Antpitta	-	-	-	-	-	-	-	-	-	-	S	-	-	-
Speckle-breasted Antpitta	-	-	X	X	S	-	X	-	-	-	-	-	-	-

-154-

	Species	SdM	Ita	SdO	RdJ	PNC	N.L	FMC	RDS	Soo	S.B	Sal	B.N	N.B	Mac
E	Black-cheeked Gnateater	X	H	X	X	-	S	X	X	X	X	X	X	-	X
	Rufous Gnateater	X	X	X	S	X	X	X	X	-	S	-	X	-	X
	"Caatinga" Gnateater	-	-	-	-	-	-	-	-	-	-	-	H	-	-
	Spotted Bamboowren	X	A	S	-	-	-	-	-	-	S	-	-	-	-
E	Slaty Bristlefront	X	X	-	S	-	-	-	-	-	H	-	-	-	-
E	Stresemann's Bristlefront	-	-	-	-	-	-	-	-	-	H	-	-	-	-
	Mouse-colored Tapaculo	X	X	X	X	-	-	-	-	-	-	-	X	-	-
E	Brasilia Tapaculo	-	-	-	-	-	?	-	-	-	-	-	-	-	-
E	White-breasted Tapaculo	S	S	-	-	?	S	-	-	-	-	-	-	-	-
E	Bahia Tapaculo	-	-	-	-	-	-	-	-	-	-	A	-	-	-
	Shrike-like Cotinga	A	A	S	A	A	A	A	-	A	-	-	-	-	-
	Swallow-tailed Cotinga	S	S	S	H	S	-	-	-	-	-	-	-	-	-
E	Black-and-gold Cotinga	X	X	X	-	-	-	-	-	-	-	-	-	-	-
E	Gray-winged (Orgaos) Cotinga	-	-	S	-	-	-	-	-	-	-	-	-	-	-
E	Hooded Berryeater	X	A	X	-	-	X	-	-	-	-	A	-	-	-
E	Black-headed Berryeater	-	-	H	-	-	-	-	-	X	X	S	H	-	S
E	Banded Cotinga	-	-	H	-	-	-	-	-	S	S	H	H	-	-
E	White-winged Cotinga	-	-	H	-	-	-	-	-	X	X	X	-	-	S
E	Buff-throated Purpletuft	S	A	A	-	-	-	-	-	-	-	-	-	-	-
	"Alagoas" Purpletuft	-	-	-	-	-	-	-	-	-	-	-	-	-	S
E	Kinglet Calyptura	-	-	-	-	H	H	-	-	-	-	-	-	-	-
	Screaming Piha	-	-	-	-	-	-	-	-	X	X	X	H	-	X
E	Cinnamon-vented Piha	A	A	H	-	S	S	A	S	S	A	-	-	-	-
	Green-backed Becard	X	X	X	H	X	X	X	X	X	-	H	X	X	X
	Chestnut-crowned Becard	X	X	X	S	X	X	X	-	X	-	-	X	H	-
	White-winged Becard	X	X	X	X	X	X	X	X	X	X	X	X	X	X
	Black-capped Becard	X	-	-	H	-	X	-	-	X	X	-	H	H	X
	Crested Becard	M	M	M	M	S	S	X	S	X	S	S	H	S	S
	Black-tailed Tityra	X	X	X	-	S	X	X	X	X	X	H	H	-	-
	Black-crowned Tityra	H	S	S	-	S	S	S	-	X	X	-	H	-	-
	Red-ruffed Fruitcrow	S	S	S	H	S	S	-	X	S	S	H	-	-	-
	Bare-throated Bellbird	X	S	S	S	-	X	-	-	X	X	X	S	H	S
	Bearded Bellbird	-	-	-	-	-	-	-	-	-	-	-	-	-	S
	Sharpbill	X	X	X	S	-	X	X	-	S	-	-	X	-	S
	Red-headed Manakin	-	-	H	-	-	S	-	-	X	X	X	-	-	X
	White-crowned Manakin	-	-	-	-	-	-	-	-	S	X	X	-	-	-
	Blue-backed Manakin	-	-	H	-	-	-	-	-	-	-	X	-	-	X
	Swallow-tailed Manakin	X	X	X	X	X	X	-	-	-	-	S	X	-	-
E	Pin-tailed Manakin	X	X	X	X	X	X	X	X	-	-	X	-	-	-
	White-bearded Manakin	X	S	X	X	-	X	X	X	X	X	X	X	-	X
	Striped Manakin	-	-	S	H	-	-	-	-	S	X	X	-	-	-
E	Wied's Tyrant-Manakin	S	S	S	-	X	X	-	S	S	-	H	H	-	-

-155-

Species	SdM	Ita	SdO	RdJ	PNC	N.L	FMC	RDS	Soo	S.B	Sal	B.N	N.B	Mac
Pale-bellied Tyrant-Manakin	-	-	-	-	X	-	S	S	-	-	H	-	-	S
E Black-capped Manakin	S	X	H	-	-	-	-	-	-	-	-	-	-	-
Wing-barred Manakin	X	-	-	-	S	X	-	-	-	-	-	-	-	-
Greenish Manakin	X	X	X	S	X	X	-	-	-	-	-	X	-	-
Thrush-like Manakin	-	-	-	-	-	-	S	X	X	X	X	X	-	S
Gray Monjita	H	X	-	S	X	-	X	X	-	-	-	S	-	-
White-rumped Monjita	X	X	-	S	X	-	X	X	-	-	-	-	-	-
White Monjita	-	-	-	-	-	-	-	-	A	-	-	X	X	-
Long-tailed Tyrant	X	X	X	S	X	X	X	X	-	-	-	X	-	-
Streamer-tailed Tyrant	H	X	H	X	-	-	X	X	-	-	-	-	-	-
Cock-tailed Tyrant	-	-	H	-	-	-	-	-	-	-	-	-	-	-
Crested Black-Tyrant	H	M	M	-	M	-	M	M	-	-	-	-	-	-
E Velvety Black-Tyrant	-	X	X	X	X	-	X	-	-	-	-	-	S	-
White-winged Black-Tyrant	-	-	-	-	A	-	-	-	-	-	-	-	S	-
Blue-billed Black-Tyrant	X	X	X	-	S	S	-	-	-	-	-	-	-	-
Spectacled Tyrant	-	-	-	A	-	-	-	-	-	-	-	-	-	-
Shear-tailed Gray Tyrant	M	M	M	M	M	M	-	M	-	-	-	-	-	-
Black-backed Water-Tyrant	-	-	-	-	-	-	M	-	-	-	-	-	M	-
Masked Water-Tyrant	S	X	X	X	X	X	X	X	X	-	X	X	X	X
White-headed Marsh-Tyrant	-	X	S	X	X	H	X	X	X	-	X	X	X	X
Vermilion Flycatcher	-	-	-	H	-	-	-	-	-	-	-	-	-	-
Yellow-browed Tyrant	X	X	X	H	X	X	X	X	X	-	-	X	X	-
Cattle Tyrant	X	X	S	S	X	X	X	X	X	X	X	X	X	H
Sirystes	X	X	H	-	X	X	X	X	X	-	-	X	-	X
Fork-tailed Flycatcher	M	M	M	M	-	-	M	M	M	M	M	-	-	-
Tropical Kingbird	X	X	X	X	X	X	X	X	X	X	X	X	X	X
White-throated Kingbird	-	-	-	-	-	-	A	-	-	-	-	-	-	-
Variegated Flycatcher	M	M	M	M	M	M	M	M	M	M	M	M	M	M
Piratic Flycatcher	M	M	M	-	-	M	M	M	M	M	-	-	-	M
Three-striped Flycatcher	S	A	-	-	-	-	-	-	S	S	S	S	-	-
Boat-billed Flycatcher	X	X	X	H	X	X	X	X	X	X	X	X	X	X
Streaked Flycatcher	M	M	M	M	M	M	M	M	M	M	H	H	H	M
Rusty-margined Flycatcher	-	-	S	-	-	S	S	-	-	S	S	S	S	-
Social Flycatcher	X	X	X	X	X	X	X	X	X	X	X	X	X	X
Great Kiskadee	X	X	X	X	X	X	X	X	X	X	X	X	X	X
Lesser Kiskadee	S	-	-	-	-	-	-	-	S	-	X	-	-	-
Bright-rumped Attila	-	-	-	-	-	-	-	-	X	X	X	-	-	H
E Gray-hooded Attila	X	X	X	X	X	X	-	X	X	X	X	X	-	-
Rufous-tailed Attila	M	M	-	-	-	-	-	-	-	-	-	-	-	-
Rufous Casiornis	-	-	-	-	-	-	S	-	-	-	-	-	S	-
E Ash-throated Casiornis	-	-	-	-	-	-	-	-	-	-	-	-	S	S
Cinereous Mourner	-	-	-	-	-	-	-	-	S	S	S	H	-	-

E	Species	SdM	Ita	SdO	RdJ	PNC	N.L	FMC	RDS	Soo	S.B	Sal	B.N	N.B	Mac
	Grayish Mourner	S	H	S	S	-	X	-	X	X	X	X	X	-	X
	Short-crested Flycatcher	S	S	S	S	X	X	X	X	X	X	X	X	-	X
	Brown-crested Flycatcher	S	-	-	-	X	S	X	S	S	S	H	X	X	H
	Swainson's Flycatcher	M	M	M	M	M	M	-	-	M	M	H	-	H	-
	Dusky-capped Flycatcher	-	-	S	H	-	S	S	S	X	S	S	-	-	S
	Olive-sided Flycatcher	M	M	-	-	-	-	-	-	-	-	-	-	-	-
	Tropical Pewee	X	X	S	S	X	X	X	-	X	-	-	X	-	X
	Euler's Flycatcher	X	X	X	X	X	X	X	X	X	-	-	X	H	X
	Fuscous Flycatcher	S	-	H	S	-	-	-	-	S	-	-	-	-	-
	"Audible" Fuscous-Flycatcher	-	A	H	-	S	-	-	-	-	-	-	S	-	-
	Sulphur-rumped Flycatcher	X	X	X	X	-	X	X	-	X	X	H	X	H	X
	Black-tailed Flycatcher	S	X	X	-	-	S	X	X	S	S	-	-	-	-
	Bran-colored Flycatcher	X	X	X	X	X	X	X	X	X	X	X	X	X	X
	Swallow Flycatcher	X	X	X	X	X	-	X	X	-	X	-	X	X	X
	Atlantic Royal-Flycatcher	S	S	H	-	-	-	-	-	-	S	-	-	-	-
	Russet-winged Spadebill	S	-	H	A	-	S	-	-	-	-	-	-	-	-
	White-throated Spadebill	X	X	X	X	X	X	X	X	-	-	-	X	-	X
	Yellow-olive Flycatcher	X	X	X	X	X	X	X	X	S	-	-	X	-	X
	Gray-crowned Flycatcher	-	S	S	-	-	S	-	-	X	X	-	H	-	-
	Yellow-breasted Flycatcher	-	-	H	X	-	-	-	X	X	X	X	X	X	X
	Olivaceous Flatbill	-	-	-	-	-	-	X	X	X	X	H	H	-	-
	Large-headed Flatbill	X	X	H	-	-	S	-	-	-	-	-	-	-	-
E	Yellow-lored Tody-Flycatcher	X	X	S	X	X	X	X	X	-	-	-	X	-	-
	Common Tody-Flycatcher	-	-	A	X	-	A	-	-	X	X	X	X	X	X
	Smoky-fronted Tody-Flycatcher	-	-	-	-	-	-	-	-	-	-	H	-	-	S
	Ochre-faced Tody-Flycatcher	X	X	X	-	X	X	X	X	-	-	-	X	-	X
	Stripe-necked Tody-Tyrant	-	-	-	-	-	-	-	-	-	-	H	-	-	-
	Pearly-vented Tody-Tyrant	-	-	-	-	-	-	S	S	-	-	-	X	X	X
E	Buff-breasted Tody-Tyrant	-	-	-	-	-	-	-	-	-	-	-	-	-	S
	White-eyed Tody-Tyrant	-	-	-	-	-	-	-	-	-	-	-	-	-	S
E	Eye-ringed Tody-Tyrant	X	A	X	X	S	S	-	-	-	-	-	-	-	-
E	Hangnest Tody-Tyrant	X	S	S	-	X	X	-	-	-	-	H	X	H	-
E	Fork-tailed Tody(Pygmy)-Tyrant	X	A	H	-	-	-	-	-	-	-	-	-	-	-
	Drab-breasted Bamboo-Tyrant	X	X	X	S	X	X	-	-	-	-	-	X	-	-
E	Brown-breasted Bamboo-Tyrant	X	X	-	-	-	-	-	-	-	-	-	-	-	-
	Eared Pygmy-Tyrant	X	X	X	X	X	X	X	X	X	X	-	H	-	-
	Southern Bristle-Tyrant	A	H	H	-	-	-	-	-	-	-	-	-	-	-
	Bay-ringed Tyrannulet	X	-	-	-	-	A	A	-	-	-	-	-	-	-
	Mottle-cheeked Tyrannulet	X	X	X	-	X	-	S	-	-	-	-	-	-	-
E	Oustalet's Tyrannulet	X	-	S	H	-	X	-	-	-	-	-	-	-	-
E	Serra do Mar Tyrannulet	X	X	X	-	-	-	-	-	-	-	-	-	-	-
	Sao Paulo Tyrannulet	S	-	-	-	-	-	-	-	-	-	-	-	-	-

	SdM	Ita	SdO	RdJ	PNC	N.L	FMC	RDS	Soo	S.B	Sal	B.N	N.B	Mac
E Alagoas Tyrannulet	-	-	-	-	-	-	-	-	-	-	-	-	-	X
Yellow Tyrannulet	X	X	X	S	X	X	X	X	X	X	X	X	-	X
Tawny-crowned Pygmy-Tyrant	S	-	S	S	-	-	-	S	-	S	-	X	X	S
Crested Doradito	-	-	-	S	-	-	-	-	-	-	-	-	-	-
E Gray-backed Tachuri	A	-	-	-	X	-	-	-	-	-	-	-	H	-
Greater Wagtail-Tyrant	-	-	-	-	-	-	-	-	-	-	-	-	H	S
Lesser Wagtail-Tyrant	-	-	-	-	-	-	-	-	-	-	-	-	X	-
White-crested Tyrannulet	X	X	X	M	X	X	X	X	-	-	-	X	X	X
Sooty Tyrannulet	X	X	H	-	S	S	-	-	-	-	-	-	-	-
Yellow-bellied Elaenia	X	X	X	X	X	X	X	X	X	X	X	X	X	X
Large Elaenia	-	-	-	A	-	-	-	-	-	-	M	M	-	-
White-crested Elaenia	M	H	-	M	-	-	-	-	-	-	-	-	H	-
Small-billed Elaenia	M	H	-	-	-	-	M	-	-	-	-	-	-	-
Olivaceous Elaenia	M	M	M	-	M	M	M	M	-	M	-	-	-	-
Plain-crested Elaenia	-	-	-	-	-	-	-	-	-	-	-	-	S	S
Lesser Elaenia	-	-	-	-	-	-	M	M	-	M	-	-	-	-
Highland Elaenia	S	S	S	H	X	S	-	S	-	S	-	-	-	-
Forest Elaenia	-	-	S	-	-	-	-	-	X	X	X	-	-	S
Gray Elaenia	X	X	X	-	S	S	S	S	S	-	H	-	-	S
Greenish Elaenia	-	A	-	-	A	-	-	-	-	-	-	-	H	H
Campo Suiriri	-	-	-	-	-	-	-	-	-	-	S	S	-	-
Southern Scrub-Flycatcher	-	-	-	-	-	-	-	-	-	-	S	S	-	-
Mouse-colored Tyrannulet	-	-	-	-	S	-	-	-	-	-	H	X	X	X
Southern Beardless-Tyrannulet	X	X	X	X	X	X	X	X	X	X	X	X	X	X
Greenish Tyrannulet	S	S	S	-	S	A	-	-	-	-	-	-	-	-
Planalto Tyrannulet	X	X	X	A	X	-	-	-	S	-	-	X	H	S
Rough-legged Tyrannulet	X	X	X	X	-	X	-	-	-	-	-	S	-	S
E Gray-capped Tyrannulet	X	X	X	-	X	X	-	-	S	-	-	-	-	-
White-lored Tyrannulet	-	-	-	-	-	-	-	-	S	S	S	-	-	-
Sepia-capped Flycatcher	X	X	X	X	X	X	X	X	X	X	-	X	H	X
Ochre-bellied Flycatcher	-	-	S	-	-	-	-	-	X	X	X	-	-	-
Gray-hooded Flycatcher	X	X	X	X	X	X	X	X	-	-	-	-	-	-
Southern Antpipit	-	-	S	H	-	-	-	-	-	-	-	-	-	-
White-winged Swallow	H	-	H	-	-	X	X	X	X	X	X	X	X	X
White-rumped Swallow	M	M	M	M	-	-	-	-	M	-	-	-	-	-
Brown-chested Martin	M	M	M	M	-	-	M	M	M	M	M	M	-	M
Purple Martin	-	-	-	-	-	-	-	-	M	-	-	-	-	-
Gray-breasted Martin	X	X	X	X	X	X	X	X	X	X	X	X	X	X
Blue-and-white Swallow	X	X	X	X	X	X	X	X	X	X	X	X	X	-
White-thighed Swallow	M	M	M	-	-	-	-	-	-	-	-	M	-	-
Tawny-headed Swallow	-	M	-	A	-	-	-	-	-	-	-	-	-	-
Southern Rough-winged Swallow	X	X	X	X	X	X	X	X	X	X	X	X	X	X

	SdM	Ita	SdO	RdJ	PNC	N.L	FMC	RDS	Soo	S.B	Sal	B.N	N.B	Mac
Bank Swallow (Sand Martin)	-	-	-	-	M	-	-	-	-	-	-	-	-	-
Barn Swallow	-	M	M	H	-	-	M	-	M	M	-	-	-	M
Cliff Swallow	-	-	-	A	-	-	-	-	-	-	-	-	-	-
Curl-crested Jay	-	-	S	-	-	-	-	-	-	-	-	-	-	-
Plush-crested Jay	-	-	S	-	-	-	-	-	-	-	-	-	-	-
E White-naped Jay-	-	-	-	-	-	-	-	-	-	-	-	-	X	H
Thrush-like Wren	-	-	-	-	-	-	-	-	X	X	-	H	-	-
Moustached Wren-	-	-	-	A	-	-	X	X	X	X	X	X	H	X
Buff-breasted Wren	-	-	-	-	-	-	A	-	-	-	-	-	-	-
E Long-billed Wren	X	-	S	X	-	-	-	-	S	-	X	X	X	X
House Wren	X	X	X	X	X	X	X	X	X	X	X	X	X	X
Black-capped Donacobius-	-	S	H	X	X	-	X	X	X	X	X	X	-	X
Tropical Mockingbird	-	-	-	-	X	-	-	-	-	-	-	-	-	-
Chalk-browed Mockingbird	X	X	S	S	X	X	X	X	X	X	X	X	X	X
Rufous-brown Solitaire	-	-	-	-	-	-	S	-	-	-	-	H	-	-
Swainson's Thrush	A	-	A	-	-	-	-	-	-	-	-	-	-	-
Yellow-legged Thrush	X	X	X	X	X	X	X	X	X	X	-	-	-	-
Eastern Slaty-Thrush	-	-	M	-	-	-	-	-	-	-	-	-	-	-
Rufous-bellied Thrush	X	X	X	X	X	X	X	X	X	X	X	X	X	X
Pale-breasted Thrush	-	S	S	S	X	S	X	X	X	S	X	-	X	X
Creamy-bellied Thrush	X	M	X	X	X	X	X	X	X	S	X	-	-	X
Cocoa Thrush	-	-	-	-	-	-	-	-	X	X	X	H	-	X
White-necked Thrush-	X	X	X	X	X	X	X	-	S	-	X	X	-	X
Long-billed Gnatwren	X	-	S	S	-	S	X	X	X	X	X	X	-	X
Tropical Gnatcatcher	-	-	-	-	-	-	-	-	-	-	H	X	X	X
Hellmayr's Pipit	-	X	X	-	-	-	-	-	-	-	-	-	-	-
Yellowish Pipit-	X	S	H	X	-	-	-	X	X	-	H	X	H	X
Rufous-browed Peppershrike	X	X	X	X	X	X	X	X	X	X	X	X	X	X
Red-eyed Vireo	X	X	X	X	-	X	X	X	X	X	X	H	X	X
Rufous-crowned Greenlet-	X	X	X	S	-	X	-	-	A	-	-	-	-	-
E Gray-eyed Greenlet	-	-	-	-	X	-	X	X	-	H	X	X	X	X
Lemon-chested Greenlet	X	S	S	X	-	-	X	-	S	-	-	X	-	-
Shiny Cowbird	X	X	X	X	X	X	X	X	X	X	H	X	X	X
Bay-winged Cowbird	-	-	-	-	-	-	-	-	-	-	S	S	X	X
Giant Cowbird	X	-	S	H	S	-	-	-	X	X	-	S	-	-
Crested Oropendola	S	H	S	S	X	X	X	X	S	X	-	S	-	-
Yellow-rumped Cacique	-	-	-	-	-	-	-	-	-	-	X	X	-	H
Red-rumped Cacique	X	X	X	A	S	-	X	X	X	X	X	H	-	H
Golden-winged Cacique	X	X	-	-	-	-	-	-	-	-	-	-	-	-
E Forbes's Blackbird	-	-	-	-	-	-	-	X	-	-	-	-	-	X
Chopi Blackbird-	H	X	H	S	X	-	X	X	X	X	-	X	X	H
Chestnut-capped Blackbird	-	S	-	S	-	-	X	X	X	-	X	X	X	X

	SdM	Ita	SdO	RdJ	PNC	N.L	FMC	RDS	Soo	S.B	Sal	B.N	N.B	Mac
Unicolored Blackbird	-	-	-	-	S	-	-	-	S	S	-	-	-	-
Epaulet Oriole	-	A	-	H	-	-	-	-	S	-	H	H	X	X
Troupial (Campo Oriole)	-	-	-	-	-	-	-	-	-	S	X	X	X	X
Yellow-rumped Marshbird	-	S	-	-	-	-	-	-	-	-	-	-	-	-
White-browed Blackbird	-	S	-	S	X	-	-	X	X	X	X	X	X	X
Bobolink	-	-	-	H	-	-	-	-	-	-	-	-	-	-
Tropical Parula	-	X	X	X	X	X	X	X	X	X	X	X	X	X
Cerulean Warbler	-	A	-	A	-	-	-	-	-	-	-	-	-	-
Blackburnian Warbler	-	-	-	A	-	-	A	-	-	-	-	-	-	-
Blackpoll Warbler	-	A	-	A	A	-	-	-	-	-	-	-	-	-
Masked Yellowthroat	-	X	X	X	X	X	X	X	X	X	-	X	-	X
Flavescent Warbler	-	-	-	-	-	S	-	-	-	S	H	X	H	X
Golden-crowned Warbler	-	X	X	X	X	X	X	X	-	X	-	X	-	X
White-bellied Warbler	-	-	-	-	-	X	-	S	-	-	-	-	-	-
White-rimmed Warbler	-	X	X	X	H	-	A	-	-	-	-	-	-	-
Neotropical River Warbler	-	X	-	H	-	S	-	-	X	-	-	-	X	-
Bananaquit	-	X	X	X	X	X	X	X	X	X	X	X	X	X
Chestnut-vented Conebill	-	X	X	S	X	X	S	X	X	X	-	X	X	X
Bicolored Conebill	-	H	-	-	S	-	-	-	-	-	X	-	H	-
Red-legged Honeycreeper	-	-	-	-	S	-	-	-	X	X	X	H	-	S
Green Honeycreeper	-	S	-	H	H	-	X	-	X	S	X	X	H	-
Blue Dacnis	-	X	X	X	X	X	X	X	X	X	X	X	X	X
E Black-legged Dacnis	-	S	A	S	-	-	A	-	-	-	-	-	-	-
Swallow-Tanager	-	H	M	M	M	X	-	S	S	S	S	-	-	-
Blue-naped Chlorophonia	-	X	X	X	S	X	X	-	-	-	S	-	-	-
Golden-rumped Euphonia	-	X	S	-	H	S	-	-	-	-	-	-	-	-
Orange-bellied Euphonia	-	-	-	S	S	-	-	-	X	X	X	-	H	-
Purple-throated Euphonia	-	-	X	X	X	X	S	-	X	X	S	X	X	X
Violaceous Euphonia	-	X	X	X	X	X	X	X	X	X	X	X	-	X
Chestnut-bellied Euphonia	-	X	X	X	X	-	X	X	X	X	X	X	-	S
Green-chinned Euphonia	-	A	A	A	H	-	-	-	-	-	-	-	-	-
Fawn-breasted Tanager	-	X	X	X	X	-	S	S	S	-	-	-	-	-
Silvery-breasted Tanager	-	-	-	H	-	-	-	-	X	X	X	H	-	-
E Seven-colored Tanager	-	-	-	-	-	-	-	-	-	-	-	-	-	X
Green-headed Tanager	-	X	X	X	X	X	X	X	X	X	X	-	-	-
Red-necked Tanager	-	X	-	X	X	-	X	X	X	-	-	-	-	X
E Gilt-edged Tanager	-	X	X	-	-	X	X	S	-	-	-	X	H	-
E Brassy-breasted Tanager	-	X	X	X	-	X	-	S	-	-	-	-	-	-
White-bellied Tanager	-	-	-	S	H	-	-	S	-	X	X	-	-	-
E Black-backed Tanager	-	M	-	-	M	-	-	-	-	-	-	-	-	-
Burnished-buff Tanager	-	X	X	X	S	X	X	X	X	A	X	X	X	X
Diademed Tanager	-	S	X	X	-	-	-	-	-	-	-	-	-	-

	SdM		SdO		PNC		FMC		Soo		Sal		N.B	
		Ita		RdJ		N.L		RDS		S.B		B.N		Mac
Sayaca Tanager	X	X	X	X	X	X	X	X	X	X	X	X	X	X
E Azure-shouldered Tanager	X	S	S	S	-	S	-	-	-	-	-	-	-	-
E Golden-chevroned Tanager	X	X	X	X	X	X	X	-	X	X	-	X	-	-
Palm Tanager	X	X	X	X	X	X	X	X	X	X	X	X	H	X
Brazilian Tanager	X	S	S	X	-	-	-	X	X	X	X	X	H	H
Silver-beaked Tanager	-	-	-	-	-	-	-	X	-	-	-	-	-	-
Lowland Hepatic-Tanager	S	S	S	H	X	-	-	-	A	-	-	-	-	S
E Olive-green Tanager	X	X	S	-	-	S	-	-	-	-	-	-	-	-
Red-crowned Ant-Tanager	X	X	X	X	-	X	X	-	X	X	-	X	-	H
White-lined Tanager	-	M	M	M	-	S	S	-	S	S	S	X	X	X
Ruby-crowned Tanager	X	X	X	X	X	X	X	X	-	-	-	-	-	-
Flame-crested Tanager	X	S	X	X	-	X	X	X	X	X	-	X	-	X
Black-goggled Tanager	X	X	X	X	X	X	X	X	-	S	-	X	-	-
Chestnut-headed Tanager	-	X	S	A	-	-	-	-	-	-	-	-	-	-
Hooded Tanager	-	A	S	A	S	S	X	X	X	-	X	X	X	X
E Rufous-headed Tanager	X	X	X	X	X	X	X	-	X	-	-	X	-	-
Guira Tanager	-	-	-	-	-	-	-	S	-	-	-	-	-	X
Yellow-backed Tanager	-	-	X	X	-	-	-	X	X	X	-	-	-	H
Orange-headed Tanager	S	X	S	X	X	-	X	X	-	X	X	X	X	X
E Scarlet-throated Tanager	-	-	-	-	-	-	-	-	-	-	H	H	S	S
E Brown Tanager	X	X	X	-	-	S	-	-	-	-	H	-	-	-
Magpie Tanager	S	X	H	-	X	H	X	X	-	X	-	-	-	-
E Cinnamon Tanager	S	X	S	S	X	X	X	X	-	S	H	X	X	X
Black-faced Tanager	-	-	-	-	S	-	-	-	S	-	-	-	-	-
Buff-throated Saltator	S	-	S	X	-	X	X	X	X	X	X	X	-	X
Grayish Saltator	-	-	-	-	-	-	-	-	-	-	-	-	H	-
Green-winged Saltator	X	X	X	X	X	X	X	X	S	S	-	X	X	-
Thick-billed Saltator	S	X	S	-	-	-	-	-	-	-	-	-	-	-
Black-throated Saltator	-	-	H	-	X	-	-	-	-	-	-	-	X	-
Yellow-green Grosbeak	-	M	S	S	-	X	X	X	X	X	X	X	-	S
Black-throated Grosbeak	X	X	M	M	-	X	X	-	S	-	-	H	-	X
E Red-cowled Cardinal	-	-	-	-	-	-	-	-	-	S	X	X	X	X
Ultramarine Grosbeak	A	S	S	S	X	-	X	X	S	-	H	X	X	X
Blue Finch	-	-	S	-	-	-	-	-	-	-	-	-	-	-
Blue-black Grassquit	X	X	X	X	X	X	X	X	X	X	X	X	X	X
Sooty Grassquit	M	-	-	S	-	S	S	-	X	-	-	-	-	X
Buffy-fronted Seedeater	M	M	M	M	-	M	-	-	A	-	-	-	-	-
Temminck's Seedeater	M	M	M	-	-	-	-	-	A	-	A	-	-	-
Plumbeous Seedeater	-	-	-	H	-	-	-	-	-	-	-	-	-	-
Rusty-collared Seedeater	-	-	-	-	S	-	-	-	S	S	-	-	-	-
Lined Seedeater	-	-	-	-	M	-	-	-	M	M	-	-	-	M
Yellow-bellied Seedeater	-	-	S	M	X	X	X	X	S	X	H	X	X	X

	Species	SdM	Ita	SdO	RdJ	PNC	N.L	FMC	RDS	Soo	S.B	Sal	B.N	N.B	Mac
E	Dubois's Seedeater	-	-	-	-	M	X	X	X	-	X	X	-	S	S
	Double-collared Seedeater	X	X	X	X	X	X	X	X	X	X	X	-	-	-
E	White-throated Seedeater	-	-	-	-	-	-	-	-	-	-	X	X	X	X
	White-bellied Seedeater	X	X	S	S	-	-	-	-	X	X	H	X	X	X
	Capped Seedeater	-	-	H	-	-	S	-	-	S	-	X	X	X	X
	"Rio de Janeiro" Seedeater	-	-	-	-	S	-	-	-	-	-	-	-	-	-
	Great-billed Seed-Finch	-	-	-	-	H	-	S	-	S	S	-	-	H	-
	Lesser Seed-Finch	A	S	S	S	-	S	-	S	S	S	S	H	-	H
	Blackish-blue Seedeater	S	S	S	-	-	-	-	-	-	-	-	-	-	-
	Stripe-tailed Yellow-Finch	-	-	-	-	-	-	-	-	-	-	-	S	-	-
	Orange-fronted Yellow-Finch	-	-	-	-	-	-	-	-	-	-	-	-	H	-
	Saffron Finch	S	S	S	X	-	-	X	X	S	S	S	S	-	X
	Misto Yellow-Finch	-	-	-	-	M	-	-	-	-	M	-	-	M	-
	Uniform Finch	X	X	X	X	M	M	-	M	-	-	-	-	-	-
	Coal-crested Finch	-	-	-	-	-	-	-	-	-	-	-	-	S	-
	Red Pileated-Finch	H	-	S	S	-	-	S	-	S	-	-	-	A	-
	Gray Pileated-Finch	-	-	S	A	S	X	X	X	X	-	H	X	X	H
	Pectoral Sparrow	-	-	-	-	-	-	X	X	X	X	X	X	-	X
	"Semipectoral" Sparrow	S	S	S	-	-	-	-	-	-	-	-	-	-	-
	Grassland Sparrow	-	-	S	S	X	-	S	-	X	X	X	X	X	X
	Rufous-collared Sparrow	X	X	X	X	X	X	X	X	X	S	-	X	X	X
	Wedge-tailed Grass-Finch	-	X	S	X	X	S	X	-	X	X	H	X	X	X
	Long-tailed Reed-Finch	S	S	-	-	-	-	-	-	-	-	-	-	-	-
E	Bay-chested Warbling-Finch	X	X	X	-	-	-	-	-	-	-	-	-	-	-
E	Cinereous Warbling-Finch	-	-	-	-	A	-	-	-	-	-	-	-	-	-
	Red-rumped(Buff-throated)W.-F.	X	X	S	-	-	-	-	-	-	-	-	-	-	-
	Great Pampa-Finch	-	X	-	-	-	-	-	-	-	-	-	-	-	-
E	Buff-throated Pampa-Finch	-	-	-	-	X	-	-	-	-	-	-	-	X	-
	Yellow-faced Siskin	-	-	-	-	-	-	-	-	-	-	-	-	A	S
	Hooded Siskin	X	X	X	M	X	-	X	-	-	-	-	-	-	-
	House Sparrow	X	X	X	X	X	X	X	-	X	X	X	X	X	X
	Common Waxbill	I	I	I	I	-	-	-	-	-	-	I	I	I	-

Amazonia and Atlantic Islands

S.L - São Luís including São Bento, Pindaré-Mirim and Rosário.

Bel - Belém. Includes the area east to Bragança and south to Igarapé-Miri.

Ama - Amapá. (Species within brackets refer to the Ilha de Marajó and Ilha Mexiana, both areas with similar fauna to Amapá, though in Pará.)

San - Santarém. Includes Alter do Chão south to Rurópolis.

ANP - Amazônia (Tapajós) National Park.

Man - Manaus. North to Balbina, east to Itacoatiara, west to Manacapuru and south to Careiro.

Tef - Tefé.

Tab - Tabatinga and Benjamin Constant. Includes east to São Paulo de Olivenca and west to Amaca Yacú National Park (Colombia) which is in parentheses.

PdN - Pico da Neblina National Park, west to Rio Içana and Rio Uaupés and east to Rio Padauari.

N.R - Northern Roraima. From the Rio Anaua Biological Reserve north to the Tepui Region. Includes the Ilha de Maraca.

R.B - Rio Branco, Acre. West to Sena Madureira, east to Plácido de Castro and south to Brasiléia. Records in parentheses are from adjacent parts of Bolivia (though species within curved brackets refer to Cruzeiro do Sul, in western Acre.)

Ron - Rondônia.

MVI - Ilha de Trindade and the Ilha de Martim Vaz.

FdN - Fernando de Noronha.

Harpy Eagle

	S.L	Bel	Ama	San	ANP	Man	Tef	Tab	PdN	N.R	R.B	Ron	MVI	FdN
Greater Rhea	-	-	-	-	-	-	-	-	-	-	-	(S)	-	-
Gray Tinamou	-	H	-	H	S	S	S	-	-	-	S	H	-	-
Great Tinamou	-	S	S	S	S	S	S	S	H	S	(H)	S	-	-
White-throated Tinamou	-	S	-	X	X	S	S	H	H	-	S	X	-	-
Cinereous Tinamou	H	X	X	H	X	S	X	X	-	X	X	X	-	-
Little Tinamou	-	X	X	X	X	X	S	X	H	X	X	X	-	-
Brown Tinamou	-	-	-	H	-	-	-	-	-	-	-	S	-	-
Undulated Tinamou	-	X	H	X	X	X	X	X	{S}	X	X	S	-	-
Rusty Tinamou	-	-	-	S	-	-	S	-	-	H	-	-	-	-
Bartlett's Tinamou	-	-	-	-	-	-	-	H	-	-	(H)	(H)	-	-
Variegated Tinamou	S	X	H	X	X	X	X	X	H	X	(X)	X	-	-
Gray-legged Tinamou	-	-	-	-	-	-	-	-	-	H	-	-	-	-
Red-legged Tinamou	-	-	-	S	-	-	-	H	-	-	-	S	-	-
Brazilian Tinamou	-	S	-	H	H	-	-	-	-	-	(S)	(S)	-	-
Barred Tinamou	-	-	-	-	-	-	-	-	-	H	-	-	-	-
Small-billed Tinamou	-	S	(S)	S	-	-	-	-	-	-	-	X	-	-
Tataupa Tinamou	-	-	-	-	-	-	-	-	-	-	-	H	-	-
Red-winged Tinamou	-	-	(S)	-	-	-	-	-	-	-	-	-	-	-
Least Grebe	-	S	(S)	-	S	S	-	-	-	-	-	S	-	-
Pied-billed Grebe	-	S	-	-	-	A	-	-	(A)	-	-	-	-	-
Black-browed Albatross	-	-	-	-	-	-	-	-	-	-	-	-	-	H
Herald (Trindade) Petrel	-	-	-	-	-	-	-	-	-	-	-	-	X	-
White-chinned Petrel	-	-	-	(A)	-	-	-	-	-	-	-	-	-	-
Cory's Shearwater	-	-	-	-	-	-	-	-	-	-	-	-	-	M
Great Shearwater	-	-	-	-	-	-	-	-	-	-	-	-	H	-
Manx Shearwater	-	-	-	-	-	-	-	-	-	-	-	-	-	M
Little Shearwater	-	-	-	-	-	-	-	-	-	-	-	-	-	A
Wilson's Storm-Petrel	-	-	-	-	-	-	-	-	-	-	-	-	-	M
Black-bellied Storm-Petrel	-	-	-	-	-	-	-	-	-	-	-	-	-	M
Leach's Storm-Petrel	-	A	H	-	-	-	-	-	-	-	-	-	-	-
Red-billed Tropicbird	-	-	-	-	-	-	-	-	-	-	-	-	M	X
White-tailed Tropicbird	-	-	-	-	-	-	-	-	-	-	-	-	-	X
Brown Pelican	-	-	M	-	-	H	-	-	-	H	-	-	-	-
Masked Booby	-	-	-	-	-	-	-	-	-	-	-	-	M	X
Red-footed Booby	-	-	-	-	-	-	-	-	-	-	-	-	X	X
Brown Booby	-	-	-	-	-	-	-	-	-	-	-	-	-	X
Neotropic Cormorant	X	-	X	X	X	M	X	X	(X)	X	-	X	-	-
Anhinga	X	X	X	X	X	X	S	X	H	X	X	X	-	-
Magnificent Frigatebird	-	M	-	H	-	-	-	-	-	-	-	-	-	X
Great Frigatebird	-	-	-	-	-	-	-	-	-	-	-	-	X	-
Lesser Frigatebird	-	-	-	-	-	-	-	-	-	-	-	-	M	-
Purple Heron	-	-	-	-	-	-	-	-	-	-	-	-	-	A

	S.L	Bel	Ama	San	ANP	Man	Tef	Tab	PdN	N.R	R.B	Ron	MVI	FdN
Grey Heron	-	-	A	-	-	-	-	-	-	-	-	-	-	-
Cocoi (White-necked) Heron	X	X	X	X	X	X	S	{X}	{X}	X	-	X	-	A
Great (Common) Egret	X	X	X	X	X	M	S	X	H	X	X	X	-	M
Snowy Egret	X	X	X	X	X	M	S	X	-	X	-	X	-	M
Little Blue Heron	X	X	X	X	-	A	-	-	{S}	X	-	-	-	-
Tricolored Heron	-	H	S	A	-	-	-	-	-	-	-	-	-	A
Striated Heron	X	X	X	X	X	X	X	X	H	X	X	X	-	A
Agami Heron	-	-	S	H	H	S	S	{S}	{S}	-	-	H	-	-
Squacco Heron	-	-	-	-	-	-	-	-	-	-	-	-	-	A
Cattle Egret	X	X	X	X	X	M	X	{X}	{X}	X	X	X	-	M
Capped Heron	X	-	X	X	X	X	X	{X}	{X}	X	X	X	-	-
Black-crowned Night-Heron	X	X	X	-	-	S	S	-	-	X	-	{X}	-	A
Yellow-crowned Night-Heron	X	X	(H)	-	-	-	-	-	{S}	-	-	-	-	-
Boat-billed Heron	-	H	H	H	-	S	S	{S}	{X}	X	-	{X}	-	-
Rufescent Tiger-Heron	X	X	X	S	S	X	S	{S}	H	X	X	X	-	-
Fasciated Tiger-Heron	-	-	-	-	-	-	-	-	{A}	-	-	-	-	-
Zigzag Heron	-	H	S	H	-	S	-	-	{S}	S	-	S	-	-
Stripe-backed Bittern	A	-	-	-	-	-	-	-	-	S	-	-	-	-
Least Bittern	H	H	H	H	-	M	-	-	-	S	-	-	-	-
Pinnated Bittern	-	-	-	-	-	-	-	-	-	S	-	-	-	-
Wood Stork	M	-	X	M	M	M	M	{M}	-	X	-	{X}	-	-
Maguari Stork	-	-	X	-	-	-	-	-	-	X	-	-	-	-
Jabiru	-	-	X	-	-	-	-	-	-	X	-	{X}	-	-
Buff-necked Ibis	-	-	X	X	-	-	-	-	-	X	-	-	-	-
Sharp-tailed Ibis	-	-	-	H	-	-	-	-	S	S	-	H	-	-
Green Ibis	X	X	X	X	X	X	X	{X}	H	X	(X)	X	-	-
Bare-faced Ibis	-	-	-	-	-	-	-	-	-	H	-	{X}	-	-
Scarlet Ibis	X	S	X	-	-	-	-	-	{A}	-	-	-	-	-
Roseate Spoonbill	X	-	X	-	-	A	A	-	-	{S}	-	{X}	-	-
Greater Flamingo	-	-	S	-	-	-	-	-	-	-	-	-	-	-
Horned Screamer	-	-	X	X	-	X	X	{X}	-	-	-	{X}	-	-
Southern Screamer	-	-	-	-	-	-	-	-	-	-	-	{X}	-	-
Fulvous Whistling-Duck	-	-	(H)	-	-	-	-	-	-	-	-	-	-	-
White-faced Whistling-Duck	M	M	M	-	-	A	-	-	-	M	-	-	-	-
Black-bellied Whistling-Duck	M	-	M	M	M	M	-	-	{M}	M	-	-	-	-
Orinoco Goose	-	-	H	H	-	-	-	-	{A}	S	-	S	-	-
White-cheeked Pintail	-	-	H	-	-	-	-	-	-	-	-	-	-	-
Northern Pintail	-	-	-	-	-	-	-	-	-	-	-	-	-	A
Blue-winged Teal	-	-	-	-	-	-	-	-	{A}	-	-	-	-	-
Brazilian Duck (Teal)	X	H	X	S	S	-	A	-	-	S	-	-	-	-
Comb Duck	M	-	(H)	-	-	-	-	-	-	-	-	-	-	-
Muscovy Duck	X	H	X	X	X	S	X	X	-	X	-	X	-	-

Species	S.L		Ama		ANP		Tef		PdN		R.B		MVI	
		Bel		San		Man		Tab		N.R		Ron		FdN
Masked Duck	-	H	-	-	-	S	S	-	-	-	-	-	-	-
King Vulture	-	H	S	S	S	S	S	{S}	{S}	X	(H)	S	-	-
Black Vulture	X	X	X	X	X	X	X	X	-	X	X	X	-	X
Turkey Vulture	X	X	X	X	X	X	X	X	{X}	X	X	X	-	-
Lesser Yellow-headed Vulture	X	X	X	X	S	S	-	-	-	-	X	-	X	-
Greater Yellow-headed Vulture	-	X	X	X	X	X	X	X	{X}	X	X	X	-	-
White-tailed Kite	-	X	X	(H)	X	-	-	-	-	-	X	-	{X}	-
Pearl Kite	-	X	-	X	X	X	S	-	{X}	-	X	-	X	-
Swallow-tailed Kite	-	-	M	M	M	M	M	M	M	H	M	M	M	-
Gray-headed Kite	-	S	S	S	H	S	S	-	{S}	H	S	S	-	-
Hook-billed Kite	-	-	-	S	H	-	S	S	{S}	{S}	{S}	-	S	-
Rufous-thighed Kite	-	-	H	-	H	-	H	-	-	-	H	-	-	-
Double-toothed Kite	-	X	H	X	X	X	X	X	{S}	X	(X)	X	-	-
Plumbeous Kite	-	-	M	M	M	M	M	M	M	{M}	M	M	M	-
Mississippi Kite	-	-	-	-	-	A	-	-	-	-	-	-	-	-
Snail Kite	-	M	M	M	M	M	M	M	{M}	-	M	-	M	-
Slender-billed Kite	-	S	S	H	-	-	S	S	{S}	-	-	-	-	-
Bicolored Hawk	-	S	-	H	-	S	-	-	-	-	S	-	-	-
Tiny Hawk	-	S	-	H	-	S	S	{S}	-	-	-	-	H	-
Gray-bellied Hawk	-	H	-	H	-	A	A	-	H	S	-	{H}	-	-
Sharp-shinned Hawk	-	-	-	-	-	A	-	-	-	-	-	-	-	-
Black-chested Buzzard-Eagle	-	-	-	-	-	-	-	-	{A}	-	-	-	-	-
White-tailed Hawk	-	S	-	X	X	S	S	-	-	-	X	-	S	-
Zone-tailed Hawk	-	-	(S)	-	-	A	-	{S}	-	S	-	-	-	-
Swainson's Hawk	-	-	-	-	-	-	-	-	{M}	-	M	-	-	-
Broad-winged Hawk	-	-	-	-	-	-	M	M	H	H	-	-	M	-
Roadside Hawk	X	X	X	X	X	X	X	X	{X}	X	X	X	-	-
Short-tailed Hawk	-	S	-	-	S	S	X	S	{S}	{S}	S	(S)	S	-
Gray-lined Hawk	-	X	X	X	X	X	X	S	{X}	-	X	X	X	-
White Hawk	-	H	-	X	X	X	S	-	A	X	X	X	-	-
Black-faced Hawk	-	-	S	-	-	S	-	-	H	S	-	-	-	-
White-browed Hawk	-	S	-	S	H	-	S	-	-	-	(H)	S	-	-
Slate-colored Hawk	-	S	H	S	S	S	S	{S}	{S}	-	-	S	-	-
Black-collared Hawk	X	-	X	X	S	X	S	{S}	{S}	X	-	{X}	-	-
Savanna Hawk	X	S	X	X	-	X	-	-	-	X	A	-	-	-
Rufous Crab-Hawk	S	-	H	-	-	-	-	-	-	-	-	-	-	-
Great Black-Hawk	X	H	X	X	S	X	X	{X}	-	X	-	X	-	-
{Solitary Eagle}	-	-	-	-	-	-	-	-	{A}	-	-	-	-	-
Crested Eagle	-	-	-	-	H	-	S	S	{S}	-	S	-	-	-
Harpy Eagle	-	H	H	H	S	S	S	S	-	S	S	S	-	-
Black-and-white Hawk-Eagle	-	A	-	H	A	S	S	{A}	-	-	S	-	-	-
Ornate Hawk-Eagle	-	-	X	H	X	X	S	{S}	H	X	-	X	-	-

Species	S.L	Bel	Ama	San	ANP	Man	Tef	Tab	PdN	N.R	R.B	Ron	MVI	FdN
Black Hawk-Eagle	-	S	(H)	-	S	S	-	S	S	S	(S)	{S}	-	-
Long-winged Harrier	-	-	X	H	-	-	-	-	-	S	-	-	-	-
Crane Hawk	X	S	H	S	-	S	S	{S}	H	S	X	S	-	-
Osprey	M	-	H	M	M	X	S	X	{M}	X	-	H	-	-
Laughing Falcon	X	X	H	X	X	S	X	X	-	X	X	X	-	-
Collared Forest-Falcon	-	-	-	-	H	-	S	S	{S}	-	S	S	-	-
{Buckley's Forest-Falcon}	-	-	-	-	-	-	-	-	{A}	-	(?)	-	-	-
Slaty-backed Forest-Falcon	-	H	-	H	H	S	S	-	H	-	-	H	-	-
Barred Forest-Falcon	-	-	X	-	S	S	S	S	-	X	-	S	-	-
Lined Forest-Falcon	-	X	X	H	H	X	X	H	H	X	-	X	-	-
Black Caracara	S	-	H	X	S	S	X	{X}	{X}	X	X	X	-	-
Red-throated Caracara	-	X	X	X	X	X	X	{X}	{X}	X	X	X	-	-
Yellow-headed Caracara	X	X	X	X	X	X	S	X	-	X	-	X	-	-
Crested Caracara	X	S	X	S	-	S	-	-	-	X	-	{X}	-	-
Peregrine Falcon	-	-	-	A	A	M	-	-	-	M	-	-	-	M
Orange-breasted Falcon	-	-	H	S	H	S	M	-	{S}	-	-	S	-	-
Bat Falcon	X	X	X	X	X	X	X	X	{X}	X	X	X	-	-
Aplomado Falcon	-	-	-	(H)	A	-	-	-	-	S	-	-	-	-
Merlin	-	-	-	-	-	A	-	-	-	A	-	-	-	-
American Kestrel	-	-	-	-	-	-	-	-	-	X	-	-	-	-
Little Chachalaca	-	-	-	X	X	-	X	-	-	X	-	-	-	-
"Buff-browed" Chachalaca	S	X	-	-	-	-	-	-	-	-	-	-	-	-
"Speckled" Chachalaca	-	-	-	-	X	-	S	H	H	-	X	X	-	-
Marail Guan	-	-	-	X	-	-	X	-	-	X	-	-	-	-
Rusty-margined Guan	-	S	H	-	H	S	-	-	-	-	-	H	-	-
Spix's Guan	-	-	-	-	-	X	X	{X}	X	X	X	X	-	-
E White-crested Guan	-	-	-	-	S	X	-	-	-	-	-	-	-	-
Blue-throated Piping-Guan	-	-	-	H	-	S	S	{X}	{X}	X	-	-	-	-
"Red-throated" Piping-Guan	-	-	H	-	H	X	-	-	-	-	-	X	-	-
Black Curassow	-	-	-	X	-	X	-	-	-	X	X	-	-	-
Bare-faced Curassow	-	-	-	-	-	-	-	-	-	-	-	H	-	-
"Natterer's" Curassow	H	S	-	-	-	-	-	-	-	-	-	-	-	-
Wattled Curassow	-	-	-	-	-	H	-	S	-	-	-	H	-	-
Razor-billed Curassow	-	-	H	-	H	X	-	S	-	-	-	H	-	-
Crestless Curassow	-	-	-	-	-	-	S	-	H	X	-	-	-	-
Nocturnal Curassow	-	-	-	-	-	-	-	{S}	H	-	-	S	-	-
Crested Bobwhite	-	-	-	X	-	-	-	-	-	X	-	-	-	-
Marbled Wood-Quail	-	H	X	X	X	X	S	{X}	H	X	X	H	-	-
Starred Wood-Quail	-	-	-	-	-	-	S	-	-	-	(H)	X	-	-
Limpkin	-	X	-	X	X	S	X	-	{H}	X	-	X	-	-
Gray-winged Trumpeter	-	-	-	S	-	X	-	S	H	S	-	-	-	-
Pale-winged Trumpeter	-	-	-	-	-	-	S	X	H	-	-	S	-	-

	S.L	Bel	Ama	San	ANP	Man	Tef	Tab	PdN	N.R	R.B	Ron	MVI	FdN
E Dark-winged Trumpeter	-	H	-	S	S	-	-	-	-	-	-	S	-	-
Clapper Rail	-	H	-	(H)	-	-	-	-	-	-	-	-	-	-
Blackish Rail	-	-	-	-	-	-	-	-	-	-	-	{S}	-	-
Spotted Rail	S	H	H	-	-	-	-	-	-	-	-	-	-	-
Uniform Crake	-	H	-	H	S	S	-	-	H	-	-	H	-	-
E Little Wood-Rail	X	X	-	-	-	-	-	-	-	-	-	-	-	-
Gray-necked Wood-Rail	X	X	X	X	S	X	X	{X}	{H}	X	X	X	-	-
Red-winged Wood-Rail	-	-	-	-	-	A	S	-	-	-	-	-	-	-
Ash-throated Crake	-	-	-	-	-	-	-	-	X	-	S	-	-	-
Yellow-breasted Crake	-	H	H	-	-	A	-	-	H	-	-	-	-	-
Gray-breasted Crake	-	X	H	H	-	S	X	X	-	X	X	-	-	-
Rufous-sided Crake	-	X	-	-	H	S	-	S	-	-	X	X	-	-
Russet-crowned Crake	X	X	X	X	X	X	-	-	X	(X)	{S}	-	-	-
Black-banded Crake	-	-	-	-	-	A	H	S	-	-	(H)	-	-	-
{Chestnut-headed Crake}	-	-	-	-	-	-	-	-	-	-	{S}	-	-	-
Paint-billed Crake	-	S	-	-	-	A	-	-	{A}	-	-	H	-	-
Common Gallinule (Moorhen)	-	H	-	-	-	-	-	-	-	-	-	-	-	-
Purple Gallinule	X	H	H	H	-	X	-	H	H	X	X	X	-	A
Azure Gallinule	-	H	M	H	-	X	-	{S}	-	M	-	-	-	-
Sungrebe	X	S	X	X	X	X	S	{X}	H	X	-	H	-	-
Sunbittern	-	S	H	H	X	S	X	X	H	X	X	H	-	-
Wattled Jacana	X	X	X	X	X	X	X	X	{X}	X	X	X	-	-
American Oystercatcher	X	X	-	-	-	-	-	-	-	-	-	-	-	-
Southern Lapwing	X	X	X	X	-	A	-	-	{S}	X	-	-	-	-
Pied Plover	-	S	(H)	H	X	X	S	X	{H}	X	(H)	X	-	-
Black-bellied Plover	M	M	M	-	-	A	-	-	-	-	-	-	-	M
American Golden Plover	M	M	M	-	-	M	M	-	H	M	H	M	-	M
Semipalmated Plover	M	M	M	-	-	A	-	-	-	-	-	-	-	M
Collared Plover	X	X	X	X	X	X	X	M	{H}	X	(H)	X	-	-
Wilson's Plover	M	M	H	-	-	-	-	-	-	-	-	-	-	-
Ruddy Turnstone	M	M	M	-	-	M	-	-	-	-	-	-	-	M
Solitary Sandpiper	M	M	M	H	M	M	M	{M}	H	M	H	M	-	-
Lesser Yellowlegs	M	M	M	H	-	M	M	-	{M}	M	H	H	-	M
Greater Yellowlegs	M	M	M	-	-	M	M	-	-	M	-	H	-	-
Spotted Sandpiper	M	M	M	M	M	M	M	{M}	H	M	-	H	-	M
Willet	M	M	M	-	-	-	-	-	-	-	-	-	-	M
Red Knot	M	M	M	-	-	-	-	-	-	-	-	-	-	M
Least Sandpiper	M	M	M	-	A	M	M	-	-	M	-	{H}	-	A
White-rumped Sandpiper	M	M	(M)	-	-	M	-	{A}	-	A	-	M	-	A
Pectoral Sandpiper	M	M	-	-	-	M	M	{M}	-	-	H	M	-	M
Semipalmated Sandpiper	M	M	M	H	-	M	-	-	-	-	-	-	-	-
Western Sandpiper	-	M	-	-	-	-	-	-	-	-	-	-	-	-

	S.L	Bel	Ama	San	ANP	Man	Tef	Tab	PdN	N.R	R.B	Ron	MVI	FdN
Sanderling	M	M	M	-	-	M	-	-	-	-	-	-	-	M
Stilt Sandpiper	-	-	-	A	-	M	-	-	-	-	-	M	-	-
Buff-breasted Sandpiper	-	-	-	-	-	-	-	-	-	M	H	H	-	-
Upland Sandpiper	M	M	-	M	-	M	-	{M}	-	M	-	M	-	-
Whimbrel	M	M	M	-	-	-	-	-	-	-	-	-	M	M
Hudsonian Godwit	A	-	-	-	-	A	-	-	-	A	-	M	-	-
Bar-tailed Godwit	-	-	-	-	-	-	-	-	-	-	-	-	-	A
Short-billed Dowitcher	M	M	M	-	-	-	-	{A}	-	-	-	-	-	A
South American Snipe	X	H	H	H	-	S	-	{S}	-	X	-	-	-	-
Giant Snipe	-	-	-	-	-	-	-	-	{S}	X	-	-	-	-
Black-necked Stilt	M	M	H	M	-	-	-	{A}	-	-	-	-	-	-
Wilson's Phalarope	-	-	-	-	-	-	-	-	-	-	-	M	-	-
Double-striped Thick-knee	-	-	A	X	-	-	-	-	-	X	-	-	-	-
Great Skua	-	-	-	(H)	-	-	-	-	-	-	-	-	-	-
Pomarine Skua (Jaeger)	-	-	-	-	A	-	-	-	-	-	-	-	-	-
Ring-billed Gull	-	-	-	-	-	-	A	-	-	-	-	-	-	-
Laughing Gull	-	-	M	-	-	A	-	-	-	-	-	-	-	-
Gray-hooded Gull	M	-	-	-	-	-	-	-	-	-	-	-	-	-
Franklin's Gull	-	-	-	-	-	-	-	-	-	-	-	-	-	A
Large-billed Tern	X	X	X	X	X	X	X	X	-	X	-	X	-	-
Gull-billed Tern	X	X	X	-	-	A	-	-	-	-	-	-	-	-
Common Tern	-	-	M	(M)	-	A	-	-	-	-	-	(M)	-	-
Roseate Tern	-	[A]	-	-	-	-	-	-	-	-	-	-	-	-
Sooty Tern	-	-	M	-	-	-	-	-	-	-	-	-	M	M
Yellow-billed Tern	X	X	X	-	S	X	X	X	{X}	X	(H)	X	-	-
Least Tern	-	-	-	(H)	-	-	-	-	-	-	-	-	-	-
Royal Tern	-	-	-	-	-	-	-	-	-	-	-	-	-	M
Brown Noddy	-	-	-	-	-	-	-	-	-	-	-	-	M	M
Black Noddy	-	-	-	-	-	-	-	-	-	-	-	-	M	X
White (Fairy) Tern	-	-	-	-	-	-	-	-	-	-	-	-	X	X
Black Skimmer	M	-	M	M	M	M	M	M	{H}	M	-	M	-	-
Band-tailed Pigeon	-	-	-	-	-	-	-	-	{S}	S	-	-	-	-
Scaled Pigeon	-	X	X	X	S	X	-	-	S	X	-	{X}	-	-
Pale-vented Pigeon	X	X	X	X	X	X	X	X	H	X	-	X	-	-
Ruddy Pigeon	-	-	H	H	X	X	X	X	H	X	-	X	-	-
Plumbeous Pigeon	-	X	X	X	X	X	X	X	-	X	-	X	-	-
Eared Dove	X	H	X	X	-	-	-	-	-	X	-	-	-	X
Scaled Dove	X	-	-	-	S	S	-	-	-	-	-	-	-	-
Common Ground-Dove	X	X	X	X	X	X	-	-	H	X	-	-	-	-
Plain-breasted Ground-Dove	X	-	-	S	-	S	S	-	{X}	X	S	{H}	-	-
Ruddy Ground-Dove	X	X	X	X	X	X	X	X	H	X	X	X	-	-
Picui Ground-Dove	X	-	-	-	-	-	-	{A}	-	-	-	-	-	-

Species	S.L	Bel	Ama	San	ANP	Man	Tef	Tab	PdN	N.R	R.B	Ron	MVI	FdN
Blue Ground-Dove	S	H	H	S	S	-	-	-	{S}	S	X	X	-	-
Long-tailed Ground-Dove	-	H	X	-	-	-	-	-	-	-	-	-	-	-
White-tipped Dove	X	X	X	X	X	X	X	-	-	X	X	X	-	-
Gray-fronted Dove	-	X	S	X	X	S	X	X	H	X	-	X	-	-
Sapphire Quail-Dove	-	-	-	-	-	-	-	A	-	-	-	-	-	-
Ruddy Quail-Dove	-	X	S	H	S	M	S	{S}	H	S	H	S	-	-
Violaceous Quail-Dove	-	H	-	-	-	-	-	-	-	-	-	{S}	-	-
Hyacinth Macaw	-	-	H	H	-	-	-	-	-	-	-	-	-	-
Blue-and-yellow Macaw	-	-	X	H	X	X	X	{X}	{S}	X	-	X	-	-
Scarlet Macaw	-	H	X	X	S	X	X	X	{X}	X	X	X	-	-
Red-and-green Macaw	-	H	H	H	X	X	S	{X}	H	X	(H)	{X}	-	-
Chestnut-fronted Macaw	-	-	X	H	X	H	X	X	-	X	X	X	-	-
Blue-headed Macaw	-	-	-	-	-	-	-	-	-	-	A	-	-	-
Blue-winged Macaw	-	S	(H)	H	-	-	-	-	-	-	-	-	-	-
Red-bellied Macaw	-	S	X	H	X	X	X	M	-	X	X	X	-	-
Red-shouldered Macaw	S	-	H	-	S	S	-	-	-	X	-	-	-	-
E Golden Parakeet	-	-	-	-	M	-	-	-	-	-	A	-	-	-
White-eyed Parakeet	X	H	X	X	X	X	X	{X}	-	X	X	X	-	-
Sun Parakeet	-	-	-	-	-	-	-	-	-	X	-	-	-	-
E Jandaya Parakeet	X	S	-	-	A	-	-	-	-	-	-	-	-	A
Dusky-headed Parakeet	-	-	-	-	-	-	-	X	-	-	X	X	-	-
Brown-throated Parakeet	-	-	-	-	A	S	-	-	H	X	-	X	-	-
Peach-fronted Parakeet	-	S	X	X	-	-	-	-	-	-	-	-	-	-
Crimson-bellied Parakeet	-	-	-	-	-	-	-	-	-	-	-	X	-	-
"Pearly" Parakeet	-	H	-	-	-	-	-	-	-	-	-	-	-	-
Painted Parakeet	-	-	X	X	X	S	X	{S}	-	X	(X)	X	-	-
Fiery-shouldered Parakeet	-	-	-	-	-	-	-	-	-	S	-	-	-	-
Maroon-tailed Parakeet	-	-	-	-	-	-	S	S	{X}	H	-	-	-	-
Rock Parakeet	-	-	-	-	-	-	-	-	-	-	S	{H}	-	-
Green-rumped Parrotlet	-	-	X	X	X	X	-	-	-	X	-	-	-	-
Blue-winged Parrotlet	X	-	-	X	-	S	X	{X}	H	-	-	S	-	-
Dusky-billed Parrotlet	-	S	H	-	-	-	X	H	H	-	S	{H}	-	-
Canary-winged Parakeet	-	-	X	X	X	X	M	M	X	-	-	X	-	-
Yellow-chevroned Parakeet	-	-	-	-	-	-	-	-	-	-	-	X	-	-
Cobalt-winged Parakeet	-	-	-	-	-	-	-	{X}	H	X	X	{X}	-	-
Golden-winged Parakeet	X	X	X	X	X	X	-	-	H	X	-	X	-	-
Tui Parakeet	-	H	X	X	-	X	S	X	-	-	X	X	-	-
Tepui Parrotlet	-	-	-	-	-	-	-	-	{S}	S	-	-	-	-
Sapphire-rumped Parrotlet	S	S	H	-	S	X	S	{X}	H	{H}	-	-	-	-
Scarlet-shouldered Parrotlet	-	S	-	-	-	-	-	-	-	-	-	{S}	-	-
Black-headed Parrot	-	-	X	-	-	S	-	{X}	H	S	-	-	-	-
White-bellied Parrot	-	X	-	X	-	A	X	X	-	-	X	X	-	-

	S.L	Bel	Ama	San	ANP	Man	Tef	Tab	PdN	N.R	R.B	Ron	MVI	FdN
Caica Parrot	-	-	H	-	-	X	-	-	-	S	-	-	-	-
Orange-cheeked Parrot	-	-	-	-	-	S	S	{X}	H	X	X	X	-	-
Vulturine Parrot	-	-	-	X	-	H	X	-	-	-	-	-	-	-
Short-tailed Parrot	-	-	-	X	X	-	M	X	{X}	-	-	-	S	-
Blue-headed Parrot	-	-	X	X	X	X	X	X	X	H	X	X	X	-
Dusky Parrot	-	-	-	-	X	X	H	S	X	-	-	-	-	-
Red-lored Parrot	-	-	-	-	-	-	-	X	S	-	-	-	-	-
{Blue-cheeked Parrot}	-	-	-	-	-	-	-	-	-	-	{S}	-	-	-
Festive Parrot	-	-	H	X	-	X	X	X	{S}	-	-	H	-	-
Turquoise-fronted Parrot	-	-	-	-	-	-	-	-	-	-	-	{H}	-	-
Yellow-crowned Parrot	-	-	(S)	H	X	-	S	{X}	{S}	X	X	X	-	-
Orange-winged Parrot	-	X	X	X	X	X	S	-	{S}	X	-	-	-	-
Mealy Parrot	-	X	X	S	X	X	X	X	{X}	X	X	X	-	A
"Kawall's" Parrot	-	-	-	A	-	-	-	-	-	-	-	-	-	-
Red-fan Parrot	-	H	X	H	S	X	X	-	S	X	-	X	-	-
Ash-colored Cuckoo	-	-	-	-	-	-	-	A	{A}	-	-	-	A	-
Dwarf Cuckoo	-	-	-	-	-	-	-	-	-	A	-	-	-	-
Black-billed Cuckoo	-	-	-	-	-	-	-	-	-	-	(A)	-	-	-
Yellow-billed Cuckoo	-	-	-	H	-	-	A	-	{M}	{A}	A	-	A	-
Pearly-breasted Cuckoo	-	H	M	-	H	-	M	M	-	{A}	-	-	-	-
Mangrove Cuckoo	-	-	-	S	-	-	-	-	-	-	-	-	-	-
Dark-billed Cuckoo	-	-	M	M	M	M	M	M	M	-	M	M	M	-
Squirrel Cuckoo	-	X	X	X	X	X	X	X	X	H	X	X	X	-
Black-bellied Cuckoo	-	-	-	-	-	S	X	X	X	H	X	X	X	-
Little Cuckoo	S	S	S	H	H	S	S	S	{S}	S	(S)	{H}	-	-
Greater Ani	-	X	X	X	X	X	M	X	X	{H}	X	(X)	X	-
Smooth-billed Ani	-	X	X	X	X	X	X	X	X	H	X	X	X	-
Hoatzin	-	X	H	X	X	X	X	S	{X}	{S}	X	-	X	-
Guira Cuckoo	-	-	X	S	X	-	-	-	-	-	-	-	-	-
Striped Cuckoo	-	X	X	X	X	X	X	-	X	-	X	(X)	X	-
Pheasant Cuckoo	-	-	-	-	-	H	-	{S}	-	-	S	{S}	-	-
Pavonine Cuckoo	-	-	-	-	-	-	A	-	-	-	H	-	S	-
Rufous-vented Ground-Cuckoo	-	-	H	-	H	-	-	-	-	-	-	(S)	H	-
Scaled Ground-Cuckoo	-	-	-	-	H	H	-	-	-	-	-	-	-	-
Rufous-winged Ground-Cuckoo	-	-	-	-	-	-	-	-	-	S	-	-	-	-
Red-billed Ground-Cuckoo	-	-	-	-	-	-	S	H	-	-	-	-	-	-
Barn Owl	-	S	-	(H)	H	-	S	-	-	-	S	S	S	-
Tropical Screech-Owl	-	X	X	(H)	X	X	S	S	{X}	{X}	X	X	-	-
Guatemalan Screech-Owl	-	-	-	-	-	-	-	-	{S}	S	-	-	-	-
Tawny-bellied Screech-Owl	-	-	-	H	-	-	X	-	{X}	H	S	-	-	-
"Austral" Screech-Owl	-	-	-	-	H	X	-	X	-	-	X	X	-	-
Crested Owl	-	-	S	H	H	-	S	-	-	-	X	X	-	-

Species	S.L	Bel	Ama	San	ANP	Man	Tef	Tab	PdN	N.R	R.B	Ron	MVI	FdN
Great Horned Owl	-	H	S	-	-	-	-	-	-	-	S	-	-	-
Spectacled Owl	X	S	H	X	X	X	X	{S}	{S}	X	H	X	-	-
Amazonian Pygmy-Owl	-	-	-	S	-	S	S	S	-	-	S	{S}	X	-
Ferruginous Pygmy-Owl	X	-	-	X	X	-	S	S	{X}	{S}	X	X	{H}	-
Burrowing Owl	-	-	-	-	-	-	-	-	-	-	X	S	-	-
Black-banded Owl	-	S	-	-	-	-	A	S	{S}	{H}	-	S	-	-
Mottled Owl	-	S	-	-	H	-	S	-	{S}	S	-	{S}	-	-
Striped Owl	-	S	-	-	S	-	-	-	-	S	-	-	-	-
Stygian Owl	-	-	-	-	S	-	A	-	-	-	-	{S}	-	-
Buff-fronted Owl	-	-	-	-	-	-	-	-	{S}	-	-	-	-	-
{Oilbird}	-	-	-	-	-	-	-	-	{S}	{S}	-	-	-	-
Great Potoo	-	-	H	H	-	-	S	X	{S}	H	S	{S}	-	-
Long-tailed Potoo	-	H	-	-	-	-	S	-	{S}	S	-	{S}	-	-
Common Potoo	H	H	H	H	X	X	S	{X}	H	X	-	{X}	-	-
E White-winged Potoo	-	-	-	-	-	-	X	-	-	-	-	-	-	-
Rufous Potoo	-	-	-	-	-	-	S	-	-	-	-	-	-	-
Short-tailed Nighthawk	-	-	-	-	X	X	X	-	{A}	H	-	(X)	X	-
Least Nighthawk	-	-	X	-	-	-	-	-	{H}	X	-	{S}	-	-
Sand-colored Nighthawk	-	-	-	H	-	M	-	M	{X}	M	(H)	M	-	-
Lesser Nighthawk	-	H	M	M	M	M	M	M	-	{H}	M	-	-	-
Common Nighthawk	-	-	-	-	-	-	M	M	-	{M}	M	-	M	-
Band-tailed Nighthawk	X	-	H	H	X	X	-	S	{X}	X	-	H	-	-
Nacunda Nighthawk	X	X	X	X	X	M	M	{M}	{H}	X	X	X	-	-
Pauraque	X	X	X	X	X	X	X	X	H	X	X	X	-	-
Ocellated Poorwill	-	-	-	-	H	-	S	-	-	-	H	S	-	-
Rufous Nightjar	-	-	-	-	S	-	S	-	-	-	-	-	-	-
Silky-tailed Nightjar	-	-	-	-	H	-	-	-	-	-	-	-	-	-
Band-winged Nightjar	-	-	-	-	-	-	-	-	{S}	{S}	-	-	-	-
White-tailed Nightjar	-	-	-	H	-	-	-	-	-	-	-	X	-	-
Spot-tailed Nightjar	-	-	-	(H)	H	-	-	A	-	{A}	X	S	-	-
Little Nightjar	-	-	-	M	-	H	-	-	-	-	-	-	M	-
Blackish Nightjar	-	-	-	X	X	X	X	X	S	{S}	H	X	-	H
{Roraiman Nightjar}	-	-	-	-	-	-	-	-	-	{S}	-	-	-	-
Ladder-tailed Nightjar	-	-	-	X	H	H	X	S	{X}	H	X	(H)	X	-
Scissor-tailed Nightjar	-	-	-	X	-	-	-	-	-	-	H	X	-	-
White-collared Swift	-	-	-	-	-	-	A	M	M	{X}	X	-	{S}	-
Tepui Swift	-	-	-	-	-	-	-	-	{S}	S	-	-	-	-
Chapman's Swift	-	-	-	M	M	-	-	M	M	{M}	-	H	M	-
Chimney Swift	-	-	-	-	-	-	A	-	-	-	-	-	-	-
Gray-rumped Swift	-	-	A	-	M	M	M	M	{M}	{S}	{S}	M	M	-
Pale-rumped Swift	-	-	-	-	-	-	-	-	-	-	-	(M)	M	-
Band-rumped Swift	-	-	-	X	X	X	X	X	X	{X}	{H}	X	-	-

-172-

	S.L	Bel	Ama	San	ANP	Man	Tef	Tab	PdN	N.R	R.B	Ron	MVI	FdN
Ashy-tailed Swift	-	-	-	-	-	-	-	-	{A}	M	-	-	-	-
Short-tailed Swift	X	X	X	X	X	X	X	X	{X}	X	X	X	-	-
White-tipped Swift	-	-	-	-	-	-	-	-	H	{S}	-	-	-	-
Lesser Swallow-tailed Swift	X	X	-	X	-	S	X	X	{S}	-	X	X	-	-
Fork-tailed Palm-Swift	X	X	X	H	X	X	-	X	{X}	X	X	X	-	-
Blue-fronted Lancebill	-	-	-	-	-	-	-	-	{S}	H	-	-	-	-
Rufous-breasted Hermit	X	X	X	-	-	S	S	{X}	{A}	-	X	X	-	-
Pale-tailed Barbthroat	-	-	X	X	S	S	S	X	X	H	X	-	H	-
"Bronze-tailed" Barbthroat	-	-	S	-	-	-	-	-	-	-	-	-	-	-
Long-tailed Hermit	-	-	-	X	-	-	-	-	X	X	X	-	-	-
"Red-billed" Long-tailed Hermit	X	-	X	X	X	X	X	-	-	-	-	X	-	-
Great-billed Hermit	-	-	-	X	-	-	-	-	-	-	-	-	-	-
White-bearded Hermit	-	-	-	-	-	-	S	X	-	-	X	X	-	-
Straight-billed Hermit	-	-	-	X	H	X	X	-	{X}	H	X	-	-	-
Needle-billed Hermit	-	-	-	-	-	-	S	X	X	-	-	X	X	-
Streak-throated Hermit	-	-	-	S	S	S	S	-	-	-	X	-	-	-
Sooty-capped Hermit	-	-	-	-	-	-	-	-	-	S	-	-	-	-
Cinnamon-throated Hermit	-	-	-	-	-	-	-	-	-	-	-	{S}	-	-
Reddish Hermit	-	-	X	X	X	X	X	X	{X}	X	X	X	X	-
Gray-chinned Hermit	-	-	-	-	-	-	-	-	{S}	S	-	-	-	-
Little Hermit	-	-	-	-	S	-	-	-	{S}	H	-	-	-	-
Gray-breasted Sabrewing	-	-	X	X	-	X	X	X	{X}	X	X	X	X	-
Rufous-breasted Sabrewing	-	-	-	-	-	-	-	-	-	S	-	-	-	-
Buff-breasted Sabrewing	-	-	-	-	-	-	-	-	S	S	-	-	-	-
Swallow-tailed Hummingbird	-	-	X	X	-	-	-	-	-	-	-	-	-	-
White-necked Jacobin	-	H	X	X	-	S	X	X	X	X	X	-	{X}	-
Brown Violetear	-	-	-	-	-	-	-	-	-	S	S	-	-	-
Sparkling Violetear	-	-	-	A	-	-	-	-	-	S	S	-	-	-
White-vented Violetear	-	-	-	-	-	-	-	-	-	-	-	S	-	-
Green-throated Mango	-	-	X	X	X	-	S	-	-	-	-	-	-	-
Black-throated Mango	-	H	M	M	M	M	-	M	M	M	H	M	M	-
Fiery-tailed Awlbill	-	-	M	M	H	M	-	-	-	-	M	-	M	-
Ruby-topaz Hummingbird	-	M	-	M	M	-	-	-	-	-	M	-	M	-
Violet-headed Hummingbird	-	-	-	-	-	-	A	-	A	-	-	-	-	-
Tufted Coquette	-	-	-	-	S	-	-	-	-	-	S	-	-	-
Dot-eared Coquette	-	-	S	(S)	-	-	-	-	-	-	-	-	-	-
Frilled Coquette	-	-	-	-	-	-	-	-	-	-	-	A	-	-
Butterfly Coquette	-	-	-	-	-	-	-	S	{A}	S	-	S	-	-
Peacock Coquette	-	-	-	-	-	-	-	-	-	S	-	-	-	-
Black-bellied Thorntail	-	-	A	-	-	A	S	S	S	S	{S}	S	-	-
Racket-tailed Coquette	-	-	-	S	-	A	S	-	-	-	-	-	-	-
Blue-chinned Sapphire	-	-	X	X	X	H	S	X	X	S	X	(H)	H	-

	S.L	Bel	Ama	San	ANP	Man	Tef	Tab	PdN	N.R	R.B	Ron	MVI	FdN
Blue-tailed Emerald-	-	-	S	-	-	-	S	S	-	X	S	S	-	-
Glittering-bellied Emerald	X	-	-	-	-	-	-	-	-	-	-	-	-	-
Fork-tailed Woodnymph	-	H	X	X	X	X	X	X	X	X	X	X	X	-
Rufous-throated Sapphire	-	-	M	M	M	M	M	-	-	H	-	{M}	-	-
White-chinned Sapphire	S	-	S	-	S	H	S	-	-	{S}	S	S	S	-
Gilded Sapphire-	-	-	-	-	-	-	-	-	-	-	-	M	-	-
Golden-tailed Sapphire	-	-	-	-	-	-	-	S	-	-	-	-	-	-
White-tailed Goldenthroat	-	H	-	X	-	-	-	-	-	X	-	-	-	-
Tepui Goldenthroat	-	-	-	-	-	-	-	-	-	S	-	-	-	-
Green-tailed Goldenthroat	-	-	S	S	-	S	X	-	-	S	S	-	H	-
Olive-spotted Hummingbird	-	-	-	-	-	-	-	S	-	-	-	-	-	-
White-chested Emerald	-	-	-	-	-	-	-	-	-	S	-	-	-	-
Versicolored Emerald	-	X	H	X	H	-	X	X	-	{S}	X	-	-	-
"Blue-headed" Emerald	-	-	-	-	-	-	-	-	-	-	-	A	-	-
Glittering-throated Emerald-	-	X	-	X	X	X	X	X	X	-	X	X	-	-
Sapphire-spangled Emerald	-	-	-	-	-	-	-	-	-	-	-	M	-	-
Plain-bellied Emerald	-	X	X	X	-	-	-	-	-	-	A	-	-	-
Green-bellied Hummingbird	-	-	-	-	-	-	-	-	{X}	S	-	-	-	-
Gould's Jewelfront	-	-	-	-	-	-	-	S	S	S	-	S	-	-
Velvet-browed Brilliant-	-	-	-	-	-	-	-	-	-	H	S	-	-	-
Black-throated Brilliant	-	-	-	-	-	-	-	-	-	S	-	-	-	-
Pink-throated Brilliant-	-	-	-	-	-	-	-	S	-	-	-	-	-	-
Crimson Topaz	-	-	S	X	-	X	-	-	-	X	-	S	-	-
Fiery Topaz	-	-	-	-	-	-	A	-	S	-	-	-	-	-
Black-eared Fairy	-	-	X	X	X	X	X	X	X	H	X	X	X	-
Horned Sungem	-	-	-	M	-	-	-	-	-	-	-	M	-	-
Long-billed Starthroat	-	-	-	M	H	M	-	-	M	M	M	M	M	-
Blue-tufted Starthroat	-	-	-	-	-	-	-	{A}	-	-	-	M	-	-
Amethyst Woodstar	-	M	M	M	-	-	-	-	-	{M}	M	-	-	-
Pavonine Quetzal	-	-	-	-	H	-	S	X	X	H	-	-	S	-
Black-tailed Trogon-	-	-	X	X	X	X	X	X	{X}	H	X	X	X	-
White-tailed Trogon-	-	H	X	X	X	X	X	X	X	H	X	X	X	-
Collared Trogon-	-	-	S	H	S	-	S	{S}	-	{S}	S	X	-	-
Masked Trogon	-	-	-	-	-	-	-	-	S	{H}	-	-	-	-
Black-throated Trogon	-	-	S	S	H	S	X	S	{X}	H	-	(H)	S	-
Blue-crowned Trogon-	-	-	-	-	-	A	S	S	{S}	-	-	X	X	-
Violaceous Trogon	-	X	X	X	X	X	X	X	X	H	X	X	X	-
Ringed Kingfisher	-	X	X	X	X	X	X	X	X	{X}	X	X	X	-
Amazon Kingfisher	-	X	X	X	X	X	X	X	X	H	X	X	X	-
Green Kingfisher	-	X	X	X	X	X	X	X	X	{X}	X	X	X	-
Green-and-rufous Kingfisher-	-	S	S	S	S	-	S	S	S	H	S	H	S	-
American Pygmy Kingfisher	-	-	X	X	X	-	X	X	X	{H}	S	X	H	-

-174-

	S.L	Bel	Ama	San	ANP	Man	Tef	Tab	PdN	N.R	R.B	Ron	MVI	FdN
Broad-billed Motmot-	-	-	-	-	-	X	-	S	S	-	S	H	H	-
Rufous Motmot	-	-	-	-	H	H	-	X	{S}	-	-	X	X	-
Blue-crowned Motmot-	-	H	X	X	H	-	X	-	{S}	H	X	H	H	-
White-eared Jacamar-	-	-	-	-	-	H	S	{X}	-	-	-	-	-	-
Chestnut Jacamar	-	-	-	-	-	-	-	-	-	-	H	{H}	-	-
White-throated Jacamar	-	-	-	-	-	-	-	H	-	-	{S}	-	-	-
Brown Jacamar	-	H	S	-	-	-	-	-	{S}	{S}	S	-	X	-
Yellow-billed Jacamar	-	-	-	X	-	-	X	-	{X}	H	-	-	-	-
Blue-cheeked Jacamar	-	-	X	-	H	X	S	X	X	-	-	(X)	X	-
Green-tailed Jacamar	-	-	-	X	X	S	S	-	-	H	X	-	-	-
White-chinned Jacamar	-	-	-	-	-	-	X	{X}	{S}	-	-	-	-	-
Bluish-fronted Jacamar	-	-	-	-	-	-	-	{S}	-	-	X	-	-	-
Rufous-tailed Jacamar	-	X	S	(H)	X	-	-	-	-	-	X	-	X	-
Bronzy Jacamar	-	-	-	S	H	S	S	S	-	H	-	{S}	S	-
"Purplish" Jacamar	-	-	-	-	-	-	-	-	{S}	-	-	-	-	-
Paradise Jacamar	-	-	X	X	X	X	X	X	X	H	X	X	X	-
Great Jacamar	-	-	H	X	X	X	X	X	X	H	X	X	X	-
White-necked Puffbird	-	-	S	X	H	X	X	X	-	H	X	(X)	X	-
Brown-banded Puffbird	-	-	-	-	H	-	-	S	-	H	-	{S}	-	-
Pied Puffbird	-	-	S	H	X	X	X	-	{X}	{S}	S	-	S	-
Chestnut-capped Puffbird	-	-	-	-	-	-	S	S	X	H	S	S	H	-
Spotted Puffbird	-	-	H	S	H	H	X	S	{X}	{S}	S	-	{S}	-
Collared Puffbird	-	-	S	S	H	S	S	S	{X}	S	-	-	S	-
Striolated Puffbird-	-	-	S	-	-	-	-	-	-	-	-	H	S	-
Spot-backed Puffbird	-	X	-	(H)	X	-	-	-	-	-	-	-	-	-
White-chested Puffbird	-	-	-	S	-	-	S	-	{S}	{S}	-	-	-	-
Semicollared Puffbird	-	-	-	-	-	-	-	S	-	-	-	(H)	{H}	-
Rufous-necked Puffbird	-	-	S	-	H	H	-	S	S	-	-	-	S	-
Lanceolated Monklet-	-	-	-	-	-	-	-	S	{S}	-	-	-	-	-
Rusty-breasted Nunlet	-	-	S	-	H	S	S	X	H	S	-	S	-	-
Fulvous-chinned Nunlet	-	-	-	-	-	-	-	-	-	-	{S}	-	-	-
Rufous-capped Nunlet	-	-	-	-	H	-	-	S	S	-	-	S	-	-
Chestnut-headed Nunlet	-	-	-	-	-	-	H	-	-	-	-	-	-	-
Black Nunbird	-	-	H	X	H	-	X	S	-	H	X	-	-	-
Black-fronted Nunbird	-	X	S	-	X	X	X	X	X	{S}	-	X	X	-
White-fronted Nunbird	-	-	X	-	X	X	S	X	X	X	-	X	X	-
Yellow-billed Nunbird	-	-	-	-	-	-	-	{S}	-	-	H	-	-	-
Swallow-wing	-	H	X	X	X	X	X	X	H	X	X	X	-	-
Scarlet-crowned Barbet	-	-	-	-	-	-	H	X	X	H	-	-	-	-
Black-girdled Barbet	-	-	-	-	-	-	-	-	-	-	-	X	-	-
Black-spotted Barbet	-	-	-	X	-	-	X	X	X	H	X	X	S	-
Brown-chested Barbet	-	-	-	-	-	S	-	-	-	-	-	-	-	-

	S.L	Bel	Ama	San	ANP	Man	Tef	Tab	PdN	N.R	R.B	Ron	MVI	FdN
Lemon-throated Barbet	-	-	-	-	-	-	X	X	-	-	-	X	-	-
Chestnut-tipped Toucanet	-	-	-	-	-	-	-	-	-	H	S	-	-	-
Emerald Toucanet	-	-	-	-	-	-	-	-	-	-	-	S	-	-
Black-necked Aracari	-	X	X	X	X	X	S	-	-	-	X	-	-	-
Chestnut-eared Aracari	-	-	-	-	-	-	S	X	X	-	X	X	-	-
Many-banded Aracari	-	-	-	-	-	-	-	{X}	H	S	-	-	-	-
Green Aracari	-	S	X	H	H	X	-	-	-	-	X	-	-	-
Lettered Aracari	-	X	X	-	S	X	S	X	{X}	-	-	H	X	-
Red-necked Aracari	-	H	X	(S)	X	X	-	-	-	-	-	X	-	-
Ivory-billed Aracari	-	-	-	-	-	-	S	S	X	H	X	-	-	-
"Brown-mandibled" Aracari	-	-	-	-	-	-	A	X	H	-	X	-	-	-
Curl-crested Aracari	-	-	-	-	-	-	S	X	S	-	X	X	-	-
Guianan Toucanet	-	-	-	X	-	-	X	-	-	-	{S}	-	-	-
Golden-collared Toucanet	-	-	-	-	-	-	X	X	-	-	(X)	-	-	-
Tawny-tufted Toucanet	-	-	-	-	-	-	-	-	-	S	-	-	-	-
Gould's Toucanet	-	X	X	-	H	X	-	-	-	-	-	X	-	-
Channel-billed Toucan	-	X	X	X	X	X	X	-	-	X	-	-	-	-
Yellow-ridged Toucan	-	-	-	-	-	-	A	S	X	H	-	(X)	X	-
Red-billed Toucan	S	X	X	S	S	X	-	-	-	X	-	-	-	-
"Cuvier's" Toucan	-	H	(H)	X	X	S	X	X	H	-	X	X	-	-
Toco Toucan	-	H	H	H	-	H	-	-	-	-	-	X	S	-
Rufous-breasted Piculet	-	-	-	-	-	-	-	{S}	-	-	H	-	-	-
Plain-breasted Piculet	-	-	-	-	-	-	-	{X}	-	-	-	-	-	-
Rusty-necked Piculet	-	-	-	-	-	-	-	-	-	-	-	{X}	-	-
White-bellied Piculet	-	-	-	-	-	-	-	-	-	S	-	-	-	-
Golden-spangled Piculet	-	-	-	X	-	-	X	-	-	{X}	X	-	-	-
Gold-fronted Piculet	-	-	-	-	-	-	S	X	{S}	-	-	X	X	-
"Bar-breasted" Piculet	-	-	-	-	-	X	-	-	-	-	-	(X)	-	-
"Banded" Piculet	-	S	-	-	H	-	-	-	-	-	-	-	-	-
Lafresnaye's Piculet	-	-	-	-	-	-	S	-	{S}	-	-	-	-	-
"Orinoco" Piculet	-	-	-	-	-	-	-	-	-	A	-	-	-	-
White-wedged Piculet	-	-	-	-	-	-	-	-	-	-	-	{H}	-	-
Lower Amazonian Piculet	-	-	X	H	X	-	-	-	-	-	-	-	-	-
Spot-breasted Woodpecker	S	-	H	S	-	X	S	X	-	X	X	{H}	-	-
Green-barred Woodpecker	-	-	(H)	-	-	-	-	-	-	-	-	-	-	-
Golden-olive Woodpecker	-	-	-	-	-	-	-	-	-	S	S	-	-	-
Yellow-throated Woodpecker	-	-	X	X	X	X	X	X	{X}	H	X	X	X	-
White-throated Woodpecker	-	-	-	-	-	-	-	-	-	-	-	-	S	-
Golden-green Woodpecker	-	S	S	H	H	X	S	X	H	-	H	X	-	-
Blond-crested Woodpecker	X	-	A	H	-	-	-	-	-	-	-	-	-	-
Chestnut Woodpecker	-	-	X	X	X	X	X	X	{X}	H	X	(H)	X	-
Waved Woodpecker	-	-	X	X	-	-	X	-	-	S	-	-	-	-

-176-

	S.L	Bel	Ama	San	ANP	Man	Tef	Tab	PdN	N.R	R.B	Ron	MVI	FdN
Scale-breasted Woodpecker	-	-	H	H	X	X	X	{X}	H	X	X	H	-	-
Cream-colored Woodpecker	-	X	X	H	X	X	X	S	{X}	{H}	X	H	X	-
Ringed Woodpecker	-	S	S	H	-	X	S	{A}	{S}	X	S	X	-	-
Lineated Woodpecker-	-	X	X	X	X	X	X	X	{S}	X	X	X	-	-
Yellow-tufted Woodpecker	-	X	X	H	X	X	X	X	H	X	X	X	-	-
White Woodpecker	X	-	H	H	-	H	-	-	-	-	-	-	-	-
Little Woodpecker	-	-	H	S	S	S	X	{X}	-	S	X	{H}	-	-
Red-stained Woodpecker	X	X	(H)	H	X	X	X	{X}	H	S	X	X	-	-
Golden-collared Woodpecker	-	-	X	-	-	X	-	-	-	X	-	-	-	-
Red-rumped Woodpecker	-	-	-	-	-	-	-	-	-	S	-	-	-	-
Crimson-crested Woodpecker	X	X	X	X	X	X	X	{X}	H	X	X	X	-	-
Red-necked Woodpecker	-	X	X	X	X	X	X	{X}	X	X	X	X	-	-
Plain-brown Woodcreeper-	-	X	X	X	X	X	X	X	H	X	X	X	-	-
White-chinned Woodcreeper	-	X	X	X	X	-	X	X	H	X	X	X	-	-
"Obidos" Woodcreeper	-	-	-	-	-	X	-	-	-	-	-	-	-	-
Long-tailed Woodcreeper-	-	H	X	H	H	X	X	H	H	S	H	H	-	-
Spot-throated Woodcreeper	-	-	X	X	X	X	X	X	H	-	-	X	-	-
Olivaceous Woodcreeper	-	-	-	X	X	X	X	X	H	X	X	X	-	-
Wedge-billed Woodcreeper	-	X	X	X	X	X	X	X	H	X	X	X	-	-
Long-billed Woodcreeper-	-	X	X	X	X	X	X	{X}	-	X	-	X	-	-
Cinnamon-throated Woodcreeper	-	X	X	X	X	X	-	X	-	X	X	X	-	-
Red-billed Woodcreeper	-	-	X	H	H	X	S	-	-	X	-	X	-	-
Bar-bellied Woodcreeper-	-	-	-	-	-	H	H	H	H	-	H	-	-	-
Strong-billed Woodcreeper	-	-	-	S	S	S	S	-	{S}	{H}	-	S	-	-
Barred Woodcreeper	-	X	X	-	-	X	X	{X}	X	X	X	H	-	-
"Concolor" Woodcreeper	-	-	-	X	-	-	-	-	-	-	-	S	-	-
Black-banded Woodcreeper	-	-	X	-	-	X	X	X	{A}	X	X	-	-	-
"Cross-barred" Woodcreeper	-	-	-	H	-	-	-	-	-	-	-	-	-	-
E Hoffmann's Woodcreeper	-	-	-	-	-	X	-	-	-	-	-	S	-	-
Straight-billed Woodcreeper-	X	X	X	X	X	X	X	X	H	X	X	H	-	-
E Zimmer's Woodcreeper	S	H	-	S	-	S	-	-	-	-	-	H	-	-
Striped Woodcreeper-	-	X	X	H	X	X	S	{X}	H	-	-	H	-	-
Ocellated Woodcreeper	-	-	-	-	S	X	X	X	{S}	-	H	-	-	-
Spix's Woodcreeper	X	X	-	X	-	-	H	H	{S}	-	S	S	-	-
Elegant Woodcreeper-	-	-	-	-	X	-	S	S	-	-	-	X	-	-
Chestnut-rumped Woodcreeper-	-	X	H	-	X	X	-	X	X	X	-	-	-	-
Buff-throated Woodcreeper	-	X	X	X	-	S	X	X	H	X	X	H	-	-
E Dusky-billed Woodcreeper	-	S	-	X	X	-	-	-	-	-	-	X	-	-
Streak-headed Woodcreeper	-	-	-	-	-	-	-	-	-	X	-	-	-	-
Narrow-billed Woodcreeper	-	H	X	X	-	-	-	-	-	-	-	-	-	-
Lineated Woodcreeper	-	H	X	H	X	X	S	-	{S}	X	(X)	X	-	-
Red-billed Scythebill	-	-	-	-	-	X	S	{S}	H	-	X	{H}	-	-

	S.L	Bel	Ama	San	ANP	Man	Tef	Tab	PdN	N.R	R.B	Ron	MVI	FdN
Curve-billed Scythebill-	-	-	-	X	X	S	-	X	{S}	H	-	-	S	-
Campo Miner-	-	-	-	-	-	-	-	-	-	-	-	-	{M}	-
Pale-legged Hornero-	-	-	X	-	-	-	-	-	-	X	X	H	-	-
Pale-billed Hornero-	-	-	-	-	-	-	-	H	-	-	-	-	-	-
Lesser Hornero -	-	-	-	X	-	S	-	{X}	-	-	-	H	-	-
E Wing-banded Hornero-	-	X	-	-	H	-	S	-	-	-	-	-	-	-
McConnell's Spinetail -	-	-	-	S	-	-	-	-	-	S	S	-	-	-
Pale-breasted Spinetail-	-	S	X	X	-	S	-	{X}	-	X	-	-	-	-
Dark-breasted Spinetail-	-	-	-	-	-	S	X	X	-	-	(X)	-	-	-
Plain-crowned Spinetail-	-	H	X	X	X	S	X	X	-	X	H	H	-	-
White-bellied Spinetail-	-	-	-	{S}	-	-	S	S	{S}	-	-	{S}	-	-
Ruddy Spinetail-	X	X	H	X	H	S	H	{X}	H	X	(X)	S	-	-
Chestnut-throated Spinetail-	-	-	-	-	-	-	-	-	-	-	-	H	-	-
Hoary-throated Spinetail	-	-	-	-	-	-	-	-	-	A	-	-	-	-
Tepui Spinetail-	-	-	-	-	-	-	-	-	{S}	S	-	-	-	-
Rusty-backed Spinetail -	-	X	-	-	X	-	X	X	{X}	-	H	-	H	-
E Scaled Spinetail	-	-	H	H	-	-	-	-	-	-	-	-	-	-
Speckled Spinetail -	-	-	H	-	-	S	-	S	{H}	-	H	H	-	-
Yellow-chinned Spinetail	-	X	X	X	X	-	X	X	{X}	-	X	-	{H}	-
Red-and-white Spinetail-	-	-	H	-	H	-	S	X	{X}	-	-	H	-	-
Plain Softtail -	-	-	-	-	-	H	-	-	-	-	-	-	-	-
Orange-fronted Plushcrown	-	-	-	-	-	-	S	-	-	-	(S)	{H}	-	-
Roraiman Barbtail -	-	-	-	-	-	-	-	-	-	S	-	-	-	-
Point-tailed Palmcreeper	-	S	H	H	X	X	-	{X}	-	{H}	{X}	-	-	-
Rufous Cacholote -	-	-	-	-	-	-	-	-	-	-	-	S	-	-
Chestnut-winged Hookbill-	-	-	-	S	S	-	S	{S}	-	-	H	S	-	-
Peruvian Recurvebill-	-	-	-	-	-	-	-	-	-	-	-	S	-	-
Striped Woodhaunter-	-	-	-	-	X	-	S	X	H	-	-	H	-	-
Rufous-rumped Foliage-gleaner	-	X	X	X	X	X	X	S	-	-	H	X	-	-
Cinnamon-rumped Foliage-gleaner-	-	X	S	H	S	S	S	X	S	-	-	H	-	-
Chestnut-winged Foliage-gleaner-	-	-	S	X	-	S	{X}	-	-	-	H	X	-	-
Rufous-tailed Foliage-gleaner -	-	S	X	S	H	S	X	S	-	-	X	X	-	-
Olive-backed Foliage-gleaner	-	X	S	S	-	X	X	X	H	S	{X}	S	-	-
Crested Foliage-gleaner-	-	-	-	-	-	-	-	-	-	-	-	S	-	-
Ruddy Foliage-gleaner -	-	-	S	-	-	S	-	-	-	S	-	-	-	-
White-throated Foliage-gleaner	-	-	-	-	-	-	-	-	S	S	-	-	-	-
Buff-throated Foliage-gleaner	-	-	X	H	X	X	X	{X}	H	-	(H)	X	-	-
Chestnut-crowned Fol.-gleaner	-	S	S	H	-	S	S	S	{S}	S	S	{H}	-	-
Brown-rumped Foliage-gleaner	-	-	-	-	-	-	-	-	-	-	{S}	-	-	-
Rufous-tailed Xenops-	-	-	-	-	S	-	X	S	{S}	H	-	S	S	-
Slender-billed Xenops -	-	-	-	-	S	-	S	A	-	-	S	-	X	-
Streaked Xenops-	-	-	-	-	-	-	-	-	S	-	-	S	-	-

	S.L	Bel	Ama	San	ANP	Man	Tef	Tab	PdN	N.R	R.B	Ron	MVI	FdN
Plain Xenops	H	X	X	X	X	X	X	X	H	X	X	X	-	-
Gray-throated Leaftosser	-	-	-	-	-	-	-	-	-	-	-	A	-	-
Tawny-throated Leaftosser	H	X	X	X	-	S	X	-	-	-	-	-	-	-
Short-billed Leaftosser	-	X	X	-	-	X	-	-	H	S	-	S	-	-
Black-tailed Leaftosser	-	S	X	H	X	S	X	{S}	S	S	(H)	H	-	-
Sharp-tailed Streamcreeper	-	-	-	-	-	-	-	-	H	{S}	-	-	-	-
Fasciated Antshrike	-	-	X	X	X	X	X	X	H	-	X	X	-	-
Bamboo Antshrike	-	-	-	-	-	-	-	-	-	-	A	A	-	-
Black-throated Antshrike	-	-	S	-	-	S	-	-	-	-	-	-	-	-
Undulated Antshrike	-	-	-	-	-	-	-	-	H	-	-	-	-	-
Great Antshrike	X	X	X	X	X	S	-	{X}	H	X	X	H	-	-
Black-crested Antshrike	-	-	-	-	-	S	H	{X}	H	X	-	-	-	-
Band-tailed Antshrike	-	-	S	-	H	S	-	-	-	-	-	-	-	-
E Glossy Antshrike	S	-	S	S	S	S	-	-	-	-	-	S	-	-
Barred Antshrike	X	H	X	S	-	S	S	X	-	X	X	X	-	-
Lined Antshrike	X	X	-	X	X	-	-	-	-	-	-	H	-	-
Blackish-gray Antshrike	-	S	S	H	H	S	-	-	H	-	-	H	-	-
Castelnau's Antshrike	-	-	-	-	-	S	-	{S}	-	-	-	-	-	-
White-shouldered Antshrike	-	X	-	S	S	S	S	{S}	H	X	S	X	-	-
Plain-winged Antshrike	-	-	-	X	X	S	X	{X}	{S}	-	X	X	-	-
Mouse-colored Antshrike	-	-	X	A	S	X	S	X	H	X	S	S	-	-
Eastern Slaty-Antshrike	-	-	X	X	X	X	-	-	-	X	-	-	-	-
Amazonian Antshrike	H	X	X	H	X	X	S	S	-	-	S	H	-	-
"Gray-capped" Antshrike	-	-	-	-	-	S	-	-	H	-	-	-	-	-
Streak-backed Antshrike	-	-	-	-	-	-	-	{S}	{S}	-	-	-	-	-
Rufous-winged Antshrike	S	-	-	-	-	-	-	-	-	-	-	{S}	-	-
Spot-winged Antshrike	-	X	-	X	X	S	S	{X}	H	X	H	X	-	-
Pearly Antshrike	-	-	-	-	-	-	X	-	H	-	-	S	-	-
Black Bushbird	-	-	-	-	A	-	-	{S}	H	-	S	-	-	-
E Rondonia Bushbird	-	-	-	-	-	-	-	-	-	-	-	S	-	-
Plain Antvireo	-	-	X	-	-	-	-	-	{X}	S	-	-	-	-
Dusky-throated Antshrike	-	-	X	-	-	X	S	X	H	X	X	-	-	-
Saturnine Antshrike	-	-	-	-	X	-	X	X	-	-	-	X	-	-
Cinereous Antshrike	H	X	X	X	X	X	X	X	X	X	X	X	-	-
Bluish-slate Antshrike	-	-	-	-	-	-	-	S	-	-	X	-	-	-
Pygmy Antwren	-	-	X	X	X	X	-	X	H	X	X	X	-	-
Short-billed Antwren	-	-	-	-	-	-	X	-	{S}	-	(S)	-	-	-
Sclater's Antwren	-	-	X	X	-	A	-	-	-	-	S	S	-	-
Yellow-throated Antwren	-	-	-	-	-	-	-	-	H	S	-	-	-	-
Streaked Antwren	-	-	X	X	X	X	S	H	{S}	{X}	H	X	H	-
Cherrie's Antwren	-	-	-	-	-	-	S	-	-	H	-	-	-	-
E Klages's Antwren	-	-	-	H	-	A	-	-	-	-	-	-	-	-

	S.L	Bel	Ama	San	ANP	Man	Tef	Tab	PdN	N.R	R.B	Ron	MVI	FdN
Rufous-bellied Antwren	-	-	-	X	-	-	X	-	-	H	X	-	-	-
Plain-throated Antwren	-	-	X	-	X	X	-	S	X	{S}	X	X	H	-
Brown-bellied Antwren	-	-	-	X	-	-	X	-	-	-	{H}	-	-	-
White-eyed Antwren	-	-	-	(H)	X	X	-	-	-	-	H	S	-	-
Stipple-throated Antwren	-	-	-	-	-	S	S	X	S	H	X	S	X	-
Ornate Antwren	-	-	-	-	-	S	X	-	S	-	-	S	H	-
Rufous-tailed Antwren	-	-	-	-	-	-	-	H	S	H	-	-	-	-
White-flanked Antwren	-	-	X	X	X	X	X	X	X	H	X	X	H	-
Long-winged Antwren	-	-	X	X	X	X	X	X	X	H	X	X	X	-
Rio Suno Antwren	-	-	-	-	-	-	-	-	{S}	-	-	S	-	-
Ihering's Antwren	-	-	-	-	-	X	-	S	-	-	-	(H)	S	-
Plain-winged Antwren	-	-	-	-	-	-	-	-	-	S	X	-	-	-
Gray Antwren	-	-	X	X	X	X	X	X	{X}	H	X	-	X	-
Leaden Antwren	-	-	-	-	-	S	-	S	-	{S}	-	-	S	-
Banded Antbird	-	-	-	-	-	H	S	A	{S}	H	-	-	{H}	-
Black-capped Antwren	-	-	-	-	-	-	-	-	-	-	-	-	S	-
Spot-tailed Antwren	-	-	-	S	-	-	-	-	-	{H}	-	-	-	-
{Dugand's Antwren}	-	-	-	-	-	-	-	-	{A}	-	-	-	-	-
Todd's Antwren	-	-	A	-	-	-	-	-	-	-	-	-	-	-
Spot-backed Antwren	-	-	-	-	-	-	X	-	-	H	-	-	-	-
Roraiman Antwren	-	-	-	-	-	-	-	-	-	H	-	-	-	-
E Pectoral Antwren	S	-	-	-	-	-	-	-	-	-	-	-	-	-
Large-billed Antwren	-	-	-	-	-	-	-	-	-	-	-	-	{S}	-
Rufous-winged Antwren	-	-	X	(H)	X	-	-	A	-	-	X	-	{S}	-
Dot-winged Antwren	-	-	-	X	-	X	S	H	{S}	-	-	X	X	-
"Lower Amazonian" Antwren	-	-	-	S	-	-	-	-	-	-	-	-	-	-
White-fringed Antwren	-	X	X	X	X	-	-	-	-	{S}	X	-	H	-
Rusty-backed Antwren	-	-	H	X	X	-	-	-	-	-	-	-	-	-
Striated Antbird	-	-	-	-	-	-	-	S	-	-	-	-	X	-
Chestnut-shouldered Antwren	-	-	-	-	-	-	-	-	H	-	-	{S}	S	-
Ash-winged Antwren	-	-	-	X	-	X	X	-	{A}	H	S	-	-	-
Gray Antbird	-	-	X	X	X	X	X	X	X	H	X	X	X	-
Dusky Antbird	-	-	-	X	-	-	X	-	-	H	X	-	-	-
E Lower Amazonian Antbird	-	-	X	X	-	-	S	-	-	-	-	-	-	-
Blackish Antbird	-	-	-	-	X	X	H	S	{S}	-	-	X	X	-
Black Antbird	-	-	-	-	-	-	-	-	S	-	-	S	-	-
Manu Antbird	-	-	-	-	-	-	-	-	-	-	-	(S)	-	-
Rio Branco Antbird	-	-	-	-	-	-	-	-	-	-	S	-	-	-
White-backed Fire-eye	-	-	H	X	-	X	-	-	-	-	-	-	{S}	-
White-browed Antbird	-	-	-	-	X	S	-	H	S	S	{H}	X	X	-
Ash-breasted Antbird	-	-	S	-	S	-	S	S	S	-	-	-	-	-
Black-faced Antbird	-	-	S	-	X	X	S	X	X	H	-	X	X	-

	S.L	Bel	Ama	San	ANP	Man	Tef	Tab	PdN	N.R	R.B	Ron	MVI	FdN
{Black-tailed Antbird}	-	-	-	-	-	-	-	{S}	-	-	-	-	-	-
Warbling Antbird	-	-	-	X	X	X	X	X	X	H	X	X	X	-
Yellow-browed Antbird	-	-	-	-	S	-	S	-	{X}	{S}	-	(S)	-	-
Black-chinned Antbird	-	-	H	X	H	-	X	S	{A}	H	X	-	H	-
Band-tailed Antbird	-	S	X	-	-	S	S	-	S	-	(H)	X	-	-
Black-and-white Antbird	-	-	-	-	-	-	-	S	{S}	-	-	H	-	-
Silvered Antbird	-	-	X	X	H	-	S	S	{X}	H	-	H	X	-
Black-headed Antbird	-	-	X	-	-	X	-	-	S	{S}	-	-	-	-
{White-lined Antbird}	-	-	-	-	-	-	-	-	-	-	{S}	-	-	-
Slate-colored Antbird	-	-	-	-	-	-	S	S	-	-	(H)	-	-	-
Spot-winged Antbird	-	-	X	X	X	X	X	X	H	H	{X}	X	-	-
Caura Antbird	-	-	-	-	-	-	-	-	H	{H}	-	-	-	-
Black-throated Antbird	-	-	X	H	H	S	S	-	H	X	(X)	{S}	-	-
"Spot-breasted" Antbird	-	-	-	-	H	-	-	-	-	-	-	-	-	-
White-bellied Antbird	-	-	X	-	-	-	-	-	-	X	-	-	-	-
Ferruginous-backed Antbird	-	-	X	-	X	X	-	-	-	{H}	-	-	-	-
{Yapacana Antbird}	-	-	-	-	-	-	-	{A}	-	-	-	-	-	-
Gray-bellied Antbird	-	-	-	-	-	-	-	-	H	-	-	-	-	-
Chestnut-tailed Antbird	-	-	-	X	S	-	X	X	-	-	X	X	-	-
Plumbeous Antbird	-	-	-	-	-	-	S	S	-	-	H	{H}	-	-
Goeldi's Antbird	-	-	-	-	-	-	-	-	-	-	H	-	-	-
White-shouldered Antbird	-	-	-	-	-	-	{X}	{S}	-	(X)	-	-	-	-
Sooty Antbird	-	-	-	-	-	-	S	H	-	-	H	-	-	-
White-plumed Antbird	-	-	X	-	-	X	-	{X}	X	X	-	-	-	-
White-throated Antbird	-	-	-	-	-	-	X	X	-	-	(H)	H	-	-
White-cheeked Antbird	-	-	-	-	-	S	-	{X}	H	-	-	-	-	-
Rufous-throated Antbird	-	-	X	-	-	X	-	-	H	X	-	-	-	-
Wing-banded Antbird	-	S	X	H	S	X	-	-	{S}	-	-	S	-	-
E Bare-eyed Antbird	-	-	-	S	-	-	-	-	-	-	-	-	-	-
Harlequin Antbird	-	-	-	S	-	-	-	-	-	-	-	-	-	-
Chestnut-crested Antbird	-	-	-	-	-	-	-	-	-	S	-	-	-	-
E White-breasted Antbird	-	-	-	-	-	-	-	-	-	-	-	-	S	-
Hairy-crested Antbird	-	-	-	-	-	-	X	S	-	-	(H)	H	-	-
Spot-backed Antbird	-	-	X	X	X	X	X	{X}	H	X	S	H	-	-
Dot-backed Antbird	-	-	-	X	S	S	-	H	H	H	-	H	-	-
Scale-backed Antbird	-	-	X	-	-	X	-	X	H	X	-	-	-	-
"Plain-backed" Antbird	-	X	-	X	X	S	X	X	-	-	(H)	H	-	-
E Pale-faced Antbird	-	-	-	-	S	-	-	-	-	-	-	-	-	-
Black-spotted Bare-eye	-	X	H	X	X	-	H	X	-	-	X	X	-	-
Reddish-winged Bare-eye	-	-	-	-	-	S	X	{X}	H	-	{S}	-	-	-
Northern Short-tailed Antthrush	-	-	-	-	-	-	-	-	{H}	X	-	-	-	-
Striated Antthrush	-	-	-	S	-	-	X	{X}	-	-	{S}	-	-	-

	S.L	Bel	Ama	San	ANP	Man	Tef	Tab	PdN	N.R	R.B	Ron	MVI	FdN
Rufous-capped Antthrush-	-	H	X	X	X	X	X	{X}	H	X	X	S	-	-
Black-faced Antthrush	-	-	X	X	X	X	X	X	X	{S}	{S}	X	X	-
Variegated Antpitta-	-	-	-	X	X	S	X	X	-	H	-	S	-	-
Scaled Antpitta	-	-	-	-	-	-	-	-	H	H	-	-	-	-
{Ochre-striped Antpitta}	-	-	-	-	-	-	-	{A}	-	-	-	-	-	-
Elusive Antpitta	-	-	-	-	-	-	-	A	-	-	-	-	-	-
Spotted Antpitta	-	-	H	S	H	H	X	X	{S}	-	-	S	-	-
Amazonian Antpitta	-	-	-	-	S	S	-	-	-	-	-	H	-	-
Thrush-like Antpitta	-	-	-	X	X	X	X	-	X	H	-	{S}	S	-
Brown-breasted Antpitta-	-	-	-	-	-	-	-	-	{S}	S	-	-	-	-
{Slate-crowned Antpitta}	-	-	-	-	-	-	-	-	-	{S}	-	-	-	-
Black-bellied Gnateater-	-	-	-	-	X	H	-	-	-	-	-	S	-	-
E Hooded Gnateater	S	S	-	-	-	-	-	-	-	-	-	-	-	-
Ash-throated Gnateater	-	-	-	-	-	-	-	-	-	-	-	H	-	-
Chestnut-belted Gnateater	-	-	X	H	H	X	X	X	H	-	-	S	-	-
Rusty-belted Tapaculo	-	-	-	-	X	-	S	X	-	-	(X)	X	-	-
Purple-throated Cotinga-	-	-	-	-	-	H	-	H	S	-	-	S	-	-
Plum-throated Cotinga	-	-	-	-	-	H	X	X	{S}	-	H	{H}	-	-
Spangled Cotinga	-	X	X	X	X	X	X	{X}	H	X	(X)	X	-	-
Purple-breasted Cotinga-	-	S	S	S	S	S	-	-	H	{H}	-	-	-	-
Pompadour Cotinga	-	-	X	-	-	X	X	-	H	X	-	S	-	-
E White-tailed Cotinga	-	S	S	-	S	S	-	-	-	-	-	-	-	-
{Red-banded Fruiteater}-	-	-	-	-	-	-	-	-	-	{S}	-	-	-	-
Dusky Purpletuft	-	-	-	-	-	S	-	-	A	-	-	-	-	-
White-browed Purpletuft-	-	X	(H)	-	X	S	S	H	H	-	X	X	-	-
Screaming Piha	-	X	X	X	X	X	X	X	X	X	X	X	-	-
Rose-collared Piha	-	-	-	-	-	-	-	-	-	S	-	-	-	-
White-naped Xenopsaris	-	-	-	-	-	-	-	-	-	S	-	S	-	-
Green-backed Becard-	-	-	-	(H)	-	-	-	-	-	{H}	-	{S}	-	-
Glossy-backed Becard	-	-	-	-	-	S	-	-	-	-	-	-	-	-
Cinereous Becard	-	X	X	X	X	X	X	X	{A}	-	X	(X)	S	-
Chestnut-crowned Becard-	-	-	-	S	S	S	S	H	{S}	-	S	S	-	-
White-winged Becard-	-	X	X	X	X	X	X	X	{S}	X	X	H	-	-
Black-capped Becard-	-	X	X	X	X	X	X	{X}	H	-	X	X	-	-
Crested Becard	-	H	(H)	-	-	-	-	-	-	-	-	{S}	-	-
Pink-throated Becard	-	S	X	X	S	X	X	H	H	X	H	X	-	-
Black-tailed Tityra-	-	X	X	H	X	X	X	{X}	S	X	(S)	{X}	-	-
Masked Tityra	-	X	H	X	X	X	S	X	X	-	-	X	X	-
Black-crowned Tityra	-	-	S	H	S	S	S	{S}	{S}	S	S	{H}	-	-
Crimson Fruitcrow	-	X	X	-	-	X	-	-	{A}	{H}	-	S	-	-
Purple-throated Fruitcrow	-	X	X	X	-	S	S	X	-	X	(X)	X	-	-
Amazonian Umbrellabird	-	-	-	-	-	S	S	{S}	H	S	S	{S}	-	-

	S.L	Bel	Ama	San	ANP	Man	Tef	Tab	PdN	N.R	R.B	Ron	MVI	FdN
Capuchinbird	-	-	X	-	-	X	-	-	H	X	-	-	-	-
Bare-necked Fruitcrow	-	S	X	H	S	S	X	X	-	X	X	X	-	-
White Bellbird	-	-	H	-	A	-	-	-	-	-	{H}	-	-	-
Bearded Bellbird	-	-	-	-	-	-	-	-	-	-	X	-	-	-
Guianan Red-Cotinga	-	-	H	X	X	H	X	-	-	-	{H}	-	-	-
Black-necked Red-Cotinga	-	-	-	H	H	S	S	{X}	H	-	-	X	-	-
Guianan Cock-of-the-Rock	-	-	H	-	-	S	-	-	S	S	-	-	-	-
Sharpbill	-	-	H	S	-	-	-	-	-	-	{S}	-	-	-
Golden-headed Manakin	-	-	X	-	-	X	-	{X}	H	X	-	-	-	-
Red-headed Manakin	-	X	-	X	X	S	X	X	-	-	X	S	-	-
Scarlet-horned Manakin	-	-	-	-	-	-	-	-	-	S	H	-	-	-
White-crowned Manakin	-	-	X	X	-	S	X	X	S	X	X	(H)	-	-
Blue-crowned Manakin	-	-	-	-	-	S	X	X	H	X	H	-	-	-
White-fronted Manakin	-	-	X	-	-	X	-	-	-	X	H	-	-	-
E Opal-crowned Manakin	-	-	X	-	X	-	-	-	-	-	-	-	-	-
Snow-capped Manakin	-	-	-	-	-	S	-	-	-	-	-	X	-	-
Crimson-hooded Manakin	-	-	H	X	H	-	H	-	-	-	-	-	-	-
Band-tailed Manakin	-	-	X	-	H	-	-	-	-	-	X	X	-	-
Wire-tailed Manakin	-	-	-	-	-	S	-	X	H	X	(H)	-	-	-
Helmeted Manakin	-	-	-	-	-	-	-	-	-	-	-	{S}	-	-
Blue-backed Manakin	-	X	X	X	X	X	S	X	X	H	X	H	H	-
White-throated Manakin	-	-	X	-	-	X	-	-	-	-	S	-	-	-
White-bearded Manakin	-	X	X	X	H	S	S	-	X	H	X	(X)	X	-
Fiery-capped Manakin	-	-	S	S	-	-	-	-	-	S	S	{S}	-	-
Striped Manakin	-	-	-	-	-	-	S	S	{S}	S	(S)	-	-	-
Black Manakin	-	-	H	-	A	-	-	-	-	H	H	{S}	-	-
Olive Manakin	-	-	-	-	-	-	-	-	-	-	H	-	-	-
Cinnamon Manakin	-	-	S	-	H	S	-	-	-	S	S	X	-	-
Flame-crested Manakin	-	-	-	H	X	-	S	-	-	-	(H)	H	-	-
Yellow-crested Manakin	-	-	H	-	A	-	-	-	-	H	-	-	-	-
Sulphur-bellied Tyrant-Manakin	-	-	-	-	-	-	-	-	-	-	H	H	-	-
Saffron-crested Tyrant-Manakin	-	-	H	-	-	X	-	-	-	H	{H}	-	-	-
Pale-bellied Tyrant-Manakin	H	-	X	H	H	-	-	-	-	-	-	{S}	-	-
Tiny Tyrant-Manakin	-	-	X	-	-	X	-	-	-	-	-	-	-	-
Dwarf Tyrant-Manakin	-	-	X	-	X	X	S	X	{S}	H	S	X	-	-
Wing-barred Manakin	-	-	S	X	S	S	X	S	S	H	S	X	X	-
Greater Manakin	-	-	H	H	-	S	S	{S}	H	-	(H)	S	-	-
Thrush-like Manakin	-	-	X	X	H	S	X	X	-	H	{H}	(S)	X	-
Gray Monjita	-	-	X	H	-	-	-	-	-	-	-	-	-	-
White-rumped Monjita	-	-	H	(H)	-	-	-	-	-	-	A	-	-	-
Little Ground-Tyrant	-	-	-	-	-	-	-	{S}	-	-	(S)	S	-	-
Long-tailed Tyrant	-	-	S	-	-	-	-	-	{A}	S	H	H	-	-

-183-

Species	S.L		Ama		ANP		Tef		PdN		R.B		MVI	
		Bel		San		Man		Tab		N.R		Ron		FdN
Riverside Tyrant	-	-	-	H	-	S	X	{X}	-	.	-	-	-	-
Rufous-tailed Tyrant	-	-	-	-	-	-	-	-	-	H	H	-	-	-
Amazonian Black-Tyrant	-	-	-	H	H	S	X	{S}	S	H	-	H	-	-
Pied Water-Tyrant	-	-	-	H	-	-	-	-	{S}	X	-	-	-	-
Black-backed Water-Tyrant	-	-	-	(H)	X	A	S	-	-	-	S	H	-	-
Masked Water-Tyrant	S	-	-	-	-	-	-	-	-	-	-	-	-	-
White-headed Marsh-Tyrant	X	X	H	X	-	X	S	{X}	-	X	-	-	-	-
Vermilion Flycatcher	-	-	-	H	-	M	M	M	{M}	M	M	M	-	-
Drab Water-Tyrant	-	-	-	H	-	S	S	-	X	H	X	H	X	-
Cattle Tyrant	S	-	-	-	-	-	-	-	-	-	-	-	-	-
Sirystes	-	-	X	-	S	X	S	{S}	-	-	X	X	-	-
Fork-tailed Flycatcher	M	M	M	M	M	M	M	{M}	H	M	M	M	-	-
Eastern Kingbird	-	-	-	-	-	-	M	M	{M}	-	M	M	-	-
Tropical Kingbird	X	X	X	X	X	X	X	X	H	X	X	X	-	-
Gray Kingbird	-	-	-	-	-	-	-	-	-	-	A	-	-	-
White-throated Kingbird	-	-	M	M	M	M	M	M	-	M	-	M	-	-
Variegated Flycatcher	X	X	X	X	X	X	X	{X}	{X}	X	X	X	-	-
Crowned Slaty Flycatcher	H	-	-	M	-	M	M	{M}	M	-	M	M	-	-
Piratic Flycatcher	-	X	X	X	X	X	M	{M}	H	X	H	M	-	-
Three-striped Flycatcher	-	-	-	S	S	S	S	-	-	-	-	-	-	-
Yellow-throated Flycatcher	-	-	-	X	-	-	X	-	-	H	X	(X)	-	-
Boat-billed Flycatcher	X	X	X	X	X	X	S	X	-	X	X	X	-	-
Sulphur-bellied Flycatcher	-	-	-	-	-	-	-	-	-	-	(A)	-	-	-
Streaked Flycatcher	M	M	H	M	M	M	M	{M}	H	M	M	M	-	-
Rusty-margined Flycatcher	X	X	X	X	X	X	-	{S}	{X}	X	X	X	-	-
Social Flycatcher	X	X	H	X	X	X	S	{X}	-	X	X	S	-	-
Gray-capped Flycatcher	-	-	-	-	-	-	S	X	{X}	-	X	H	-	-
Dusky-chested Flycatcher	-	-	X	-	X	S	X	{X}	H	S	H	S	-	-
Sulphury Flycatcher	S	S	X	H	-	X	S	{X}	H	X	(S)	-	-	-
Great Kiskadee	X	X	X	X	X	X	X	X	H	X	X	X	-	-
Lesser Kiskadee	X	X	X	X	X	X	X	{X}	-	X	H	X	-	-
Bright-rumped Attila	-	X	X	X	X	X	X	X	H	X	-	X	-	-
White-eyed Attila	-	H	-	H	-	S	-	{S}	-	-	(H)	{S}	-	-
Citron-bellied Attila	-	-	-	-	-	-	S	-	H	-	-	(S)	-	-
Cinnamon Attila	S	X	X	H	S	S	S	{X}	{S}	X	(H)	{H}	-	-
Rufous-tailed Attila	-	-	-	H	-	-	-	-	-	-	-	{M}	-	-
Rufous Casiornis	-	-	-	-	-	-	-	-	-	-	-	S	-	-
E Ash-throated Casiornis	H	H	-	H	-	-	-	-	-	-	-	-	-	-
Cinereous Mourner	-	H	X	H	X	X	X	X	H	X	(H)	H	-	-
Grayish Mourner	-	X	X	X	X	X	X	{X}	H	X	X	X	-	-
Pale-bellied Mourner	-	-	-	H	H	-	A	-	-	-	-	-	-	-
Short-crested Flycatcher	X	X	X	X	X	X	X	X	H	X	X	X	-	-

	S.L	Bel	Ama	San	ANF	Man	Tef	Tab	PdN	N.R	R.B	Ron	MVI	FdN
Brown-crested Flycatcher	X	H	X	X	-	-	-	-	-	X	H	S	-	-
Swainson's Flycatcher	M	H	M	H	-	M	M	{M}	H	H	H	M	-	-
Dusky-capped Flycatcher	-	S	S	S	S	X	-	H	{X}	X	S	X	-	-
Olive-sided Flycatcher	-	-	-	-	-	M	-	-	-	A	-	-	-	-
Eastern Wood-Pewee	-	-	-	-	-	M	M	-	-	-	-	M	-	-
Tropical Pewee	-	H	H	H	-	-	-	-	-	S	-	-	-	-
White-throated Pewee	-	-	H	-	-	-	-	-	-	-	-	-	-	-
Blackish Pewee	A	-	-	-	-	-	-	-	-	-	-	-	-	-
{Smoke-colored Pewee}	-	-	-	-	-	-	-	-	-	{S}	-	-	-	-
Alder Flycatcher	-	-	-	A	-	A	-	-	-	-	-	{H}	-	-
Euler's Flycatcher	-	H	S	H	S	S	S	-	-	S	(S)	S	-	-
Audible Fuscous-Flycatcher	H	-	H	H	-	S	S	{X}	H	S	S	H	-	-
Ruddy-tailed Flycatcher	-	X	X	H	X	X	X	{X}	H	X	(H)	S	-	-
Sulphur-rumped Flycatcher	-	S	X	H	X	X	X	S	H	X	H	S	-	-
Black-tailed Flycatcher	H	-	-	X	-	-	-	-	-	X	-	-	-	-
Bran-colored Flycatcher	H	X	X	-	-	-	-	{A}	-	X	X	{S}	-	-
Roraiman Flycatcher	-	-	-	-	-	-	-	-	S	{H}	-	-	-	-
Cliff Flycatcher	-	-	-	-	-	-	-	-	H	{S}	-	-	-	-
Amazonian Royal-Flycatcher	-	S	S	H	H	S	S	{S}	H	-	-	{S}	-	-
White-crested Spadebill	-	S	S	H	-	S	S	-	H	S	-	S	-	-
White-throated Spadebill	H	-	-	-	-	-	-	-	H	H	-	-	-	-
Golden-crowned Spadebill	-	-	S	H	S	S	S	{S}	H	H	-	H	-	-
Cinnamon-crested Spadebill	-	X	S	H	H	S	-	-	H	-	-	S	-	-
Brownish Flycatcher (Twistwing)	-	-	-	-	-	S	S	{S}	H	-	(S)	-	-	-
Yellow-olive Flycatcher	-	S	S	-	S	S	S	-	H	X	S	{H}	-	-
Yellow-margined Flycatcher	-	H	X	X	-	X	X	-	H	S	X	X	-	-
Gray-crowned Flycatcher	-	S	X	H	X	X	S	X	H	X	(X)	X	-	-
Yellow-breasted Flycatcher	X	X	X	X	X	-	-	-	-	X	-	X	-	-
"Upper Amazonian" Flycatcher	-	-	-	-	A	H	{X}	-	-	-	-	-	-	-
Olivaceous Flatbill	-	X	X	H	S	X	S	{X}	-	-	S	-	-	-
Rufous-tailed Flatbill	-	X	X	H	X	X	X	{S}	H	X	H	H	-	-
Dusky-tailed Flatbill	-	-	-	-	-	-	-	-	-	-	-	{H}	-	-
Large-headed Flatbill	-	-	-	-	-	-	S	{H}	-	-	-	S	-	-
Yellow-browed Tody-Flycatcher	-	X	-	H	X	X	X	{S}	H	-	X	S	-	-
Painted Tody-Flycatcher	-	-	X	-	-	X	-	-	H	X	-	-	-	-
Common Tody-Flycatcher	X	-	X	X	-	-	-	-	-	X	-	-	-	-
Spotted Tody-Flycatcher	X	X	X	X	X	X	S	X	H	X	X	S	-	-
Smoky-fronted Tody-Flycatcher	A	-	-	A	-	-	-	-	-	{A}	-	-	-	-
Ruddy Tody-Flycatcher	-	-	-	-	-	-	-	-	-	S	-	-	-	-
Rusty-fronted Tody-Flycatcher	-	-	-	H	-	S	S	X	{S}	-	(X)	{H}	-	-
Slate-headed Tody-Flycatcher	H	H	-	-	-	-	-	-	-	X	-	-	-	-
Black-and-white Tody-Tyrant	-	-	-	-	-	-	-	{S}	-	-	-	-	-	-

	S.L	Bel	Ama	San	ANP	Man	Tef	Tab	PdN	N.R	R.B	Ron	MVI	FdN
"Tricolored" Tody-Tyrant	-	-	-	-	-	-	-	-	-	-	-	S	-	-
Black-chested Tyrant	-	-	S	-	-	H	-	-	-	S	-	-	-	-
Snethlage's Tody-Tyrant	-	-	-	-	X	X	S	S	-	-	-	-	X	-
Zimmer's Tody-Tyrant	-	-	-	-	H	-	-	-	-	-	-	{S}	-	-
Stripe-necked Tody-Tyrant	X	-	-	X	X	-	-	-	-	-	-	{S}	-	-
Johanne's Tody-Tyrant	-	-	-	-	-	-	-	-	S	-	S	S	-	-
Pearly-vented Tody-Tyrant	-	-	-	-	-	-	-	-	-	{S}	-	-	-	-
E Pelzeln's Tody-Tyrant	-	-	-	-	-	-	-	-	H	-	-	-	-	-
White-eyed Tody-Tyrant	-	-	-	X	-	-	S	-	S	-	-	-	-	-
"White-bellied" Tody-Tyrant	-	-	-	H	-	S	X	-	-	-	-	H	-	-
Boat-billed Tody-Tyrant	-	-	A	-	-	S	-	-	-	-	-	-	-	-
Flammulated Bamboo-Tyrant	-	-	-	-	-	-	-	-	-	-	(S)	S	-	-
Double-banded Pygmy-Tyrant	-	-	X	-	-	X	-	-	H	-	-	-	-	-
"Golden-scaled" Pygmy-Tyrant	-	-	-	-	-	-	S	-	-	(S)	-	-	-	-
Long-crested Pygmy-Tyrant	-	-	-	-	-	-	-	-	-	-	(S)	-	-	-
Helmeted Pygmy-Tyrant	-	-	H	X	X	X	X	X	X	{A}	H	X	-	-
Pale-eyed Pygmy-Tyrant	-	-	-	-	-	-	-	-	-	S	-	-	-	-
Short-tailed Pygmy-Tyrant	-	-	X	X	H	X	X	S	{S}	{S}	X	X	X	-
Chapman's Tyrann.(Bristle-Tyr.)	-	-	-	-	-	-	-	-	S	{S}	-	-	-	-
Olive-green Tyrannulet	-	-	-	-	-	-	S	-	-	-	-	-	-	-
Black-fronted Tyrannulet	-	-	-	-	-	-	-	-	S	{H}	-	-	-	-
Yellow Tyrannulet	-	-	-	X	-	X	H	-	-	H	H	-	-	-
Tawny-crowned Pygmy-Tyrant	-	-	-	-	-	-	-	-	-	-	-	{S}	-	-
Rufous-sided Pygmy-Tyrant	-	-	-	-	-	-	-	-	-	-	-	{S}	-	-
Bearded Tachuri	-	-	-	S	-	-	-	-	-	-	S	-	-	-
Lesser Wagtail-Tyrant	-	-	-	-	H	-	S	X	{S}	-	-	-	-	-
River Tyrannulet	-	-	-	-	S	-	S	X	{S}	-	-	-	{H}	-
White-crested Tyrannulet	-	-	-	-	-	-	-	-	-	-	-	{H}	-	-
Pale-tipped Tyrannulet	-	-	H	-	H	H	S	-	-	H	X	-	{S}	-
Plain Tyrannulet	-	-	-	-	-	-	-	-	-	-	S	S	-	-
White-throated Tyrannulet	-	-	-	-	-	-	-	-	H	H	-	-	-	-
Yellow-bellied Elaenia	-	X	X	X	X	-	H	-	-	-	X	X	X	-
Large Elaenia	-	-	-	-	-	-	A	M	{M}	-	-	(H)	H	-
E Noronha Elaenia	-	-	-	-	-	-	-	-	-	-	-	-	-	X
Small-billed Elaenia	-	-	-	-	H	-	M	M	{M}	-	M	(M)	H	-
{Slaty Elaenia}	-	-	-	-	-	-	-	-	{A}	-	-	-	-	-
Brownish Elaenia	-	-	-	-	H	-	H	S	{S}	-	-	-	{H}	-
Plain-crested Elaenia	-	-	S	X	X	S	-	-	-	-	X	-	S	-
Lesser Elaenia	-	-	-	-	H	-	-	M	-	-	H	M	-	{M}
Rufous-crowned Elaenia	-	-	-	H	-	A	-	-	H	S	-	-	-	-
{Great Elaenia}	-	-	-	-	-	-	-	-	-	{S}	-	-	-	-
Sierran Elaenia	-	-	-	-	-	-	-	-	{S}	H	-	-	-	-

Species	S.L	Bel	Ama	San	ANP	Man	Tef	Tab	PdN	N.R	R.B	Ron	MVI	FdN
Forest Elaenia	H	X	X	X	X	X	X	X	H	X	(X)	X	-	-
Gray Elaenia	-	-	-	-	-	S	S	-	S	-	(S)	S	-	-
Yellow-crowned Elaenia	-	S	S	H	-	A	S	-	-	S	(H)	H	-	-
Greenish Elaenia	-	-	-	-	H	A	-	-	-	A	-	{S}	-	-
Campo Suiriri	-	-	X	X	-	-	-	-	-	-	-	-	-	-
Southern Scrub-Flycatcher	-	S	(H)	S	-	S	-	-	S	X	H	S	-	-
"Northern" Scrub-Flycatcher	-	-	(A)	-	-	-	-	-	-	-	-	-	-	-
Mouse-colored Tyrannulet	-	X	X	X	X	X	S	X	-	X	X	S	-	-
Southern Beardless-Tyrannulet	X	X	X	X	X	S	S	{X}	H	X	X	{S}	-	-
Planalto Tyrannulet	-	-	-	-	-	-	-	-	-	-	-	{S}	-	-
Sooty-headed Tyrannulet	-	-	-	S	-	-	S	-	-	H	S	-	-	-
Slender-footed Tyrannulet	X	X	X	X	X	X	X	{X}	H	S	X	X	-	-
Yellow-crowned Tyrannulet	X	X	X	X	X	X	X	X	H	X	X	X	-	-
White-lored Tyrannulet	-	S	X	H	X	X	X	-	S	X	(X)	X	-	-
Sepia-capped Flycatcher	-	-	X	-	-	-	-	{S}	{S}	S	(X)	{X}	-	-
Ochre-bellied Flycatcher	X	X	X	H	X	X	X	{X}	H	X	X	X	-	-
McConnell's Flycatcher	-	-	X	X	X	H	X	S	-	{S}	H	S	-	-
Ringed Antpipit	-	X	X	H	X	X	X	{X}	H	{S}	X	X	-	-
White-winged Swallow	X	X	X	X	X	X	X	X	H	X	X	X	-	-
Brown-chested Martin	M	-	X	H	X	X	X	X	-	X	X	X	-	-
Purple Martin	-	-	-	H	M	M	M	-	{M}	M	-	{H}	-	-
Gray-breasted Martin	X	X	X	X	X	X	X	X	H	X	X	X	-	M
Southern Martin	-	-	-	-	-	A	-	M	H	-	(A)	-	-	-
Blue-and-white Swallow	-	-	M	-	-	-	A	M	{M}	{M}	X	M	-	-
White-banded Swallow	-	H	X	-	X	S	X	{X}	{X}	X	X	X	-	-
Black-collared Swallow	-	-	S	-	S	S	-	-	H	S	-	X	-	-
White-thighed Swallow	-	-	H	-	-	-	X	S	{S}	{A}	-	H	-	-
Tawny-headed Swallow	-	-	-	-	-	-	-	-	-	M	-	-	-	-
Southern Rough-winged Swallow	X	X	-	X	X	X	X	X	{A}	X	X	X	-	-
Bank Swallow (Sand Martin)	-	M	M	-	-	M	M	M	-	{M}	M	-	-	-
Barn Swallow	M	M	H	H	M	M	M	{M}	M	M	H	M	-	A
Cliff Swallow	-	-	-	-	A	-	A	-	-	-	-	-	-	-
Purplish Jay	-	-	-	-	-	-	-	-	-	-	-	{A}	-	-
Violaceous Jay	-	-	-	-	-	-	-	X	S	H	X	(S)	-	-
Curl-crested Jay	-	-	-	-	-	-	-	-	-	-	-	{S}	-	-
Azure-naped Jay	-	-	-	-	-	-	-	-	-	H	-	-	-	-
Cayenne Jay	-	-	-	-	H	-	-	S	-	H	-	-	-	-
White-naped Jay	H	-	-	-	-	-	-	-	-	-	-	-	-	-
Bicolored Wren	-	-	-	-	-	-	-	-	-	X	-	-	-	-
Thrush-like Wren	X	-	-	X	X	S	X	X	-	-	H	X	-	-
Tooth-billed Wren	-	-	-	S	-	-	-	-	-	-	-	X	-	-
Grass Wren	-	-	-	-	-	-	-	-	-	-	-	S	-	-

	S.L	Bel	Ama	San	ANP	Man	Tef	Tab	PdN	N.R	R.B	Ron	MVI	FdN
Moustached Wren	X	S	-	-	X	-	X	-	-	-	X	X	-	-
Coraya Wren	-	X	X	X	S	X	-	X	H	X	-	-	-	-
Buff-breasted Wren	-	-	X	X	X	X	S	{X}	H	X	X	H	-	-
Fawn-breasted Wren	-	-	-	-	-	-	-	-	-	-	-	{X}	-	-
E Gray Wren	-	-	-	-	-	-	-	-	S	-	-	-	-	-
House Wren	X	X	X	X	X	X	X	X	H	X	X	X	-	-
Tepui Wren	-	-	-	-	-	-	-	-	S	S	-	-	-	-
White-breasted Wood-Wren	-	-	-	X	-	-	S	-	-	S	X	-	-	-
Southern Nightingale-Wren	-	X	-	X	X	-	X	X	{S}	-	X	X	-	-
Flutist Wren	-	-	-	-	-	-	-	-	{X}	S	-	-	-	-
Wing-banded Wren	-	-	-	X	-	-	X	-	-	S	X	-	-	-
Musician Wren	-	-	-	X	X	X	X	X	{S}	X	S	H	-	-
Black-capped Donacobius	X	X	X	X	-	X	X	X	-	X	X	X	-	-
Tropical Mockingbird	S	S	-	-	-	-	-	-	-	X	-	-	-	-
Chalk-browed Mockingbird	-	-	X	H	-	-	-	-	-	S	-	-	-	-
Guyanan Solitaire	-	-	-	-	-	-	-	-	-	S	-	-	-	-
Veery	-	-	-	H	-	M	-	M	{A}	M	-	M	-	-
Gray-cheeked Thrush	-	-	-	-	A	M	A	-	H	M	-	-	-	-
Swainson's Thrush	-	-	-	-	-	A	-	-	H	-	-	-	-	-
Yellow-legged Thrush	-	-	-	-	-	-	-	-	-	S	-	-	-	-
Pale-eyed Thrush	-	-	-	-	-	-	-	-	H	S	-	-	-	-
Black-hooded Thrush	-	-	-	-	-	-	-	-	S	S	-	-	-	-
Pale-breasted Thrush	X	X	X	X	X	X	-	-	-	X	-	-	-	-
Creamy-bellied Thrush	H	M	-	-	-	-	-	-	-	-	M	H	-	-
Black-billed Thrush	-	-	-	X	-	X	S	X	{S}	X	X	X	-	-
Lawrence's Thrush	-	-	-	-	-	-	-	S	{X}	S	S	S	-	-
Cocoa Thrush	X	X	X	H	X	S	-	-	{H}	X	-	-	-	-
Hauxwell's Thrush	-	-	-	-	-	-	S	H	{S}	-	H	X	-	-
Bare-eyed Thrush	X	X	X	H	-	-	-	-	-	X	-	-	-	-
White-necked Thrush	-	X	X	H	X	X	X	{X}	H	X	-	H	-	-
Collared Gnatwren	-	-	-	X	-	-	X	-	{X}	H	X	-	-	-
Long-billed Gnatwren	-	-	X	H	-	X	X	X	{S}	{H}	X	(X)	H	-
Tropical Gnatcatcher	X	X	X	X	S	S	S	S	-	X	X	-	-	-
Guianan Gnatcatcher	-	-	S	-	H	-	S	X	-	S	S	-	S	-
Yellowish Pipit	-	-	X	X	X	X	-	-	-	-	X	-	S	-
Rufous-browed Peppershrike	X	X	X	X	X	X	-	-	{S}	X	X	{X}	-	-
Slaty-capped Shrike-Vireo	-	-	X	X	X	X	X	-	-	X	(X)	X	-	-
Red-eyed Vireo	X	X	X	X	X	X	X	H	H	X	X	X	-	-
{Yellow-green Vireo}	-	-	-	-	-	-	-	{M}	-	-	-	-	-	-
E Noronha Vireo	-	-	-	-	-	-	-	-	-	-	-	-	-	X
Black-whiskered Vireo	-	-	-	H	M	M	-	-	{A}	-	-	M	-	-
Lemon-chested Greenlet	-	-	-	H	-	S	S	S	S	-	H	H	-	-

Species	S.L	Bel	Ama	San	ANP	Man	Tef	Tab	PdN	N.R	R.B	Ron	MVI	FdN
Gray-chested Greenlet	-	X	X	-	X	X	S	-	H	-	-	H	-	-
Ashy-headed Greenlet	X	X	(H)	X	-	S	-	-	-	X	-	{H}	-	-
Tepui Greenlet	-	-	-	-	-	-	-	-	{S}	S	-	-	-	-
Buff-cheeked Greenlet	-	-	X	H	X	X	S	-	H	X	-	X	-	-
Brown-headed Greenlet	-	-	-	-	-	-	S	-	H	X	-	X	-	-
Dusky-capped Greenlet	-	-	-	X	-	A	X	X	H	-	X	S	-	-
Tawny-crowned Greenlet	-	-	-	X	-	-	X	X	-	H	S	{S}	-	-
"Red-fronted" Greenlet	-	X	-	H	X	-	-	-	-	-	-	H	-	-
Shiny Cowbird	-	X	S	X	X	X	X	-	{S}	-	X	X	-	-
Giant Cowbird	X	X	H	X	X	X	S	X	-	X	X	X	-	-
Band-tailed Oropendola	-	-	-	-	-	-	-	-	{X}	-	-	-	-	-
Casqued Oropendola	-	-	-	-	-	-	-	-	{A}	-	-	-	-	-
Crested Oropendola	X	H	X	X	X	X	S	X	-	X	X	X	-	-
Green Oropendola	-	X	X	H	X	X	X	{S}	S	X	(H)	S	-	-
Russet-backed Oropendola	-	-	-	-	-	-	S	X	X	-	X	-	-	-
Para (Amazonian) Oropendola	X	X	-	-	-	-	-	-	-	-	-	-	-	-
"Olive" Oropendola	-	-	-	H	X	S	-	X	H	-	X	X	-	-
Yellow-rumped Cacique	X	X	X	X	X	X	X	X	H	X	X	X	-	-
Red-rumped Cacique	-	X	X	H	X	X	-	{X}	H	X	-	S	-	-
Solitary Cacique	-	-	H	X	-	X	S	X	-	S	(X)	{H}	-	-
Carib Grackle	-	-	H	-	-	-	-	-	-	-	-	-	-	-
Golden-tufted Grackle	-	-	-	-	-	-	-	-	-	S	-	-	-	-
Velvet-fronted Grackle	-	-	-	-	-	S	S	S	-	S	X	X	-	-
Chopi Blackbird	-	S	-	-	-	-	-	-	-	-	-	H	-	-
Chestnut-capped Blackbird	-	X	H	S	-	-	-	-	-	-	-	-	-	-
Unicolored Blackbird	-	H	-	H	-	-	-	-	-	-	-	-	-	-
Yellow-hooded Blackbird	-	-	H	X	H	-	S	S	X	-	X	(X)	-	-
Epaulet Oriole	H	H	X	H	X	X	S	{S}	-	-	(X)	X	-	-
Moriche Oriole	-	-	-	-	-	X	S	{X}	H	X	-	-	-	-
Troupial (Campo Oriole)	-	X	S	-	-	-	-	-	-	-	-	-	-	-
"Orange-backed" Troupial	-	I	-	X	-	X	S	X	-	X	X	{H}	-	-
Yellow Oriole	-	-	H	-	-	-	-	-	-	X	-	-	-	-
Oriole Blackbird	-	-	X	S	S	S	S	X	-	X	-	-	-	-
Red-breasted Blackbird	X	X	X	H	X	X	-	{X}	{X}	X	X	X	-	-
White-browed Blackbird	-	-	-	-	-	-	-	-	-	-	-	S	-	-
Eastern Meadowlark	-	-	S	X	-	-	-	-	-	S	X	-	-	-
Bobolink	-	-	-	-	-	-	-	-	-	-	-	A	-	-
{Tennessee Warbler}	-	-	-	-	-	-	-	-	{A}	-	-	-	-	-
Tropical Parula	-	-	-	-	-	-	-	-	{S}	X	-	{X}	-	-
Yellow Warbler	-	-	-	H	-	-	A	-	-	A	H	-	-	-
Blackburnian Warbler	-	-	A	-	-	-	A	-	-	{H}	A	-	-	-
{Bay-breasted Warbler}	-	-	-	-	-	-	-	-	{A}	-	-	-	-	-

	S.L	Bel	Ama	San	ANP	Man	Tef	Tab	PdN	N.R	R.B	Ron	MVI	FdN
Blackpoll Warbler	-	-	A	(H)	A	-	M	-	-	H	M	-	-	-
{Mourning Warbler}	-	-	-	-	-	-	-	-	-	{A}	-	-	-	-
Masked Yellowthroat	-	-	X	X	H	X	X	-	-	-	X	X	-	-
Rose-breasted Chat	-	-	S	X	X	X	S	-	-	H	S	-	H	-
Canada Warbler	-	-	-	-	-	-	-	-	-	-	A	-	-	-
American Redstart	-	-	-	-	H	-	A	-	-	H	M	-	-	-
Slate-throated Redstart	-	-	-	-	-	-	-	-	-	H	S	-	-	-
Tepui Redstart	-	-	-	-	-	-	-	-	{S}	H	-	-	-	-
Two-banded Warbler	-	-	-	-	-	-	-	-	S	S	-	-	-	-
Golden-crowned Warbler	-	-	-	-	-	-	-	-	-	S	-	{X}	-	-
Neotropical River Warbler	-	-	X	X	H	X	X	-	-	{X}	X	-	{X}	-
Buff-rumped Warbler	-	-	-	-	-	-	X	H	{S}	-	{X}	X	-	-
Bananaquit	-	X	X	X	X	X	X	S	-	X	X	-	{X}	-
Chestnut-vented Conebill	-	X	-	(H)	-	-	S	-	{A}	-	S	-	{X}	-
Bicolored Conebill	-	X	X	H	X	-	S	S	{S}	-	-	-	S	-
Pearly-breasted Conebill	-	-	-	-	-	S	S	-	-	-	-	-	-	-
Scaled Flower-piercer	-	-	-	-	-	-	-	-	H	-	-	-	-	-
Greater Flower-piercer	-	-	-	-	-	-	-	-	-	S	-	-	-	-
Short-billed Honeycreeper	-	-	-	H	-	-	S	X	S	H	-	(X)	S	-
Purple Honeycreeper	-	-	X	X	H	X	X	X	{X}	X	X	X	X	-
Red-legged Honeycreeper	-	H	X	X	H	X	X	X	-	H	X	-	H	-
Green Honeycreeper	-	-	X	X	-	X	X	X	X	X	X	X	X	-
Blue Dacnis	X	X	X	X	X	X	X	X	H	X	X	X	-	-
Black-faced Dacnis	-	-	S	X	-	X	X	X	{X}	{X}	X	X	X	-
Yellow-bellied Dacnis	-	-	-	-	H	S	S	X	X	H	-	H	S	-
White-bellied Dacnis	-	-	-	-	-	-	-	-	{H}	-	-	-	-	-
Swallow-Tanager	-	-	-	-	-	-	A	M	-	H	S	-	X	-
Blue-naped Chlorophonia	-	-	-	-	-	-	-	-	H	S	-	-	-	-
Golden-rumped Euphonia	-	-	-	-	-	-	-	-	-	{H}	-	S	-	-
Orange-bellied Euphonia	-	-	-	-	-	-	S	{X}	X	X	(H)	X	-	-
White-vented Euphonia	-	-	S	H	H	-	X	X	{X}	{S}	S	X	X	-
Finsch's Euphonia	-	-	-	-	-	-	-	-	-	X	-	-	-	-
Purple-throated Euphonia	-	S	S	X	S	X	X	H	-	{A}	X	(X)	H	-
Thick-billed Euphonia	-	-	-	-	-	-	-	-	-	-	-	X	-	-
"Black-tailed" Euphonia	-	-	-	S	-	S	X	{X}	-	-	X	-	-	-
Violaceous Euphonia	-	X	X	X	X	X	H	-	-	-	X	-	-	-
Rufous-bellied Euphonia	-	-	-	S	X	S	X	{X}	H	-	X	X	-	-
Golden-sided Euphonia	-	-	X	X	-	-	X	-	-	{S}	-	-	-	-
{Bronze-green Euphonia}	-	-	-	-	-	-	-	-	-	-	-	{S}	-	-
White-lored Euphonia	-	-	-	X	-	H	X	X	{S}	H	X	X	X	-
Plumbeous Euphonia	-	-	-	H	-	-	S	-	S	S	-	-	-	-
Fawn-breasted Tanager	-	-	-	-	-	-	A	-	-	S	-	-	-	-

	S.L	Bel	Ama	San	ANP	Man	Tef	Tab	PdN	N.R	R.B	Ron	MVI	FdN
Opal-rumped Tanager-	-	-	X	X	S	-	X	X	H	H	X	X	X	-
Opal-crowned Tanager	-	-	-	-	-	-	-	-	S	-	-	X	-	-
Paradise Tanager	-	-	X	--	X	X	X	X	H	X	X	X	-	-
Green-and-gold Tanager	-	-	-	-	-	-	X	X	{X}	-	X	-	-	-
Spotted Tanager-	-	-	X	X	X	H	X	-	-	H	S	-	-	-
Speckled Tanager	-	-	-	-	-	-	-	-	-	H	S	-	-	-
Dotted Tanager	-	-	-	-	S	H	S	-	-	-	-	-	-	-
Yellow-bellied Tanager	-	-	-	-	-	-	-	-	S	H	X	X	-	-
Blue-necked Tanager-	-	-	-	-	-	-	-	-	-	-	-	{S}	-	-
Masked Tanager	-	-	-	-	-	-	-	S	-	H	S	X	X	-
Turquoise Tanager	-	-	X	X	X	X	X	X	X	S	X	X	X	-
Bay-headed Tanager	-	-	X	X	-	S	S	H	S	H	S	(S)	H	-
Burnished-buff Tanager	-	-	-	X	X	-	-	-	-	{A}	X	-	{S}	-
Black-headed Tanager	-	-	-	-	-	-	-	-	-	{S}	H	-	-	-
Blue-gray Tanager	X	X	X	X	X	X	X	X	H	X	X	X	-	-
Sayaca Tanager	-	-	-	-	-	-	-	-	-	-	-	-	S	-
Palm Tanager	X	X	X	X	X	X	X	X	H	X	X	X	-	-
Silver-beaked Tanager	X	X	X	X	X	X	X	X	H	X	X	X	-	-
Masked Crimson Tanager	-	-	-	-	H	S	S	X	X	-	-	X	H	-
Lowland Hepatic-Tanager-	-	-	-	X	H	-	-	-	-	-	X	-	-	-
"Highland" Hepatic-Tanager	-	-	-	-	-	-	-	-	-	H	-	-	-	-
Summer Tanager	-	-	-	-	A	A	M	A	-	H	M	-	H	-
Scarlet Tanager-	-	-	-	-	-	-	A	-	-	-	-	-	-	-
White-winged Tanager	-	-	-	-	-	-	-	-	-	S	-	-	-	-
Blue-backed Tanager-	-	-	-	-	-	S	S	-	-	S	-	-	-	-
Red-crowned Ant-Tanager-	-	-	-	X	X	S	S	X	H	-	X	X	-	-
Fulvous Shrike-Tanager	-	-	-	X	-	-	S	-	{X}	S	-	-	-	-
White-winged Shrike-Tanager-	-	-	-	X	S	-	X	H	-	-	-	Y	-	-
White-lined Tanager-	X	X	X	H	X	-	-	-	-	-	-	-	-	-
Flame-crested Tanager	-	-	X	X	X	X	X	X	{X}	H	X	X	X	-
Fulvous-crested Tanager-	-	-	X	X	H	X	X	X	S	H	X	(X)	S	-
Red-shouldered Tanager	-	-	-	X	-	-	A	-	-	{S}	X	-	X	-
Yellow-crested Tanager	-	-	-	-	-	-	-	-	S	-	-	(X)	-	-
White-shouldered Tanager	-	-	X	X	X	X	S	-	-	X	X	X	-	-
Olive-backed Tanager	-	-	-	-	-	-	-	-	-	S	-	-	-	-
Gray-headed Tanager-	-	-	X	H	H	-	S	S	S	-	-	(H)	H	-
White-rumped Tanager	-	-	X	-	-	-	-	-	-	-	-	X	-	-
Hooded Tanager	X	H	(H)	X	-	S	X	{X}	-	X	H	{H}	-	-
Guira Tanager	-	-	X	X	X	-	S	-	{S}	-	X	H	{X}	-
Yellow-backed Tanager	-	-	-	X	-	-	X	X	H	H	X	(X)	X	-
Orange-headed Tanager	-	S	-	-	-	-	A	S	{X}	-	-	-	H	-
White-banded Tanager	-	-	-	S	-	-	-	-	-	-	-	-	-	-

	S.L	Bel	Ama	San	ANP	Man	Tef	Tab	PdN	N.R	R.B	Ron	MVI	FdN
Red-billed Pied Tanager	-	-	X	X	X	X	X	X	-	-	-	{X}	X	-
Magpie Tanager	-	-	-	-	X	-	X	X	-	X	X	X	-	-
Black-faced Tanager	-	X	X	X	-	A	-	{A}	{S}	X	X	X	-	-
Buff-throated Saltator	-	X	X	X	X	X	X	S	X	H	X	X	X	-
Grayish Saltator	-	X	X	H	X	X	X	S	X	-	X	X	X	-
Black-throated Saltator	-	-	-	-	-	-	-	-	-	-	-	{S}	-	-
Yellow-green Grosbeak	-	-	X	X	-	S	X	-	-	H	-	-	-	-
Yellow-shouldered Grosbeak	-	-	-	-	-	-	-	-	-	-	-	H	X	-
Red-and-black Grosbeak	-	-	H	X	H	-	-	-	-	{S}	-	-	-	-
Slate-colored Grosbeak	-	-	H	X	H	X	X	-	{X}	H	X	X	X	-
E Red-cowled Cardinal	-	-	I	-	-	-	-	-	-	-	-	-	-	X
Red-capped Cardinal	S	-	H	X	X	X	X	X	-	X	X	X	-	-
Black-backed Grosbeak	-	-	-	-	-	-	-	-	-	-	-	{H}	-	-
Blue-black Grosbeak	-	X	X	X	X	X	X	S	S	S	X	X	X	-
Dickcissel	-	-	-	-	-	-	-	-	-	A	-	-	-	-
Blue Finch	-	-	-	-	-	-	-	-	-	-	-	{S}	-	-
Blue-black Grassquit	-	X	X	X	X	X	X	X	H	X	X	X	-	-
Sooty Grassquit	-	-	-	-	-	-	-	-	-	{H}	-	-	-	-
White-naped Seedeater	-	-	-	-	-	-	-	-	-	H	-	-	-	-
Slate-colored Seedeater	-	S	X	-	-	-	-	-	-	X	-	{S}	-	-
Gray Seedeater	-	-	-	-	-	-	-	-	-	S	-	-	-	-
Plumbeous Seedeater	S	-	X	-	-	-	-	-	-	X	-	{H}	-	-
Wing-barred Seedeater	-	X	X	H	X	X	S	X	{S}	-	-	-	-	-
Rusty-collared Seedeater	-	-	-	-	A	-	-	-	-	-	-	-	-	-
Lined Seedeater	-	-	M	H	M	M	M	M	{M}	H	M	-	H	-
Lesson's Seedeater	-	-	-	-	-	-	M	M	{M}	H	-	H	-	-
Black-and-white Seedeater	-	-	-	-	-	-	-	-	-	-	-	(A)	-	-
Yellow-bellied Seedeater	-	X	X	X	-	-	-	-	{S}	H	X	-	-	-
Double-collared Seedeater	-	-	-	H	-	-	-	-	{M}	-	-	M	M	-
White-bellied Seedeater	S	-	(H)	-	-	-	-	-	-	-	-	-	-	X
Black-and-tawny Seedeater	-	-	-	-	-	-	-	-	-	-	-	{M}	-	-
Capped Seedeater	-	X	-	S	-	-	-	-	-	-	-	-	-	-
Ruddy-breasted Seedeater	-	-	X	X	H	-	-	-	-	H	X	-	-	-
Tawny-bellied Seedeater	-	-	-	-	-	-	-	-	-	-	-	{M}	-	-
Dark-throated Seedeater	-	-	-	-	-	-	-	-	-	-	-	{S}	-	-
Chestnut-bellied Seedeater	-	-	X	X	X	X	X	X	H	X	X	X	-	-
Rufous-rumped Seedeater	-	-	-	-	-	-	-	-	-	-	-	{M}	-	-
Large-billed Seed-Finch	-	-	X	-	-	-	-	-	-	H	X	-	-	-
Great-billed Seed-Finch	-	-	S	X	H	-	-	-	-	-	-	S	-	-
Lesser Seed-Finch	-	X	X	H	X	X	X	X	H	X	X	X	X	-
Paramo Seedeater	-	-	-	-	-	-	-	-	{S}	S	-	-	-	-
Stripe-tailed Yellow-Finch	-	-	-	-	-	-	-	-	-	X	-	{S}	-	-

	S.L	Bel	Ama	San	ANP	Man	Tef	Tab	PdN	N.R	R.B	Ron	MVI	FdN
Orange-fronted Yellow-Finch	-	-	H	X	H	X	X	-	-	-	-	-	-	-
Saffron Finch	-	-	-	-	-	-	-	-	-	-	-	-	X	X
Grassland Yellow-Finch	-	-	-	H	H	-	-	-	-	X	-	-	-	-
{Slaty Finch}	-	-	-	-	-	-	-	-	{A}	-	-	-	-	-
Coal-crested Finch	-	-	-	-	-	-	-	-	-	-	-	{S}	-	-
Red Pileated-Finch	-	S	-	-	-	-	-	-	-	-	-	{S}	-	-
Tepui Brush-Finch	-	-	-	-	-	-	-	-	H	S	-	-	-	-
Pectoral Sparrow	X	X	X	H	X	X	-	-	X	X	X	X	-	-
Black-striped Sparrow	-	-	-	-	-	-	-	-	-	X	-	-	-	-
Grassland Sparrow	X	-	X	X	-	-	-	-	-	X	-	{X}	-	-
Yellow-browed Sparrow	-	-	X	-	H	X	X	X	X	H	-	X	X	-
Rufous-collared Sparrow	-	-	X	X	-	-	-	-	H	X	-	-	-	-
Wedge-tailed Grass-Finch	H	-	X	-	-	-	-	-	-	X	-	-	-	-
Black-masked Finch	-	-	-	(H)	-	-	-	-	-	-	-	-	-	-
Hooded Siskin	-	-	-	-	-	-	-	-	-	-	-	S	-	-
{Olivaceous Siskin}	-	-	-	-	-	-	-	-	-	-	-	{A}	-	-
House Sparrow	X	X	X	-	-	X	-	-	-	-	-	-	X	X
Common Waxbill	-	I	-	-	-	I	-	-	-	-	-	-	-	-

Brazilian Checklist

Here, a full checklist of the birds of Brazil is given. Alternative name forms are in brackets. Species within inverted commas, are either, subspecies of the previous bird, or in some cases are colour morphs that have been considered as separate species in the past. Species totally within parenthesis have not been located within Brazil, though have been recorded very close to the border and could be expected to occur, within Brazil, in the future. The appearance of an "E" prior to the English name, denotes an endemic, though includes species that have occurred as a non-breeding accidental in a neighbouring country.

The three site group lists are:-

Those species that have not been located at any of the 42 sites have the state listed where they have been recorded. Where possible the state name is given in full but has been abbreviated where necessary:-

E. Santo	= Espírito Santo
Mato	= Mato Grosso
Minas G.	= Minas Gerais
RdJ	= Rio de Janeiro
S. Cat	= Santa Catarina
S.P	= São Paulo

Atol das Rocas is 220 kms off Rio Grande de Norte.

Long-tailed Cinclodes

Common name	Scientific name	1	2	3
Rockhopper Penguin	Eudyptes chrysocome (crestatus)	X	-	-
Macaroni Penguin	Eudyptes chrysolophus	X	-	-
Magellanic Penguin	Spheniscus magellanicus	X	X	-
Greater Rhea	Rhea americana	X	X	X
Gray Tinamou	Tinamus tao	-	-	X
Solitary Tinamou	Tinamus solitarius	X	X	-
Great Tinamou.	Tinamus major	-	-	X
White-throated Tinamou	Tinamus guttatus	-	-	X
Cinereous Tinamou	Crypturellus cinereus	X	-	X
Little Tinamou	Crypturellus soui	X	X	X
Brown Tinamou	Crypturellus obsoletus	X	X	X
Undulated Tinamou	Crypturellus undulatus	X	X	X
Rusty Tinamou	Crypturellus brevirostris	-	-	X
Bartlett's Tinamou	Crypturellus bartletti	-	-	X
Variegated Tinamou	Crypturellus variegatus	-	X	X
E Yellow-legged Tinamou	Crypturellus noctivagus	X	X	-
Gray-legged Tinamou	Crypturellus duidae	-	-	X
Red-legged(Red-footed) Tinamou	Crypturellus erythropus	-	-	X
Brazilian Tinamou	Crypturellus strigulosus	X	X	X
Barred Tinamou	Crypturellus casiquiare	-	-	X
Small-billed Tinamou	Crypturellus parvirostris	X	X	X
Tataupa Tinamou	Crypturellus tataupa	X	X	X
Red-winged Tinamou	Rhynchotus rufescens	X	X	X
White-bellied Nothura	Nothura boraquira	-	X	-
E Lesser Nothura	Nothura minor	X	-	-
Spotted Nothura	Nothura maculosa	X	X	-
Dwarf Tinamou	Taoniscus nanus	X	-	-
Least Grebe	Tachybaptus (Podiceps) dominicus	X	X	X
White-tufted Grebe	Podiceps (Rollandia) rolland	X	-	-
Great Grebe	Podiceps major	X	-	-
Pied-billed Grebe	Podilymbus podiceps	X	X	X
Wandering Albatross	Diomedea exulans	X	-	-
Royal Albatross	Diomedea epomophora	X	-	-
Black-browed Albatross	Diomedea melanophris	X	X	X
Shy (White-capped) Albatross	Diomedea cauta	X	-	-
Yellow-nosed Albatross	Diomedea chlororhynchos	X	X	-
Gray-headed Albatross	Diomedea chrysostoma	-	X	-
Sooty Albatross	Phoebetria fusca	1 dead Sao Paulo		
Light-mantled Sooty Albatross	Phoebetria palpebrata	X	-	-
Southern Giant-Petrel	Macronectes giganteus	X	X	-

Common Name	Scientific Name			
Southern (Antarctic) Fulmar-	Fulmarus glacialoides	X	-	-
Cape (Pintado) Petrel	Daption capense-	X	X	-
Gray-faced (Great-winged) Petrel	Pterodroma macroptera	Rio Grande do Sul		
Kerguelen Petrel	Pterodroma brevirostris-	X	X	-
White-headed Petrel-	Pterodroma lessonii-	X	-	-
Atlantic (Hooded) Petrel	Pterodroma incerta	X	X	-
Black-capped Petrel-	Pterodroma hasitata-	Oceanic off Brazil		
Herald (Trindade) Petrel	Pterodroma arminjoniana-	-	-	X
Soft-plumaged Petrel	Pterodroma mollis	X	-	-
Blue Petrel-	Halobaena caerulea	Rio de Janeiro		
Broad-billed Prion	Pachyptila vittata	X	-	-
"Antarctic (Dove)" Prion	Pachyptila "desolata"	X	X	-
Thin-billed(Slender-billed)Prion	Pachyptila belcheri-	X	X	-
Gray (Brown) Petrel-	Procellaria (Adamastor) cinerea-	X	-	-
White-chinned Petrel (Shoemaker)	Procellaria aequinoctialis	X	X	X
Cory's Shearwater	Calonectris (Puffinus) diomedea-	X	X	X
Great Shearwater	Puffinus gravis-	X	X	X
Sooty Shearwater	Puffinus griseus	X	X	-
Manx Shearwater-	Puffinus puffinus	X	X	X
Little Shearwater	Puffinus assimilis -	-	-	X
Wilson's Storm-Petrel	Oceanites oceanicus-	X	X	X
White-bellied Storm-Petrel	Fregatta grallaria	Espirito Santo		
Black-bellied Storm-Petrel	Fregatta tropica	-	-	X
Leach's Storm-Petrel	Oceanodroma leucorhoa	-	X	X
Magellanic Diving-Petrel	Pelecanoides magellani	X	-	-
Red-billed Tropicbird	Phaethon aethereus -	-	-	X
White-tailed Tropicbird-	Phaethon lepturus	-	-	X
Brown Pelican	Pelecanus occidentalis -	-	X	X
Australasian Gannet-	Sula (Morus) serrator	Santa Catarina		
Masked (Blue-faced) Booby	Sula dactylatra-	-	X	X
Red-footed Booby	Sula sula -	-	X	X
Brown Booby-	Sula leucogaster	-	X	X
Neotropic (Olivaceous) Cormorant	Phalacrocorax brasilianus(olivaceus)	X	X	X
Anhinga (American Darter)	Anhinga anhinga-	X	X	X
Magnificent Frigatebird-	Fregata magnificens-	X	X	X
Great Frigatebird	Fregata minor	-	-	X
Lesser Frigatebird	Fregata ariel	-	-	X
Purple Heron	Ardea purpurea	-	-	X
Grey Heron	Ardea cinerea	-	-	X
Cocoi (White-necked) Heron	Ardea cocoi-	X	X	X
Great (Common) Egret	Casmerodius albus	X	X	X
Snowy Egret-	Egretta thula	X	X	X
Little Blue Heron	Egretta (Florida) caerulea -	X	X	X
Tricolored Heron	Egretta (Hydranassa) tricolor	X	-	X

Common Name	Scientific Name			
Striated Heron - - - -	- Butorides striatus - - - -	X	X	X
Agami (Chestnut-bellied) Heron	- Agamia agami - - - -	X	-	X
Squacco Heron - - - -	- Ardeola ralloides - - - -	-	-	X
Cattle Egret - - - -	- Bubulcus ibis - - - -	X	X	X
Whistling Heron- - - -	- Syrigma sibilatrix - - -	X	X	-
Capped Heron - - - -	- Pilherodius pileatus - - -	X	X	X
Black-crowned Night-Heron -	- Nycticorax nycticorax - - -	X	X	X
Yellow-crowned Night-Heron -	- Nyctanassa (Nycticorax) violacea	X	X	X
Boat-billed Heron - - -	- Cochlearius cochlearius- - -	X	X	X
Rufescent Tiger-Heron - -	- Tigrisoma lineatum - - -	X	X	X
Fasciated Tiger-Heron - -	- Tigrisoma fasciatum- - -	X	-	X
Zigzag Heron - - -	- Zebrilus undulatus - - -	-	-	X
Stripe-backed Bittern - -	- Ixobrychus involucris - -	X	X	X
Least Bittern - - -	- Ixobrychus exilis - - -	X	X	X
Pinnated Bittern - - -	- Botaurus pinnatus - - -	X	X	X
Wood Stork - - -	- Mycteria americana - - -	X	-	X
Maguari Stork - - -	- Ciconia (Euxenura) maguari - -	X	X	X
Jabiru - - - -	- Jabiru mycteria- - - -	X	X	X
Plumbeous Ibis - - -	- Theristicus(Harpiprion)caerulescens	X	-	-
Buff-necked Ibis - - -	- Theristicus caudatus - - -	X	-	X
Sharp-tailed Ibis - - -	- Cercibis oxycerca - - -	-	-	X
Green Ibis - - - -	- Mesembrinibis cayennensis - -	X	-	X
Bare-faced (Whispering) Ibis	- Phimosus infuscatus- - -	X	X	X
Scarlet Ibis - - -	- Eudocimus ruber- - - -	-	X	X
White-faced Ibis - - -	- Plegadis chihi - - -	X	X	-
Roseate Spoonbill - - -	- Ajaia ajaja- - - -	X	X	X
Greater Flamingo - - -	- Phoenicopterus ruber - -	-	-	X
Chilean Flamingo - - -	- Phoenicopterus chilensis - -	X	-	-
Andean Flamingo- - -	- Phoenicopterus andinus - -	X	-	-
Horned Screamer- - -	- Anhima cornuta - - -	X	X	X
Southern Screamer - - -	- Chauna torquata- - -	X	-	X
Fulvous Whistling-Duck - -	- Dendrocygna bicolor- - -	X	X	X
White-faced Whistling-Duck -	- Dendrocygna viduata- - -	X	X	X
Black-bellied Whistling-Duck	- Dendrocygna autumnalis - -	X	X	X
Coscoroba Swan - - -	- Coscoroba coscoroba- - -	X	-	-
Black-necked Swan - - -	- Cygnus melancorypha- - -	X	-	-
Orinoco Goose - - -	- Neochen jubata - - -	X	-	X
Speckled Teal - - -	- Anas flavirostris - - -	X	-	-
Chiloe (Southern) Wigeon -	- Anas sibilatrix- - -	X	-	-
White-cheeked (Bahama) Pintail	- Anas bahamensis- - -	X	X	X
Northern Pintail - - -	- Anas acuta - - - -	-	-	X
Yellow-billed Pintail - -	- Anas georgica - - -	X	-	X
Silver Teal- - - -	- Anas versicolor- - -	X	-	-
Blue-winged Teal - - -	- Anas discors - - -	X	-	X

Common Name	Scientific Name			
Cinnamon Teal	Anas cyanoptera	X		
Red Shoveler	Anas platalea	X	X	
Ringed Teal	Anas (Callonetta) leucophrys	X		
Rosy-billed Pochard	Netta peposaca	X		
Southern Pochard	Netta erythrophthalma	X	X	
Brazilian Duck (Teal)	Amazonetta brasiliensis	X	X	X
Comb Duck	Sarkidiornis melanotos	X	X	X
Muscovy Duck	Cairina moschata	X	X	X
Brazilian Merganser	Mergus octosetaceus	X		
Lake Duck	Oxyura vittata	X		
Masked Duck	Oxyura dominica	X	X	X
Black-headed Duck	Heteronetta atricapilla	X		
Andean Condor	Vultur gryphus	X		
King Vulture	Sarcoramphus papa	X	X	X
Black Vulture	Coragyps atratus	X	X	X
Turkey Vulture	Cathartes aura	X	X	X
Lesser Yellow-headed Vulture	Cathartes burrovianus	X	X	X
Greater Yellow-headed Vulture	Cathartes melambrotus	X	-	X
White-tailed Kite	Elanus leucurus (caeruleus)	X	X	X
Pearl Kite	Gampsonyx swainsonii	X	X	X
Swallow-tailed Kite	Elanoides forficatus	X	X	X
E White-collared Kite	Leptodon forbesi	-	X	-
Gray-headed Kite	Leptodon cayanensis	X	X	X
Hook-billed Kite	Chondrohierax uncinatus	X	X	X
Rufous-thighed Kite	Harpagus diodon	X	X	X
Double-toothed Kite	Harpagus bidentatus	-	X	X
Plumbeous Kite	Ictinia plumbea	X	X	X
Mississippi Kite	Ictinia mississippiensis	-	-	X
Snail Kite	Rostrhamus sociabilis	X	X	X
Slender-billed Kite	Rostrhamus(Helicolestes) hamatus	X	-	X
Bicolored Hawk	Accipiter bicolor	X	X	X
Tiny Hawk	Accipiter superciliosus	X	X	X
Gray-bellied Hawk (Goshawk)	Accipiter poliogaster	X	X	X
Sharp-shinned Hawk	Accipiter striatus	X	X	X
Black-chested Buzzard-Eagle	Geranoaetus melanoleucus	X	X	X
White-tailed Hawk	Buteo albicaudatus	X	X	X
Zone-tailed Hawk	Buteo albonotatus	X	X	X
Swainson's Hawk	Buteo swainsoni	X	-	X
Broad-winged Hawk	Buteo platypterus	X	X	X
Roadside Hawk	Buteo magnirostris	X	X	X
White-rumped Hawk	Buteo leucorrhous	X	X	-
Short-tailed Hawk	Buteo brachyurus	X	X	X
Gray-lined (Gray) Hawk	Asturina (Buteo) nitida	X	X	X
Harris's (Bay-winged) Hawk	Parabuteo unicinctus	X	X	-

Common Name	Scientific Name			
White Hawk	Leucopternis albicollis	X	-	X
Mantled Hawk	Leucopternis polionota	X	X	-
E White-necked Hawk	Leucopternis lacernulata	X	X	-
Black-faced Hawk	Leucopternis melanops	-	-	X
White-browed Hawk	Leucopternis kuhli	-	-	X
Slate-colored Hawk	Leucopternis schistacea	-	-	X
Black-collared Hawk	Busarellus nigricollis	X	X	X
Savanna Hawk	Buteog.(Heterospizias) meridionalis	X	X	X
Rufous Crab-Hawk	Buteogallus aequinoctialis	-	-	X
Great Black-Hawk	Buteogallus urubitinga	X	X	X
{Solitary Eagle}	{Harpyhaliaetus solitarius}	-	-	{X}
Crowned Eagle	Harpyhaliaetus coronatus	X	X	-
Crested Eagle	Morphnus guianensis	X	X	X
Harpy Eagle	Harpia harpyja	X	X	X
Black-and-white Hawk-Eagle	Spizastur melanoleucus	X	X	X
Ornate Hawk-Eagle	Spizaetus ornatus	X	X	X
Black Hawk-Eagle	Spizaetus tyrannus	X	X	X
Cinereous Harrier	Circus cinereus	X	-	-
Long-winged Harrier	Circus buffoni	X	X	X
Crane Hawk	Geranospiza caerulescens	X	X	X
Osprey	Pandion haliaetus	X	X	X
Laughing Falcon	Herpetotheres cachinnans	X	X	X
Collared Forest-Falcon	Micrastur semitorquatus	X	X	X
{Buckley's Forest-Falcon}	{Micrastur buckleyi}	-	-	{X}
Slaty-backed Forest-Falcon	Micrastur mirandollei	-	-	X
Barred Forest-Falcon	Micrastur ruficollis	X	X	X
Lined Forest-Falcon	Micrastur gilvicollis	-	X	X
Black (Yellow-throated) Caracara	Daptrius ater	-	-	X
Red-throated Caracara	Daptrius americanus	X	-	X
Yellow-headed Caracara	Milvago chimachima	X	X	X
Chimango Caracara	Milvago chimango	X	-	-
Crested Caracara	Polyborus (Caracara) plancus	X	X	X
Peregrine Falcon	Falco peregrinus	X	X	X
Orange-breasted Falcon	Falco deiroleucus	X	-	X
Bat Falcon	Falco rufigularis	X	X	X
Aplomado Falcon	Falco femoralis	X	X	X
Merlin	Falco columbarius	-	-	X
American Kestrel	Falco sparverius	X	X	X
Little (Variable) Chachalaca	Ortalis motmot	-	-	X
"Buff-browed" Chachalaca	Ortalis "superciliaris"	-	-	X
"Speckled" Chachalaca	Ortalis "guttata"	X	X	X
Chaco Chachalaca	Ortalis canicollis	X	-	-
Marail Guan	Penelope marail	-	-	X
Rusty-margined Guan	Penelope superciliaris	X	X	X

Common Name	Scientific Name			
Dusky-legged Guan	Penelope obscura	X	X	-
Spix's Guan	Penelope jacquacu	-	-	X
E White-browed Guan	Penelope jacucaca	-	X	-
E Chestnut-bellied Guan	Penelope ochrogaster	X	-	-
E White-crested Guan	Penelope pileata	-	-	X
Blue-throated(Common)Piping-Guan	Pipile (Aburria) cumanensis(pipile)	X	-	X
"Red-throated" Piping-Guan	Pipile (Aburria)"cujubi"	X	-	X
Black-fronted Piping-Guan	Pipile (Aburria) jacutinga	X	X	-
Black Curassow	Crax alector	-	-	X
E Red-billed Curassow	Crax blumenbachii	-	X	..
Bare-faced Curassow	Crax fasciolata	X	-	X
"Natterer's" Curassow	Crax "pinima"	-	-	X
Wattled Curassow	Crax globulosa	-	-	X
E Alagoas Curassow	Mitu mitu	-	X	-
Razor-billed Curassow	Mitu (Crax) tuberosa	-	-	X
Crestless(LesserRazor-b)Curassow	Mitu (Crax) tomentosa	-	-	X
Nocturnal (Rufous) Curassow	Nothocrax urumutum	-	-	X
Crested Bobwhite	Colinus cristatus	-	-	X
Marbled Wood-Quail	Odontophorus gujanensis	X	-	X
Spot-winged Wood-Quail	Odontophorus capueira	X	X	-
Starred Wood-Quail	Odontophorus stellatus	-	-	X
Limpkin	Aramus guarauna	X	X	X
Gray-winged (Common) Trumpeter	Psophia crepitans	-	-	X
Pale-winged (White-w.)Trumpeter	Psophia leucoptera	-	-	X
E Dark-winged (Green-w.)Trumpeter	Psophia viridis	-	-	X
Clapper Rail	Rallus longirostris	-	X	X
Plumbeous Rail	Rallus (Pardirallus) sanguinolentus	X	X	-
Blackish Rail	Rallus (Pardirallus) nigricans	X	X	X
Spotted Rail	Pardirallus (Rallus) maculatus	X	-	X
Uniform Crake	Amaurolimnas (Eulabeor.) concolor	X	X	X
E Little Wood-Rail	Aramides (Eulabeornis) mangle	-	X	X
Gray-necked Wood-Rail	Aramides (Eulabeornis) cajanea	X	X	X
Giant Wood-Rail	Aramides (Eulabeornis) ypecaha	X	-	-
Slaty-breasted Wood-Rail	Aramides (Eulabeornis) saracura	X	X	-
Red-winged Wood-Rail	Aramides (Eulabeornis) calopterus	-	-	X
Ash-throated Crake	Porzana albicollis	X	X	X
Yellow-breasted Crake	Porzana flaviventer	X	X	X
Gray-breasted Crake	Laterallus exilis	X	X	X
Rufous-faced Crake	Laterallus xenopterus	X	-	-
Rufous-sided Crake	Laterallus melanophaius	X	X	X
Red-and-white Crake	Laterallus leucopyrrhus	X	X	-
Russet-crowned Crake	Anurolimnas (Laterallus) viridis	X	X	X
Black-banded Crake	Anurolimnas (Laterallus) fasciatus	-	-	X
{Chestnut-headed Crake}	{Anurolimnas castaneiceps}	-	-	{X}

Common name	Scientific name	Sao	Paulo	& RGdS
Ocellated Crake	Micropygia (Coturn.) schomburgkii	X	-	-
Speckled (Darwin's) Crake	Coturnicops notata			
Paint-billed Crake	Neocrex (Porzana) erythrops	X	X	X
Spot-flanked Gallinule	Porphyriops(Gallinula) melanops	X	X	-
Common Gallinule (Moorhen)	Gallinula chloropus	X	X	X
Purple Gallinule	Porphyrula (Gallin.) martinica	X	X	X
Azure Gallinule	Porphyrula (Gallin.) flavirostris	X	-	X
Red-gartered Coot	Fulica armillata	X	X	-
White-winged Coot	Fulica leucoptera	X	-	-
Red-fronted Coot	Fulica rufifrons	X	-	-
Sungrebe (American Finfoot)	Heliornis fulica	X	X	X
Sunbittern	Eurypyga helias	X	-	X
Red-legged Seriema	Cariama cristata	X	X	-
Wattled Jacana	Jacana jacana	X	X	X
South American Painted-Snipe	Nycticryphes(Rostratula)semicollaris	X	X	-
American Oystercatcher	Haematopus palliatus	X	X	X
Southern (Chilean) Lapwing	Vanellus chilensis	X	X	X
Pied Plover (Lapwing)	Hoploxypterus (Vanellus) cayanus	X	X	X
Black-bellied (Grey) Plover	Pluvialis squatarola	X	X	X
American Golden Plover	Pluvialis dominica	X	-	X
Semipalmated Plover	Charadrius semipalmatus	X	X	X
Two-banded Plover	Charadrius falklandicus	X	-	-
Collared Plover	Charadrius collaris	X	X	X
Wilson's (Thick-billed) Plover	Charadrius wilsonia	-	-	-
Rufous-chested Plover (Dotterel)	Charadrius (Zonibyx) modestus	X	-	-
Tawny-throated Dotterel	Oreopholus (Eudromias) ruficollis	X	-	-
Ruddy Turnstone	Arenaria interpres	X	X	X
Solitary Sandpiper	Tringa solitaria	X	X	X
Lesser Yellowlegs	Tringa flavipes	X	X	X
Greater Yellowlegs	Tringa melanoleuca	X	X	X
Spotted Sandpiper	Actitis macularia	X	X	X
Willet	Catoptrophorus semipalmatus	X	X	X
Red Knot	Calidris canutus	X	X	X
Least Sandpiper	Calidris minutilla	X	X	X
Baird's Sandpiper	Calidris bairdii	X	-	-
White-rumped Sandpiper	Calidris fuscicollis	X	X	X
Pectoral Sandpiper	Calidris melanotos	X	X	X
Semipalmated Sandpiper	Calidris pusilla	X	X	X
Western Sandpiper	Calidris mauri	-	-	X
Sanderling	Calidris alba	X	X	X
Stilt Sandpiper	Micropalama (Calidris) himantopus	X	-	X
Buff-breasted Sandpiper	Tryngites subruficollis	X	-	X
Ruff	Philomachus pugnax	X	-	-
Upland Sandpiper	Bartramia longicauda	X	-	X

English name	Scientific name			
Whimbrel	Numenius phaeopus	X	X	X
Eskimo Curlew	Numenius borealis	X	-	-
Hudsonian Godwit	Limosa haemastica	X	-	X
Bar-tailed Godwit	Limosa lapponica	-	-	X
Short-billed Dowitcher	Limnodromus griseus	X	X	X
South American Snipe	Gallinago paraguaiae (gallinago)	X	X	X
Giant Snipe	Gallinago undulata	X	X	X
Black-necked (Common) Stilt	Himantopus mexicanus (himantopus)	-	-	X
"White-backed" Stilt	Himantopus "melanurus"	X	X	-
Red (Grey) Phalarope	Phalaropus fulicarius	Mato Grosso		
Wilson's Phalarope	Steganopus (Phalaropus) tricolor	X	-	X
Double-striped Thick-knee	Burhinus bistriatus	-	-	X
Collared (Common) Pratincole	Glareola pratincola	Atol das Rocas		
Least Seedsnipe	Thinocorus rumicivorus	X	-	-
Snowy Sheathbill	Chionis alba	X	X	-
Great Skua	Catharacta skua	-	X	X
"Antarctic (Southern)" Skua	Catharacta "antarctica"	X	X	-
Chilean Skua	Catharacta chilensis	X	X	-
South Polar (McCormick's) Skua	Catharacta maccormicki	X	-	-
Pomarine Skua (Jaeger)	Stercorarius pomarinus	X	-	X
Arctic Skua (Parasitic Jaeger)	Stercorarius parasiticus	X	X	-
Long-tailed Skua (Jaeger)	Stercorarius longicaudus	X	X	-
Olrog's Gull	Larus atlanticus	X	-	-
Ring-billed Gull	Larus delawarensis	-	-	X
Kelp Gull	Larus dominicanus	X	X	-
Laughing Gull	Larus atricilla	-	-	X
Gray-hooded Gull	Larus cirrocephalus	X	X	X
Franklin's Gull	Larus pipixcan	-	-	X
Brown-hooded Gull	Larus maculipennis	X	X	-
Black Tern	Chlidonias niger	X	-	-
Large-billed Tern	Phaetusa (Sterna) simplex	X	X	X
Gull-billed Tern	Sterna (Gelochelidon) nilotica	X	X	X
South American Tern	Sterna hirundinacea	X	X	-
Common Tern	Sterna hirundo	X	X	X
Arctic Tern	Sterna paradisaea	X	-	-
Antarctic Tern	Sterna vittata	S.Cat. & RdJ		
Forster's Tern	Sterna forsteri	Pernambuco		
Trudeau's (Snowy-crowned) Tern	Sterna trudeaui	X	X	-
Roseate Tern	Sterna dougallii	-	X	X
Sooty Tern	Sterna fuscata	-	X	X
Yellow-billed (Amazon) Tern	Sterna superciliaris	X	X	X
Least Tern	Sterna antillarum	-	X	X
Royal Tern	Sterna maxima	X	X	X
Sandwich Tern	Sterna sandvicensis	-	X	-

English Name	Scientific Name			
"Cayenne" Tern	Sterna "eurygnatha"	X	X	-
Brown (Common) Noddy	Anous stolidus	-	X	X
Black (White-capped) Noddy	Anous minutus (tenuirostris)	-	-	X
White (Fairy) Tern	Gygis alba	-	-	X
Black Skimmer	Rynchops niger	X	X	X
Rock Pigeon (Feral Rock Dove)	Columba livia	throughout Brazil		
Band-tailed Pigeon	Columba fasciata	-	-	X
Scaled Pigeon	Columba speciosa	X	X	X
Picazuro Pigeon	Columba picazuro	X	X	-
Spot-winged Pigeon	Columba maculosa	X	-	-
Pale-vented Pigeon	Columba cayennensis	X	X	X
Ruddy Pigeon	Columba subvinacea	-	-	X
Plumbeous Pigeon	Columba plumbea	X	X	X
Eared Dove	Zenaida auriculata	X	X	X
Scaled Dove	Columbina (Scardafella) squammata	X	X	X
E Blue-eyed Ground-Dove	Columbina cyanopis	X	-	-
Common Ground-Dove	Columbina passerina	X	X	X
Plain-breasted Ground-Dove	Columbina minuta	X	X	X
Ruddy Ground-Dove	Columbina talpacoti	X	X	X
Picui Ground-Dove	Columbina picui	X	X	X
Blue Ground-Dove	Claravis pretiosa	X	X	X
Purple-winged Ground-Dove	Claravis godefrida	X	X	-
Long-tailed Ground-Dove	Uropelia campestris	X	-	X
White-tipped Dove	Leptotila verreauxi	X	X	X
Gray-fronted Dove	Leptotila rufaxilla	X	X	X
Sapphire Quail-Dove	Geotrygon saphirina	-	-	X
Ruddy Quail-Dove	Geotrygon montana	X	X	X
Violaceous Quail-Dove	Geotrygon violacea	X	X	X
Hyacinth Macaw	Anodorhynchus hyacinthinus	X	-	X
E Indigo (Lear's) Macaw	Anodorhynchus leari	-	X	-
Glaucous Macaw	Anodorhynchus glaucus	RGdS & Parana		
E Little Blue (Spix's) Macaw	Cyanopsitta spixii	-	X	-
Blue-and-yellow Macaw	Ara ararauna	X	-	X
Scarlet Macaw	Ara macao	X	-	X
Red-and-green(Green-winged)Macaw	Ara chloroptera	X	X	X
Chestnut-fronted Macaw	Ara severa	X	-	X
Blue-headed Macaw	Ara couloni	-	-	X
Blue-winged (Illiger's) Macaw	Ara (Propyrrhura) maracana	X	X	X
Golden-collared Macaw	Ara (Propyrrhura) auricollis	X	-	-
Red-bellied Macaw	Ara (Orthopsittaca) manilata	X	-	X
Red-shouldered Macaw	Ara (Diopsittaca) nobilis	X	X	X
E Golden Parakeet (Conure)	Guaruba (Aratinga) guarouba	-	-	X
Blue-crowned Parakeet (Conure)	Aratinga acuticauda	X	X	-
White-eyed Parakeet (Conure)	Aratinga leucophthalmus	X	X	X

Common Name	Scientific Name			
Sun Parakeet (Conure)	Aratinga solstitialis	-	-	X
E Jandaya (Flaming) Parakeet (Con.)	Aratinga jandaya	-	-	X
E Golden-capped Parakeet (Conure)	Aratinga auricapilla	X	X	-
Dusky-headed Parakeet (Conure)	Aratinga weddellii	-	-	X
Brown-throated Parakeet (Conure)	Aratinga pertinax	-	-	X
E Caatinga (Cactus) Parakeet (Con.)	Aratinga cactorum	-	X	-
Peach-fronted Parakeet (Conure)	Aratinga aurea	X	X	X
Black-hooded Parakeet(Nanday C.)	Nandayus nenday	X	-	-
E Ochre-marked(Blue-chested)Parak.	Pyrrhura cruentata	-	X	-
Reddish-bellied Par.(Maroon-b C)	Pyrrhura frontalis	X	X	-
Blaze-winged Parakeet (Conure)	Pyrrhura devillei	X	-	-
Crimson-bellied Parakeet (Conure)	Pyrrhura perlata (rhodogaster)	-	-	X
"Pearly" Parakeet (Conure)	Pyrrhura "lepida"	-	-	X
Green-cheeked Parakeet (Conure)	Pyrrhura molinae	X	-	-
Painted Parakeet (Conure)	Pyrrhura picta	-	-	X
Maroon-faced Par.(White-eared C)	Pyrrhura leucotis	-	X	-
Fiery-shouldered Parakeet (Con.)	Pyrrhura egregia	-	-	X
Maroon-tailed Parakeet (Conure)	Pyrrhura melanura	-	-	X
Rock Parakeet (Black-capped Co.)	Pyrrhura rupicola	-	-	X
Monk Parakeet	Myiopsitta monachus	X	X	-
Green-rumped Parrotlet	Forpus passerinus	-	-	X
Blue-winged Parrotlet	Forpus xanthopterygius	X	X	X
Dusky-billed(Sclater's)Parrotlet	Forpus sclateri	-	-	X
E Plain Parakeet	Brotogeris tirica	X	X	-
Canary-winged Parakeet	Brotogeris versicolurus	-	-	X
Yellow-chevroned Parakeet	Brotogeris chiriri	X	X	-
Cobalt-winged Parakeet	Brotogeris cyanoptera	-	-	X
Golden-winged Parakeet	Brotogeris chrysopterus	-	-	X
Tui Parakeet	Brotogeris sanctithomae	-	-	X
Tepui Parrotlet	Nannopsittaca panychlora	-	-	X
Sapphire-rumped Parrotlet	Touit purpurata	-	-	X
E Black-eared(Brown-backed)Par.let	Touit melanonota	-	X	-
Scarlet-shouldered Parrotlet	Touit huetii	-	-	X
E Golden-tailed Parrotlet	Touit surda	-	X	-
Black-headed Parrot (Caique)	Pionites melanocephala	-	-	X
White-bellied Parrot (Caique)	Pionites leucogaster	-	-	X
Red-capped (Pileated) Parrot	Pionopsitta pileata	X	X	-
Caica Parrot	Pionopsitta caica	-	-	X
Orange-cheeked (Barraband's)Par.	Pionopsitta barrabandi	-	-	X
E Vulturine Parrot	Gypopsitta (Pionopsitta) vulturina	-	-	X
Short-tailed Parrot	Graydidascalus brachyurus	-	-	X
Blue-headed Parrot	Pionus menstruus	X	X	X
Scaly-headed Parrot	Pionus maximiliani	X	X	-
Dusky Parrot	Pionus fuscus	-	-	X

Common Name	Scientific Name			
Red-spectacled Parrot (Amazon)	Amazona pretrei-	X	-	-
Red-lored (Diademed) Parrot-	Amazona autumnalis diadema	-	-	X
E Red-tailed Parrot (Amazon) -	Amazona brasiliensis	S.P.south to RGdS		
E Red-browed Parrot (Amazon) -	Amazona rhodocorytha	-	X	-
{Blue-cheeked Parrot} -	{Amazona dufresniana}	-	-	{X}
Festive Parrot (Amazon)	Amazona festiva-	-	-	X
Yellow-faced Parrot (Amazon)	Amazona xanthops	X	-	-
Turquoise-fronted Parrot (Amaz.)	Amazona aestiva-	X	X	X
Yellow-crowned Parrot (Amazon)	Amazona ochrocephala	-	-	X
Orange-winged Parrot (Amazon)	Amazona amazonica	X	X	X
Mealy Parrot (Amazon) -	Amazona farinosa	-	X	X
"Kawall's" Parrot (Amazon) -	Amazona "kawalli"	-	-	X
Vinaceous-breasted Parrot (Amaz)	Amazona vinacea-	X	X	-
Red-fan (Hawk-headed) Parrot	Deroptyus accipitrinus -	-	-	X
E Blue (Purple)-bellied Parrot	Triclaria malachitacea -	X	X	-
Ash-colored Cuckoo -	Coccyzus cinereus	X	-	X
Dwarf Cuckoo -	Coccyzus pumilus	-	-	X
Black-billed Cuckoo-	Coccyzus erythropthalmus	-	-	X
Yellow-billed Cuckoo	Coccyzus americanus-	X	X	X
Pearly-breasted Cuckoo -	Coccyzus julieni (euleri)	X	X	X
Mangrove Cuckoo-	Coccyzus minor	-	-	X
Dark-billed Cuckoo -	Coccyzus melacoryphus	X	X	X
Squirrel Cuckoo-	Piaya cayana	X	X	X
Black-bellied Cuckoo	Piaya melanogaster -	-	-	X
Little Cuckoo -	Piaya minuta	X	-	X
Greater Ani-	Crotophaga major	X	X	X
Smooth-billed Ani	Crotophaga ani -	X	X	X
Hoatzin-	Opisthocomus hoazin-	X	-	X
Guira Cuckoo -	Guira guira	X	X	X
Striped Cuckoo -	Tapera naevia -	X	X	X
Pheasant Cuckoo-	Dromococcyx phasianellus	X	X	X
Pavonine Cuckoo-	Dromococcyx pavoninus	X	X	X
Rufous-vented Ground-Cuckoo	Neomorphus geoffroyi	-	X	X
E Scaled Ground-Cuckoo -	Neomorphus squamiger	-	-	X
Rufous-winged Ground-Cuckoo	Neomorphus rufipennis	-	-	X
Red-billed Ground-Cuckoo -	Neomorphus pucheranii	-	-	X
Barn Owl -	Tyto alba -	X	X	X
Tropical Screech-Owl -	Otus choliba	X	X	X
Variable (Black-capped) Scr.-Owl	Otus atricapillus -	X	X	-
"Long-tufted" Screech-Owl -	Otus "sanctaecatarinae"-	X	-	-
"Guatemalan" Screech-Owl -	Otus "guatemalae roraimae" -	-	-	X
Tawny-bellied Screech-Owl -	Otus watsonii -	-	-	X
"Austral" Screech-Owl -	Otus "usta"-	-	-	X
Crested Owl-	Lophostrix cristata-	-	-	X

Common Name	Scientific Name			
Great Horned Owl - - -	- Bubo virginianus - - - -	X	X	X
Spectacled Owl - - - -	- Pulsatrix perspicillata- - -	X	X	X
Tawny-browed Owl - - -	- Pulsatrix koeniswaldiana - -	X	X	-
Least Pygmy-Owl- - - -	- Glaucidium minutissimum- - -	-	X	-
Amazonian Pygmy-Owl- - -	- Glaucidium hardyi - - - -	-	-	X
Ferruginous Pygmy-Owl - -	- Glaucidium brasilianum - - -	X	X	X
Burrowing Owl - - - -	- Speotyto (Athene) cunicularia -	X	X	X
Black-banded Owl - - -	- Strix (Ciccaba) huhula - -	X	X	X
Mottled Owl- - - -	- Strix (Ciccaba) virgata- -	X	X	X
Rusty-barred Owl - - -	- Strix hylophila- - - -	X	X	-
Striped Owl- - - -	- Asio (Rhinoptynx) clamator - -	X	X	X
Stygian Owl- - - -	- Asio stygius - - - -	X	-	X
Short-eared Owl- - -	- Asio flammeus - - - -	X	-	-
Buff-fronted Owl - - -	- Aegolius harrisii - - -	X	-	X
{Oilbird} - - - -	- {Steatornis caripensis}- - -	-	-	{X}
Great Potoo- - - -	- Nyctibius grandis - - -	X	X	X
Large-tailed Potoo - -	- Nyctibius aethereus- - -	X	X	-
"Long-tailed" Potoo- -	- Nyctibius "longicaudatus" - -	-	-	X
Common Potoo - - -	- Nyctibius griseus - - -	X	X	X
E White-winged Potoo - -	- Nyctibius leucopterus - - -	-	X	X
Rufous Potoo - - -	- Nyctibius bracteatus - - -	-	-	X
Short-tailed(Semi-col.)Nighthawk	Lurocalis semitorquatus- -	X	X	X
Least Nighthawk- - -	- Chordeiles pusillus- - -	X	X	X
Sand-colored Nighthawk - -	- Chordeiles rupestris - -	X	-	X
Lesser Nighthawk - -	- Chordeiles acutipennis - -	X	X	X
Common Nighthawk - -	- Chordeiles minor - - -	X	X	X
Band-tailed Nighthawk - -	- Nyctiprogne leucopyga - -	X	-	X
Nacunda Nighthawk - -	- Podager nacunda- - - -	X	X	X
Pauraque - - - -	- Nyctidromus albicollis - -	X	X	X
Ocellated Poorwill - -	- Nyctiphrynus ocellatus - -	X	X	X
Rufous Nightjar- - -	- Caprimulgus rufus - - -	X	X	X
Silky-tailed Nightjar - -	- Caprimulgus sericocaudatus - -	-	-	X
Band-winged Nightjar - -	- Caprimulgus longirostris - -	X	X	X
White-tailed Nightjar - -	- Caprimulgus cayennensis- -	-	-	X
White-winged Nightjar - -	- Caprimulgus candicans - -	X	-	-
Spot-tailed Nightjar - -	- Caprimulgus maculicaudus - -	X	X	X
Little Nightjar- - -	- Caprimulgus parvulus - -	X	X	X
Blackish Nightjar - -	- Caprimulgus nigrescens - -	-	-	X
{Roraiman Nightjar}- -	- {Caprimulgus whitelyi} - -	-	-	{X}
E Pygmy Nightjar - - -	- Caprimulgus hirundinaceus - -	-	X	-
Ladder-tailed Nightjar -	- Hydropsalis climacocerca -	X	-	X
Scissor-tailed Nightjar-	- Hydropsalis brasiliana - -	X	X	X
Long-trained Nightjar -	- Macropsalis creagra- - -	X	X	X
Sickle-winged Nightjar -	- Eleothreptus anomalus - -	X	-	-

White-collared Swift - -	- Streptoprocne zonaris - - -	X	X	X
Biscutate Swift- - - -	- Streptoprocne biscutata- - -	X	X	-
Tepui Swift- - - -	- Cypseloides phelpsi- - - -	-	-	X
Great Dusky Swift - -	- Cypseloides (Aerornis) senex -	X	-	-
Sooty Swift- - - -	- Cypseloides fumigatus - - -	X	X	-
Chapman's Swift- - -	- Chaetura chapmani - - - -	-	-	X
Chimney Swift - - -	- Chaetura pelagica - - - -	-	-	X
Gray-rumped Swift - -	- Chaetura cinereiventris- - -	X	X	X
Pale-rumped Swift - -	- Chaetura egregia - - - -	-	-	X
Band-rumped Swift - -	- Chaetura spinicauda- - - -	-	X	X
Ashy-tailed Swift - -	- Chaetura andrei- - - -	X	X	X
Short-tailed Swift - -	- Chaetura brachyura - - -	X	-	X
White-tipped Swift - -	- Aeronautes montivagus - - -	-	-	X
Lesser Swallow-tailed Swift-	- Panyptila cayennensis - -	X	X	X
Fork-tailed Palm-Swift - -	- Reinarda (Tachornis) squamata -	X	X	X
Blue-fronted Lancebill - -	- Doryfera johannae - - - -	-	-	X
E Saw-billed Hermit - -	- Ramphodon naevius - - -	-	X	-
E Hook-billed Hermit - -	- Ramphodon (Glaucis) dohrnii-	-	X	-
Rufous-breasted Hermit - -	- Glaucis hirsuta- - - -	X	X	X
Pale-tailed Barbthroat - -	- Threnetes leucurus - - -	-	-	X
"Bronze-tailed" Barbthroat -	- Threnetes "loehkeni" - -	-	-	X
Long-tailed Hermit - - -	- Phaethornis superciliosus - -	-	-	X
"Red-billed" Long-tailed Hermit-	Phaethornis "ochraceiventris" -	X	-	X
"Klabin Farm"Long-tailed Hermit-	Phaethornis "margarettae" -	-	X	-
Great-billed Hermit- - -	- Phaethornis malaris- - -	-	-	X
Scale-throated Hermit - -	- Phaethornis eurynome - -	X	X	-
"Black-billed" Hermit - -	- Phaethornis "nigrirostris" -	-	X	-
White-bearded Hermit - -	- Phaethornis hispidus - -	X	-	X
Straight-billed Hermit - -	- Phaethornis bourcieri - -	-	-	X
Needle-billed Hermit - -	- Phaethornis philippi - -	-	-	X
Dusky-throated Hermit - -	- Phaethornis squalidus - -	-	X	-
"Streak-throated" Hermit -	- Phaethornis "rupurumii"- -	-	-	X
Planalto Hermit- - -	- Phaethornis pretrei- - -	X	X	-
Sooty-capped Hermit- -	- Phaethornis augusti- - -	-	-	X
Buff-bellied Hermit- -	- Phaethornis subochraceus -	X	-	-
Cinnamon-throated Hermit	- Phaethornis nattereri - -	X	-	X
"Maranhao" Hermit - -	- Phaethornis "maranhaoensis"-	Maranhao.		
E Broad-tipped Hermit- -	- Phaethornis gounellei - -	-	X	-
Reddish Hermit - - -	- Phaethornis ruber - -	X	X	X
Gray-chinned Hermit- -	- Phaethornis griseogularis -	-	-	X
Little Hermit - - -	- Phaethornis longuemareus -	-	-	X
E Minute Hermit - - -	- Phaethornis idaliae- -	-	X	-
Gray-breasted Sabrewing-	- Campylopterus largipennis -	-	X	X
Rufous-breasted Sabrewing -	- Campylopterus hyperythrus -	-	-	X

Buff-breasted Sabrewing-	- - Campylopterus duidae - - - - -			X
Swallow-tailed Hummingbird -	- Eupetomena macroura- - - -	X	X	X
White-necked Jacobin - -	- Florisuga mellivora- - - -			X
Black Jacobin - - - -	- Melanotrochilus (Florisuga) fuscus	X	X	-
Brown Violetear- - -	- Colibri delphinae - - - -			X
Sparkling Violetear- -	- Colibri coruscans - - - -			X
White-vented Violetear - -	- Colibri serrirostris - - -	X	X	X
Green-throated Mango - -	- Anthracothorax viridigula - -	X	-	X
Black-throated Mango - -	- Anthracothorax nigricollis -	X	X	X
Fiery-tailed Awlbill - -	- Avocettula recurvirostris - -			X
Ruby-topaz Hummingbird - -	- Chrysolampis mosquitus - -	X	X	X
Violet-headed Hummingbird -	- Klais guimeti - - - -			X
Black-breasted Plovercrest -	- Stephanoxis lalandi- - -		X	-
"Violet-crested" Plovercrest	- Stephanoxis "loddigesii" -	X	-	
Tufted Coquette- - - -	- Lophornis ornata - - -			X
Dot-eared Coquette - - -	- Lophornis gouldii - -	X	-	X
E Frilled Coquette - - -	- Lophornis magnifica- -	X	X	X
Festive Coquette - - -	- Lophornis chalybea - -	X	-	
"Butterfly" Coquette - -	- Lophornis "verreauxii" - -			X
Peacock Coquette - - -	- Lophornis pavonina - -			X
Black-bellied Thorntail- -	- Popelairia langsdorffi -	X	X	
Racket-tailed Coquette - -	- Discosura longicauda -	X	X	
Blue-chinned Sapphire - -	- Chlorestes notatus - -	X	X	X
Blue-tailed Emerald- - -	- Chlorostilbon mellisugus	X	-	X
Glittering-bellied Emerald -	- Chlorostilbon aureoventris -	X	X	X
Fork-tailed (Common) Woodnymph	- Thalurania furcata - -	X	X	X
E Long-tailed Woodnymph - -	- Thalurania watertonii -	X	-	
Violet-capped Woodnymph-	- Thalurania glaucopis -	X	X	
Rufous-throated Sapphire -	- Hylocharis sapphirina -	X	X	X
White-chinned Sapphire - -	- Hylocharis cyanus - -	X	X	X
Gilded Sapphire (Hummingbird)	- Hylocharis chrysura- -	X	-	X
Golden-tailed Sapphire -	- Chrysuronia oenone - -			X
White-throated Hummingbird -	- Leucochloris albicollis-	X	X	
White-tailed Goldenthroat -	- Polytmus guainumbi - -	X	X	X
Tepui Goldenthroat - - -	- Polytmus milleri - -			X
Green-tailed Goldenthroat -	- Polytmus theresiae -			X
Olive-spotted Hummingbird -	- Leucippus chlorocercus -			X
Many-spotted Hummingbird -	- Taphrospilus hypostictus -	X	-	-
White-bellied Hummingbird -	- Amazilia chionogaster -	X	-	-
White-chested Emerald - -	- Amazilia chionopectus -			X
Versicolored Emerald -	- Amazilia versicolor- -	X	X	X
"Blue-headed" Emerald - -	- Amazilia "rondoniae" -			X
Glittering-throated Emerald-	- Amazilia fimbriata - -	X	X	X
"Big"Glittering-throated Emerald	Amazilia "tephrocephala" -	X	X	-

Common Name	Scientific Name			
Sapphire-spangled Emerald - -	- Amazilia lactea- - - - -	X	X	X
Plain-bellied Emerald - -	- Amazilia leucogaster - - -	X	X	X
Green-bellied Hummingbird -	- Amazilia viridigaster - - -	-	-	X
E Sombre Hummingbird - - -	- Aphantochroa cirrhochloris - -	X	X	-
E Brazilian Ruby - - - -	- Clytolaema rubricauda - - -	-	X	-
Gould's Jewelfront - - -	- Heliodoxa (Polyplancta) aurescens	-	-	X
Velvet-browed Brilliant- -	- Heliodoxa xanthogonys - - -	-	-	X
Black-throated Brilliant -	- Heliodoxa schreibersii - - -	-	-	X
Pink-throated Brilliant-	- Heliodoxa gularis - - - -	-	-	X
Crimson Topaz - - - -	- Topaza pella - - - - -	-	-	X
Fiery Topaz- - - - -	- Topaza pyra- - - - - -	-	-	X
E Hooded Visorbearer - - -	- Augastes lumachellus - - -	-	X	-
E Hyacinth Visorbearer - -	- Augastes scutatus - - - -	X	X	-
Black-eared Fairy - - -	- Heliothryx aurita - - - -	-	X	X
Horned Sungem - - - -	- Heliactin cornuta (bilophum) -	X	-	X
Long-billed Starthroat - -	- Heliomaster longirostris - -	X	X	X
E Stripe-breasted Starthroat -	- Heliomaster squamosus - - -	X	X	-
Blue-tufted Starthroat - -	- Heliomaster furcifer - - -	X	-	X
Amethyst Woodstar - - -	- Calliphlox amethystina - - -	X	X	X
Pavonine Quetzal - - -	- Pharomachrus pavoninus - - -	-	-	X
Black-tailed Trogon- - -	- Trogon melanurus - - - -	X	-	X
White-tailed Trogon- - -	- Trogon viridis - - - - -	X	X	X
Collared Trogon- - - -	- Trogon collaris- - - - -	-	X	X
Masked Trogon - - - -	- Trogon personatus - - - -	-	-	X
Black-throated Trogon - -	- Trogon rufus - - - - ..	X	X	X
Surucua Trogon - - - -	- Trogon surrucura - - - -	X	-	-
"Brazilian(Orange-breasted)"Trog.Trogon "aurantius" - - - -		X	X	-
Blue-crowned Trogon- - -	- Trogon curucui - - - - -	X	X	X
Violaceous Trogon - - -	- Trogon violaceus - - - -	-	-	X
Ringed Kingfisher - - -	- Megaceryle (Ceryle) torquata- -	X	X	X
Amazon Kingfisher - - -	- Chloroceryle amazona - - -	X	X	X
Green Kingfisher - - -	- Chloroceryle americana - -	X	X	X
Green-and-rufous Kingfisher-	- Chloroceryle inda - - - -	X	X	X
American Pygmy Kingfisher -	- Chloroceryle aenea - - - -	X	X	X
Broad-billed Motmot- - -	- Electron platyrhynchum - - -	-	-	X
Rufous-capped Motmot - -	- Baryphthengus ruficapillus - -	X	X	-
Rufous Motmot - - - -	- Baryphthengus martii - - -	-	-	X
Blue-crowned Motmot- - -	- Momotus momota - - - - -	X	X	X
White-eared Jacamar- - -	- Galbalcyrhynchus leucotis - -	-	-	X
Chestnut (Purus) Jacamar -	- Galbalcyrhynchus purusianus- -	-	-	X
White-throated Jacamar - -	- Brachygalba albogularis- - -	-	-	X
Brown (Black-billed) Jacamar	- Brachygalba lugubris - - -	X	-	X
E Three-toed Jacamar - - -	- Jacamaralcyon tridactyla - -	X	X	-
Yellow-billed Jacamar - -	- Galbula albirostris- - - -	-	-	X

E	Common Name	Scientific Name			
	Blue-cheeked(B.-necked) Jacamar	Galbula cyanicollis	-	-	X
	Green-tailed Jacamar	Galbula galbula	-	-	X
	White-chinned Jacamar	Galbula tombacea	-	-	X
	Bluish-fronted Jacamar	Galbula cyanescens	-	-	X
	Rufous-tailed Jacamar	Galbula ruficauda	X	X	X
	Bronzy Jacamar	Galbula leucogastra	-	-	X
	"Purplish" Jacamar	Galbula "chalcothorax"	-	-	X
	Paradise Jacamar	Galbula dea	-	-	X
	Great Jacamar	Jacamerops aurea	-	-	X
	White-necked Puffbird	Notharchus (Bucco) macrorhynchos	-	-	X
	"Buff-bellied" Puffbird	Notharchus (Bucco) "swainsoni"	X	X	-
	Brown-banded Puffbird	Notharchus (Bucco) ordii	-	-	X
	Pied Puffbird	Notharchus (Bucco) tectus	X	-	X
	Chestnut-capped Puffbird	Bucco macrodactylus	-	-	X
	Spotted Puffbird	Bucco tamatia	-	-	X
	Collared Puffbird	Bucco capensis	-	-	X
	White-eared Puffbird	Nystalus (Bucco) chacuru	X	X	-
	Striolated Puffbird	Nystalus (Bucco) striolatus	-	-	X
	Spot-backed Puffbird	Nystalus (Bucco) maculatus	X	X	X
E	Crescent-chested Puffbird	Malacoptila striata	X	X	-
	White-chested Puffbird	Malacoptila fusca	-	-	X
	Semicollared Puffbird	Malacoptila semicincta	-	-	X
	Rufous-necked Puffbird	Malacoptila rufa	-	-	X
	Lanceolated Monklet	Micromonacha lanceolata	-	-	X
	Rusty-breasted Nunlet	Nonnula rubecula	X	X	X
	Fulvous-chinned Nunlet	Nonnula sclateri	-	-	X
	Rufous-capped Nunlet	Nonnula ruficapilla	X	-	X
E	Chestnut-headed Nunlet	Nonnula amaurocephala	-	-	X
	Black Nunbird	Monasa atra	-	-	X
	Black-fronted Nunbird	Monasa nigrifrons	X	-	X
	White-fronted Nunbird	Monasa morphoeus	-	X	X
	Yellow-billed Nunbird	Monasa flavirostris	-	-	X
	Swallow-wing	Chelidoptera tenebrosa	X	X	X
	Scarlet-crowned Barbet	Capito aurovirens	-	-	X
	Black-girdled Barbet	Capito dayi	-	-	X
	Black-spotted Barbet	Capito niger	-	-	X
E	Brown-chested Barbet	Capito brunneipectus	-	-	X
	Lemon-throated Barbet	Eubucco richardsoni	-	-	X
	Chestnut-tipped Toucanet	Aulacorhynchus derbianus	-	-	X
	Emerald Toucanet	Aulacorhynchus prasinus	-	-	X
	Black-necked Aracari	Pteroglossus aracari	X	X	X
	Chestnut-eared Aracari	Pteroglossus castanotis	X	-	X
	Many-banded Aracari	Pteroglossus pluricinctus	-	-	X
	Green Aracari	Pteroglossus viridis	-	-	X

	Common Name	Scientific Name			
	Lettered Aracari	Pteroglossus inscriptus	X	-	X
	Red-necked Aracari	Pteroglossus bitorquatus	-	-	X
	Ivory-billed Aracari	Pteroglossus azara (flavirostris)	-	-	X
	"Brown-mandibled" Aracari	Pteroglossus "mariae"	-	-	X
	Curl-crested Aracari	Pteroglossus beauharnaesii	-	-	X
	Guianan Toucanet	Selenidera culik	-	-	X
	Golden-collared Toucanet	Selenidera reinwardtii	-	-	X
	Tawny-tufted Toucanet	Selenidera nattereri	-	-	X
	Spot-billed Toucanet	Selenidera maculirostris	X	X	-
	Gould's Toucanet	Selenidera gouldii	-	-	X
	Saffron Toucanet	Baillonius bailloni	X	X	-
	Red-breasted Toucan	Ramphastos dicolorus	X	X	-
	Channel-billed Toucan	Ramphastos vitellinus	X	X	X
	Yellow-ridged Toucan	Ramphastos culminatus	X	-	X
	Red-billed Toucan	Ramphastos tucanus	-	-	X
	"Cuvier's" Toucan	Ramphastos "cuvieri"	X	-	X
	Toco Toucan	Ramphastos toco	X	-	X
	Rufous-breasted Piculet	Picumnus rufiventris	-	-	X
E	Tawny Piculet	Picumnus fulvescens	-	X	-
E	Ochraceous Piculet	Picumnus limae	Ceara & Paraiba		
	Mottled Piculet	Picumnus nebulosus	X	-	-
	Plain-breasted Piculet	Picumnus castelnau	-	-	{X}
	Rusty-necked Piculet	Picumnus fuscus	-	-	{X}
	White-bellied Piculet	Picumnus spilogaster	-	-	X
	Golden-spangled Piculet	Picumnus exilis	-	X	X
	Gold-fronted Piculet	Picumnus aurifrons	-	-	X
	"Bar-breasted" Piculet	Picumnus "borbae"	-	-	X
	"Banded" Piculet	Picumnus "transfasciatus"	-	-	X
	Lafresnaye's Piculet	Picumnus lafresnayi	-	-	X
	"Orinoco" Piculet	Picumnus "pumilus"	-	-	X
	White-wedged Piculet	Picumnus albosquamatus	X	-	X
	White-barred Piculet	Picumnus cirratus	X	X	-
	"Pilcomayo" Piculet	Picumnus "pilcomayensis"	X	-	-
	"Lower Amazonian" Piculet	Picumnus "macconnelli"	-	-	X
	Ochre-collared Piculet	Picumnus temminckii	X	-	-
E	Varzea Piculet	Picumnus varzeae	Para & Amazonas		
E	Spotted Piculet	Picumnus pygmaeus	-	X	-
	Campo Flicker	Colaptes campestris	X	X	-
	"Field" Flicker	Colaptes "campestroides"	X	-	-
	Spot-breasted Woodpecker	Colaptes (Chrysoptilus) punctigula	-	-	X
	Green-barred Woodpecker(Flicker)	Colaptes (Chrysopt.) melanochloros	X	X	X
	Golden-olive Woodpecker	Piculus rubiginosus	-	-	X
	Yellow-throated Woodpecker	Piculus flavigula	X	X	X
	White-throated(Rufous-winged) W.	Piculus leucolaemus	X	-	X

Common Name	Scientific Name			
Golden-green Woodpecker- -	- Piculus chrysochloros - - -	X	X	X
White (Yellow)-browed Woodpecker	Piculus aurulentus - - - -	X	X	-
Blond-crested Woodpecker -	- Celeus flavescens - - -	X	X	X
Pale-crested Woodpecker-	- Celeus lugubris- - - -	X	-	-
Chestnut Woodpecker- - -	- Celeus elegans - - - -	-	-	X
Waved Woodpecker - - -	- Celeus undatus - - - -	-	-	X
Scale-breasted Woodpecker -	- Celeus grammicus - - -	-	-	X
Cream-colored Woodpecker	- Celeus flavus - - - -	X	X	X
Rufous-headed Woodpecker	- Celeus spectabilis - - -	Piaui		
Ringed Woodpecker - - -	- Celeus torquatus - - -	X	X	X
Lineated Woodpecker- - -	- Dryocopus lineatus - - -	X	X	X
Helmeted Woodpecker- - -	- Dryocopus galeatus - - -	X	X	-
Yellow-tufted Woodpecker -	- Melanerpes cruentatus - - -	X	-	X
Yellow-fronted Woodpecker -	- Melanerpes flavifrons - - -	X	X	-
White Woodpecker - - -	- Melanerpes (Leuconerpes) candidus	X	X	X
White-fronted Woodpecker -	- Melanerpes (Trichopicus) cactorum	X	-	-
White-spotted Woodpecker -	- Veniliornis spilogaster- - -	X	X	-
Little Woodpecker - - -	- Veniliornis passerinus - - -	X	X	X
Red-stained Woodpecker - -	- Veniliornis affinis- - -	X	X	X
Golden-collared Woodpecker -	- Veniliornis cassini- - - -	-	-	X
E Yellow-eared Woodpecker- -	- Veniliornis maculifrons- - -	X	X	-
Red-rumped Woodpecker - -	- Veniliornis kirkii - - - -	-	-	X
Checkered Woodpecker - -	- Picoides (Dendrocopos) mixtus -	X	-	-
Crimson-crested Woodpecker -	- Campephilus(Phloeo.) melanoleucos	X	X	X
Cream-backed Woodpecker-	- Camp.(Phloeoceastes) leucopogon-	X	-	-
Red-necked Woodpecker - -	- Campephilus (Phloeo.) rubricollis	X	-	X
Robust Woodpecker - - -	- Campephilus (Phloeo.) robustus -	X	X	-
Plain-brown Woodcreeper- -	- Dendrocincla fuliginosa- - -	X	-	X
Thrush-like(Plain-winged)Woodcr.	Dendrocincla turdina - - -	X	X	-
White-chinned Woodcreeper -	- Dendrocincla merula- - - -	-	-	X
"Obidos (Noisy)" Woodcreeper	- Dendrocincla "obidensis" - - -	-	-	X
Long-tailed Woodcreeper-	- Deconychura longicauda - - -	-	-	X
Spot-throated Woodcreeper -	- Deconychura stictolaema- - -	-	-	X
Olivaceous Woodcreeper - -	- Sittasomus griseicapillus - -	X	X	X
Wedge-billed Woodcreeper -	- Glyphorynchus spirurus - - -	X	X	X
Long-billed Woodcreeper- -	- Nasica longirostris- - - -	X	-	X
Cinnamon-throated Woodcreeper	- Dendrexetastes rufigula- - -	-	-	X
Red-billed Woodcreeper - -	- Hylexetastes perrotii - - -	-	-	X
Bar-bellied Woodcreeper- -	- Hylexetastes stresemanni - -	-	-	X
Strong-billed Woodcreeper -	- Xiphocolaptes promeropirhynchus-	-	-	X
White-throated Woodcreeper -	- Xiphocolaptes albicollis - -	X	X	-
E Moustached Woodcreeper- -	- Xiphocolaptes falcirostris - -	-	X	-
"Snethlage's" Woodcreeper -	- Xiphocolaptes "franciscanus" -	Minas Gerais		
Great Rufous Woodcreeper -	- Xiphocolaptes major- - - -	X	-	-

Common Name	Scientific Name			
Barred Woodcreeper	Dendrocolaptes certhia		X	X
"Concolor" Woodcreeper	Dendrocolaptes "concolor"			X
Black-banded Woodcreeper	Dendrocolaptes picumnus			X
"Pale-billed" Woodcreeper	Dendrocolaptes "pallescens"	X		
"Cross-barred" Woodcreeper	Dendrocolaptes "transfasciatus"			X
E Hoffmann's Woodcreeper	Dendrocolaptes hoffmannsi			X
Planalto Woodcreeper	Dendrocolaptes platyrostris	X	X	
Straight-billed Woodcreeper	Xiphorhynchus picus	X	X	X
E Zimmer's Woodcreeper	Xiphorhynchus necopinus			X
Striped Woodcreeper	Xiphorhynchus obsoletus			X
Ocellated Woodcreeper	Xiphorhynchus ocellatus			X
Spix's Woodcreeper	Xiphorhynchus spixii			X
Elegant Woodcreeper	Xiphorhynchus elegans			X
Chestnut-rumped Woodcreeper	Xiphorhynchus pardalotus			X
Buff-throated Woodcreeper	Xiphorhynchus guttatus	X	X	X
E Dusky-billed Woodcreeper	Xiphorhynchus eytoni			X
Streak-headed Woodcreeper	Lepidocolaptes souleyetii			X
Narrow-billed Woodcreeper	Lepidocolaptes angustirostris	X	X	X
Scaled Woodcreeper	Lepidocolaptes squamatus	X	X	
Lesser Woodcreeper	Lepidocolaptes fuscus	X	X	
Lineated Woodcreeper	Lepidocolaptes albolineatus			X
Scimitar-billed Woodcreeper	Drymornis bridgesii	X		
Red-billed Scythebill	Campylorhamphus trochilirostris	X	X	X
Black-billed Scythebill	Campylorhamphus falcularius	X	X	
Curve-billed Scythebill	Campylorhamphus procurvoides			X
Campo Miner	Geobates (Geositta) poeciloptera	X		X
Common Miner	Geositta cunicularia	X		
Bar-winged Cinclodes	Cinclodes fuscus	X		
E Long-tailed Cinclodes	Cinclodes pabsti	X		
Rufous Hornero	Furnarius rufus	X	X	
Pale-legged Hornero	Furnarius leucopus	X	X	X
Pale-billed (Bay) Hornero	Furnarius torridus			X
Lesser Hornero	Furnarius minor			X
E Wing-banded Hornero	Furnarius figulus	X	X	X
Curve-billed Reedhaunter	Limnornis curvirostris	X		
Straight-billed Reedhaunter	Limnornis(Limnoctites) rectirostris	X		
Wren-like Rushbird	Phleocryptes melanops	X		
E Striolated Tit-Spinetail	Leptasthenura striolata	X		
Tufted Tit-Spinetail	Leptasthenura platensis	X		
Araucaria Tit-Spinetail	Leptasthenura setaria	X	X	
Chotoy Spinetail	Schoeniophylax phryganophila	X		
E Itatiaia (Brazilian) Spinetail	Oreophylax (Schizoeaca) moreirae		X	
Rufous-capped Spinetail	Synallaxis ruficapilla	X	X	
E Pinto's (Plain) Spinetail	Synallaxis infuscata		X	

Sooty-fronted Spinetail	Synallaxis frontalis	X	X	-
Cabanis's Spinetail	Synallaxis cabanisi	Mato Grosso		
McConnell's Spinetail	Synallaxis macconnelli	-	-	X
Chicli (Spix's) Spinetail	Synallaxis spixi	X	X	-
Cinereous-breasted Spinetail	Synallaxis hypospodia	X	X	-
Pale-breasted Spinetail	Synallaxis albescens	X	X	X
Dark-breasted Spinetail	Synallaxis albigularis	-	-	X
Plain-crowned Spinetail	Synallaxis gujanensis	-	-	X
"White-lored(Ochre-breasted)"Sp.	Synallaxis "albilora"	X	-	-
White-bellied Spinetail	Synallaxis propinqua	-	-	X
Gray-bellied Spinetail	Synallaxis cinerascens	X	X	-
Ruddy Spinetail	Synallaxis rutilans	-	-	X
Chestnut-throated Spinetail	Synallaxis cherriei	-	-	X
E Red-shouldered(Reiser's) Spine.-	Gyalophylax(Synallaxis) hellmayri	-	X	-
Hoary-throated(Rio Branco)Spine.	Poecilurus (Synallaxis) kollari	-	-	X
Ochre-cheeked Spinetail	Poecilurus (Synallaxis) scutata	X	X	-
Sulphur-bearded Spinetail	Cranioleuca(Certhiaxis)sulphurifera	X	-	-
E Gray-headed Spinetail	Cranioleuca (Certh.) semicinerea	X	X	-
Stripe-crowned Spinetail	Cranioleuca (Certh.) pyrrhophia-	X	-	-
Olive Spinetail	Cranioleuca (Certhiaxis) obsoleta	X	-	-
Tepui Spinetail	Cranioleuca (Certhiaxis) demissa	-	-	X
Rusty-backed Spinetail	Cranioleuca (Certhiaxis) vulpina	X	-	X
E Pallid (Pale-browed) Spinetail	Cranioleuca (Certhiaxis) pallida	X	X	-
E Scaled Spinetail	Cranioleuca (Certhiaxis) muelleri	-	-	X
Speckled Spinetail	Cranioleuca (Certh.) gutturata	-	-	X
Yellow-chinned(-throated) Spine.	Certhiaxis (Cranio.) cinnamomea-	X	X	X
Red-and-white Spinetail	Certhiaxis (Cranio.) mustelina	-	-	X
Bay-capped Wren-Spinetail	Spartonoica maluroides	X	-	-
Hudson's Canastero	Asthenes (Thripophaga) hudsoni	X	-	-
Short-billed Canastero	Asthenes (Thripophaga) baeri	X	-	-
E Cipo Canastero	Asthenes luizae-	X	-	-
E Striated Softtail	Thripophaga macroura	-	X	-
Plain Softtail	Thripophaga(Phacellod.) fusciceps	-	-	X
Canebrake Groundcreeper-	Clibanornis(Ph.) dendrocolaptoides	X	X	-
Rufous-fronted (Plain) Thornbird	Phacellodomus rufifrons-	X	X	-
E Red-eyed Thornbird	Phacellodomus erythrophthalmus	-	X	-
"Rufous-breasted" Thornbird-	Phacellodomus "ferrugineigula" -	X	X	-
Greater (Yellow-eyed) Thornbird-	Phacellodomus ruber-	X	-	-
Freckle-breasted Thornbird	Phacellodomus striaticollis-	X	-	-
Lark-like Brushrunner	Coryphistera alaudina	X	-	-
Firewood-gatherer	Anumbius annumbi	X	X	-
Orange-fronted Plushcrown	Metopothrix aurantiacus-	-	-	X
Roraiman Barbtail	Roraimia (Margarornis) adusta	-	-	X
Point-tailed Palmcreeper	Berlepschia rikeri	X	-	X

	English Name	Scientific Name			
	Rufous Cacholote	Pseudoseisura cristata	X	X	X
	Brown Cacholote	Pseudoseisura lophotes	X	-	-
	Chestnut-winged Hookbill	Ancistrops (Philydor) strigilatus	-	-	X
E	Pale-browed Treehunter	Cichlocolaptes (Philydor) leucophrys	X	-	
	Peruvian Recurvebill	Simoxenops (Philydor) ucayalae	-	-	X
	Striped Woodhaunter (Fol-glean.)	Hyloctistes (Philydor) subulatus	-	-	X
E	White-collared Foliage-gleaner	Anabazenops (Philydor) fuscus	-	X	-
	Buff-browed Foliage-gleaner	Syndactyla (Ph.) rufosuperciliata	X	X	-
	White-browed Foliage-gleaner	Philydor (Anabacerthia) amaurotis	X	X	-
	Black-capped Foliage-gleaner	Philydor atricapillus	X	X	-
E	Alagoas Foliage-gleaner	Philydor novaesi	-	X	-
	Rufous-rumped Foliage-gleaner	Philydor erythrocercus	X	-	X
	Cinnamon-rumped Foliage-gleaner	Philydor pyrrhodes	-	-	X
	Russet-mantled Foliage-gleaner	Philydor dimidiatus	X	-	-
	Ochre-breasted Foliage-gleaner	Philydor lichtensteini	X	X	-
	Buff-fronted Foliage-gleaner	Philydor rufus	X	X	-
	Chestnut-winged Foliage-gleaner	Philydor erythropterus	-	-	X
	Rufous-tailed Foliage-gleaner	Philydor ruficaudatus	-	-	X
	White-eyed Foliage-gleaner	Automolus leucophthalmus	X	X	-
	Olive-backed Foliage-gleaner	Automolus infuscatus	-	-	X
	Crested Foliage-gleaner	Automolus dorsalis	-	-	X
	Ruddy Foliage-gleaner	Automolus rubiginosus	-	-	X
	White-throated Foliage-gleaner	Automolus roraimae (albigularis)	-	-	X
	Buff-throated Foliage-gleaner	Automolus ochrolaemus	-	-	X
	Chestnut-crowned Foliage-gleaner	Automolus rufipileatus	-	-	X
	Brown-rumped Foliage-gleaner	Automolus melanopezus	-	-	X
	Chestnut-capped Foliage-gleaner	Hylocryptus (Autom.) rectirostris	X	X	-
	Sharp-billed Treehunter	Heliobletus (Xenops) contaminatus	X	X	-
	Rufous-tailed Xenops	Xenops milleri	-	-	X
	Slender-billed Xenops	Xenops tenuirostris	-	-	X
	Streaked Xenops	Xenops rutilans	X	X	X
	Plain Xenops	Xenops minutus	X	X	X
E	Great Xenops	Megaxenops parnaguae	-	X	-
	Rufous-breasted Leaftosser (L.s)	Sclerurus scansor	X	X	-
	Gray-throated Leaft.(Leafscraper)	Sclerurus albigularis	-	-	X
	Tawny-throated Leaftosser (L.s)	Sclerurus mexicanus	-	X	X
	Short-billed Leaftosser (Leafs.)	Sclerurus rufigularis	-	-	X
	Black-tailed Leaftosser (Leafs.)	Sclerurus caudacutus	-	X	X
	Sharp-tailed Streamcreeper	Lochmias nematura	X	X	X
	Fasciated Antshrike	Cymbilaimus lineatus	-	-	X
	Bamboo Antshrike	Cymbilaimus sanctaemariae	-	-	X
	Spot-backed Antshrike	Hypoedaleus guttatus	X	X	-
	Giant Antshrike	Batara cinerea	X	X	-
	Large-tailed Antshrike	Mackenziaena leachii	X	X	-

English Name	Scientific Name			
Tufted Antshrike - - -	- Mackenziaena severa- - - -	X	X	-
Black-throated Antshrike -	- Frederickena viridis - - -	-	-	X
Undulated Antshrike- - -	- Frederickena unduligera- - -	-	-	X
Great Antshrike- - - -	- Taraba major - - - -	X	X	X
Black-crested Antshrike- -	- Sakesphorus canadensis - - -	-	-	X
E Silvery-cheeked Antshrike -	- Sakesphoros cristatus - - -	-	X	-
Band-tailed Antshrike - -	- Sakesphorus melanothorax - -	-	-	X
E Glossy Antshrike - - -	- Sakesphorus luctuosus - -	X	-	X
White-bearded Antshrike- -	- Biatas nigropectus - - -	X	X	-
Barred Antshrike - - -	- Thamnophilus doliatus - -	X	X	X
Lined (Chestnut-backed)Antshrike	Thamnophilus palliatus - -	-	X	X
Blackish-gray Antshrike- -	- Thamnophilus nigrocinereus - -	-	-	X
Castelnau's Antshrike - -	- Thamnophilus cryptoleucus - -	-	-	X
White-shouldered Antshrike -	- Thamnophilus aethiops - -	-	X	X
Plain-winged(Black-capped)Antsh.	Thamnophilus schistaceus - -	-	-	X
Mouse-colored Antshrike- -	- Thamnophilus murinus - -	-	-	X
Eastern Slaty-Antshrike- -	- Thamnophilus punctatus - -	X	X	X
Amazonian Antshrike- - -	- Thamnophilus amazonicus- -	X	-	X
"Gray-capped" Antshrike- -	- Thamnophilus "cinereiceps" - -	-	-	X
Streak-backed Antshrike- -	- Thamnophilus insignis - -	-	-	X
Variable Antshrike - - -	- Thamnophilus caerulescens - -	X	X	-
Rufous-winged Antshrike- -	- Thamnophilus torquatus - -	X	X	X
Rufous-capped Antshrike- -	- Thamnophilus ruficapillus - -	X	X	-
Spot-winged Antshrike - -	- Pygiptila stellaris- - -	-	-	X
Pearly Antshrike - - -	- Megastictus margaritatus - -	-	-	X
Black Bushbird - - - -	- Neoctantes niger - - -	-	-	X
E Rondonia Bushbird - - -	- Clytoctantes atrogularis - -	-	-	X
Spot-breasted Antvireo - -	- Dysithamnus stictothorax - -	X	X	-
Plain Antvireo - - - -	- Dysithamnus mentalis - -	X	X	X
E Rufous-backed Antvireo - -	- Dysithamnus xanthopterus - -	-	X	-
E Plumbeous Antvireo (Antshrike)	- Dysitham.(Thamnomanes) plumbeus-	-	X	-
Dusky-throated Antshrike -	- Thamnomanes (Dysitham.)ardesiacus	-	-	X
Saturnine Antshrike- - -	- Thamnomanes (Dysitham.)saturninus	-	-	X
Cinereous Antshrike- - -	- Thamnomanes caesius- - -	X	X	X
Bluish-slate Antshrike - -	- Thamnomanes schistogynus - -	-	-	X
Pygmy Antwren - - - -	- Myrmotherula brachyura - -	-	-	X
Short-billed Antwren - -	- Myrmotherula obscura - -	-	-	X
Sclater's Antwren - - -	- Mrymotherula sclateri - -	-	-	X
Yellow-throated Antwren- -	- Myrmotherula ambigua - -	-	-	X
Streaked Antwren - - -	- Myrmotherula surinamensis - -	X	-	X
Cherrie's Antwren - - -	- Myrmotherula cherriei - -	-	-	X
E Klages's Antwren - - -	- Myrmotherula klagesi - -	-	-	X
Rufous-bellied Antwren - -	- Myrmotherula guttata - -	-	-	X
Plain-throated Antwren - -	- Myrmotherula hauxwelli - - -	-	-	X

	Common name	Scientific name			
E	Star-throated Antwren	Myrmotherula gularis	-	X	-
	Brown-bellied Antwren	Myrmotherula gutturalis	-	-	X
	White-eyed Antwren	Myrmotherula leucophthalma	-	-	X
	Stipple-throated Antwren	Myrmotherula haematonota	-	-	X
	Ornate Antwren	Myrmotherula ornata	-	-	X
	Rufous-tailed Antwren	Myrmotherula erythrura	-	-	X
	White-flanked Antwren	Myrmotherula axillaris	X	X	X
E	Rio de Janeiro Antwren	Myrmotherula fluminensis	-	X	-
	Long-winged Antwren	Myrmotherula longipennis	X	-	X
	Rio Suno Antwren	Myrmotherula sunensis	-	-	X
E	Salvadori's Antwren	Myrmotherula minor	-	X	-
	Ihering's Antwren	Myrmotherula iheringi	-	-	X
E	Unicolored Antwren	Myrmotherula unicolor	X	X	-
E	Bititinga Antwren	Myrmotherula snowi	-	X	-
	Plain-winged Antwren	Myrmotherula behni	-	-	X
E	Band-tailed Antwren	Myrmotherula urosticta	-	X	-
	Gray (Menetries's) Antwren	Myrmotherula menetriesii	-	-	X
	Leaden Antwren	Myrmotherula assimilis	-	-	X
	Banded Antbird	Dichrozona cincta	-	-	X
	Stripe-backed Antbird	Myrmochilus strigilatus	X	X	-
	Black-capped Antwren	Herpsilochmus atricapillus	X	X	X
E	Pileated (White-browed) Antwren	Herpsilochmus pileatus	-	X	-
	Spot-tailed Antwren	Herpsilochmus sticturus	-	-	X
	{Dugand's Antwren}	{Herpsilochmus dugandi}	-	-	{X}
	Todd's Antwren	Herpsilochmus stictocephalus	-	-	X
	Spot-backed Antwren	Herpsilochmus dorsimaculatus	-	-	X
	Roraiman Antwren	Herpsilochmus roraimae	-	-	X
	Pectoral Antwren	Herpsilochmus pectoralis	-	X	X
	Large-billed Antwren	Herpsilochmus longirostris	X	X	X
	Rufous-winged Antwren	Herpsilochmus rufimarginatus	X	X	X
	Dot-winged Antwren	Microrhopias quixensis	-	-	X
	"Lower Amazonian" Antwren	Microrhopias "emiliae"	-	-	X
	Narrow-billed Antwren	Formicivora iheringi	-	X	-
	White-fringed Antwren	Formicivora grisea	X	X	X
	Black-bellied Antwren	Formicivora melanogaster	X	X	-
	Serra Antwren (Antbird)	Formicivora serrana	X	X	-
	"Restinga" Antwren	Formicivora "littoralis"	-	X	-
	Black-hooded Antwren	Formicivora erythronotos	-	X	-
	Rusty-backed Antwren	Formicivora rufa	X	X	X
	Ferruginous Antbird	Drymophila ferruginea	-	X	-
	Bertoni's(Rufous-necked) Antbird	Drymophila rubricollis	X	X	-
	Rufous-tailed Antbird	Drymophila genei	-	X	-
	Ochre-rumped Antbird	Drymophila ochropyga	-	X	-
	Striated Antbird	Drymophila devillei	X	-	X

Dusky-tailed Antbird	Drymophila malura	X	X	-
E Scaled Antbird	Drymophila squamata	-	X	-
Streak-capped Antwren	Terenura maculata	X	X	-
E Alagoas (Orange-bellied) Antwren	Terenura sicki	-	X	-
Chestnut-shouldered Antwren	Terenura humeralis	-	-	X
Ash-winged Antwren	Terenura spodioptila	-	-	X
Gray Antbird	Cercomacra cinerascens	-	-	X
E Rio de Janeiro Antbird	Cercomacra brasiliana	-	X	-
Dusky Antbird	Cercomacra tyrannina	-	-	X
E Lower Amazonian Antbird	Cercomacra laeta	-	X	X
Blackish Antbird	Cercomacra nigrescens	-	-	X
E Bananal Antbird	Cercomacra ferdinandi	X	-	-
Black Antbird	Cercomacra serva	-	-	X
Manu Antbird	Cercomacra manu	-	-	X
Rio Branco Antbird	Cercomacra carbonaria	-	-	X
Mato Grosso Antbird	Cercomacra melanaria	X	-	-
White-backed Fire-eye	Pyriglena leuconota	X	X	X
E Fringe-backed Fire-eye	Pyriglena atra	-	X	-
White-shouldered Fire-eye	Pyriglena leucoptera	X	X	-
E Slender Antbird	Rhopornis ardesiaca	-	X	-
White-browed Antbird	Myrmoborus leucophrys	-	-	X
Ash-breasted Antbird	Myrmoborus lugubris	-	-	X
Black-faced Antbird	Myrmoborus myotherinus	X	-	X
{Black-tailed Antbird}	{Myrmoborus melanurus}	-	-	{X}
Warbling Antbird	Hypocnemis cantator	X	-	X
Yellow-browed Antbird	Hypocnemis hypoxantha	-	-	X
Black-chinned Antbird	Hypocnemoides melanopogon	-	-	X
Band-tailed Antbird	Hypocnemoides maculicauda	X	-	X
Black-and-white Antbird	Myrmochanes hemileucus	-	-	X
Silvered Antbird	Sclateria naevia	-	-	X
Black-headed Antbird	Percnostola rufifrons	-	-	X
{White-lined Antbird}	{Percnost.lophotes (macrolopha)}	-	-	{X}
Slate-colored Antbird	Percn.(Schistocichla) schistacea	-	-	X
Spot-winged Antbird	Percn.(Schistocichla) leucostigma	-	-	X
Caura Antbird	Percn.(Schistocichla) caurensis	-	-	X
Black-throated Antbird	Myrmeciza (Schistocichla)atrothorax	X	-	X
"Spot-breasted" Antbird	Myrmeciza (Schistoc.)"stictothorax"	-	-	X
White-bellied Antbird	Myrmeciza longipes	-	-	X
Ferruginous-backed Antbird	Myrmeciza ferruginea	-	-	X
E Scalloped(Rufous-tailed)Antbird	Myrmeciza ruficauda	-	X	-
E Squamate Antbird	Myrmeciza squamosa	X	X	-
E White-bibbed Antbird	Myrmeciza loricata	-	X	-
{Yapacana Antbird}	{Myrmeciza disjuncta}	-	-	{X}
Gray-bellied Antbird	Myrmeciza pelzelni	-	-	X

Common Name	Scientific Name					
Chestnut-tailed Antbird	Myrmeciza hemimelaena	-	-	-	-	X
Plumbeous Antbird	Myrmeciza hyperythra	-	-	-	-	X
Goeldi's Antbird	Myrmeciza goeldii	-	-	-	-	X
White-shouldered Antbird	Myrmeciza melanoceps	-	-	-	-	X
Sooty Antbird	Myrmeciza fortis	-	-	-	-	X
White-plumed Antbird	Pithys albifrons	-	-	-	-	X
White-throated Antbird	Gymnopithys salvini-	-	-	-	-	X
White-cheeked(Bicolored) Antbird	Gymnopithys leucaspis	-	-	-	-	X
Rufous-throated Antbird	Gymnopithys rufigula	-	-	-	-	X
Wing-banded Antbird	Myrmornis torquata	-	-	-	-	X
Bare-eyed (Santarem) Antbird	Rhegmatorhina gymnops	-	-	-	-	X
Harlequin Antbird	Rhegmatorhina berlepschi	-	-	-	-	X
Chestnut-crested Antbird	Rhegmatorhina cristata	-	-	-	-	X
White-breasted Antbird	Rhegmatorhina hoffmannsi	-	-	-	-	X
Hairy-crested Antbird	Rhegmatorhina melanosticta	-	-	-	-	X
Spot-backed Antbird	Hylophylax naevia	-	-	-	-	X
Dot-backed Antbird	Hylophylax punctulata	-	-	-	-	X
Scale-backed Antbird	Hylophylax poecilonota	-	-	-	-	X
"Plain-backed" Antbird	Hylophylax "griseiventris"	-	-	X	-	X
Pale-faced Antbird (Bare-eye)	Skutchia (Phlegopsis) borbae	-	-	-	-	X
Black-spotted Bare-eye	Phlegopsis nigromaculata	-	-	-	-	X
Reddish-winged Bare-eye	Phlegopsis erythroptera-	-	-	-	-	X
Short-tailed Antthrush	Chamaeza campanisona	-	-	X	X	-
"Northern Short-tailed"Antthrush	Chamaeza "olivacea"-	-	-	-	-	X
Such's Antthrush	Chamaeza meruloides-	-	-	-	X	-
Striated (Noble) Antthrush	Chamaeza nobilis	-	-	-	-	X
Brazilian(Rufous-tailed) Antthr.	Chamaeza ruficauda	-	-	X	X	-
Rufous-capped Antthrush	Formicarius colma	-	-	X	X	X
Black-faced Antthrush	Formicarius analis	-	-	-	-	X
Variegated Antpitta	Grallaria varia	-	-	X	X	X
Scaled Antpitta	Grallaria guatimalensis-	-	-	-	-	X
{Ochre-striped Antpitta}	{Grallaria dignissima}	-	-	-	-	{X}
Elusive Antpitta	Grallaria eludens	-	-	-	-	X
Spotted Antpitta	Hylopezus macularius	-	-	-	-	X
Amazonian Antpitta	Hylopezus berlepschi	-	-	-	-	X
White-browed Antpitta	Hylopezus ochroleucus	-	-	-	X	-
Speckle-breasted Antpitta	Hylopezus nattereri-	-	-	X	X	-
Thrush-like Antpitta	Myrmothera campanisona	-	-	-	-	X
Brown-breasted (Tepui) Antpitta-	Myrmothera simplex	-	-	-	-	X
{Slate-crowned Antpitta}	{Grallaricula nana}-	-	-	-	-	{X}
Black-bellied Gnateater-	Conopophaga melanogaster	-	-	-	-	X
Hooded Gnateater	Conopophaga roberti-	-	-	-	-	X
Black-cheeked Gnateater-	Conopophaga melanops	-	-	-	X	-
Ash-throated Gnateater	Conopophaga peruviana	-	-	-	-	X

	English Name	Scientific Name			
	Chestnut-belted Gnateater	Conopophaga aurita	-	-	X
	Rufous(Silvery-tufted)Gnateater	Conopophaga lineata	X	X	-
	"Caatinga" Gnateater	Conopophaga "cearae"	-	X	-
	Rusty-belted Tapaculo	Liosceles thoracicus	-	-	X
	Collared Crescentchest	Melanopareia torquata	X	-	-
	Spotted Bamboowren	Psilorhamphus guttatus	X	X	-
E	Slaty Bristlefront	Merulaxis ater	-	X	-
E	Stresemann's Bristlefront	Merulaxis stresemanni	-	X	-
	Mouse-colored Tapaculo	Scytalopus speluncae	X	X	-
E	Brasilia Tapaculo	Scytalopus novacapitalis	X	X	-
E	White-breasted Tapaculo	Scytalopus indigoticus	X	X	-
E	Bahia Tapaculo	Scytalopus psychopompus	-	X	-
	Shrike-like Cotinga	Laniisoma elegans	-	X	-
	Swallow-tailed Cotinga	Phibalura flavirostris	X	X	-
E	Black-and-gold Cotinga	Tijuca atra	-	X	-
E	Gray-winged (Orgaos) Cotinga	Tijuca condita	-	X	-
E	Hooded Berryeater	Carpornis cucullatus	X	X	-
E	Black-headed Berryeater	Carpornis melanocephalus	-	X	-
	Purple-throated Cotinga	Porphyrolaema porphyrolaema	-	-	X
	Plum-throated Cotinga	Cotinga maynana	-	-	X
	Spangled Cotinga	Cotinga cayana	-	-	X
	Purple-breasted Cotinga	Cotinga cotinga	-	-	X
E	Banded Cotinga	Cotinga maculata	-	X	-
	Pompadour Cotinga	Xipholena punicea	-	-	X
E	White-tailed Cotinga	Xipholena lamellipennis	-	-	X
E	White-winged Cotinga	Xipholena atropurpurea	-	X	-
	{Red-banded Fruiteater}	{Pipreola whitelyi}	-	-	{X}
E	Buff-throated Purpletuft	Iodopleura pipra	-	X	-
	"Alagoas(White-rumped)"Purpletuft	Iodopleura "leucopygia"	-	X	-
	Dusky Purpletuft	Iodopleura fusca	-	-	X
	White-browed Purpletuft	Iodopleura isabellae	-	-	X
E	Kinglet Calyptura	Calyptura cristata	-	X	-
	Screaming Piha	Lipaugus vociferans	X	X	X
E	Cinnamon-vented Piha	Lipaugus lanioides	X	X	-
	Rose-collared Piha	Lipaugus streptophorus	-	-	X
	White-naped Xenopsaris	Xenopsaris albinucha	X	-	X
	Green-backed Becard	Pachyramphus viridis	X	X	X
	Glossy-backed Becard	Pachyramphus surinamus	-	-	X
	Cinereous Becard	Pachyramphus rufus	-	-	X
	Chestnut-crowned Becard	Pachyramphus castaneus	X	X	X
	White-winged Becard	Pachyramphus polychopterus	X	X	X
	Black-capped Becard	Pachyramphus marginatus	X	X	X
	Crested (Plain) Becard	Pachyramphus validus (Platy. rufus)	X	X	X
	Pink-throated Becard	Pachyramphus (Platypsaris) minor	-	-	X

Common Name	Scientific Name			
Black-tailed Tityra	Tityra cayana	X	X	X
Masked Tityra	Tityra semifasciata	X	-	X
Black-crowned Tityra	Tityra inquisitor	X	X	X
Crimson Fruitcrow	Haematoderus militaris	-	-	X
Purple-throated Fruitcrow	Querula purpurata	X	-	X
Red-ruffed Fruitcrow	Pyroderus scutatus	X	X	-
Amazonian Umbrellabird	Cephalopterus ornatus	X	-	X
Capuchinbird	Perissocephalus tricolor	-	-	X
Bare-necked Fruitcrow	Gymnoderus foetidus	X	-	X
White Bellbird	Procnias alba	-	-	X
Bare-throated Bellbird	Procnias nudicollis	X	X	-
Bearded Bellbird	Procnias averano	-	X	X
Guianan Red-Cotinga	Phoenicircus carnifex	-	-	X
Black-necked Red-Cotinga	Phoenicircus nigricollis	-	-	X
Guianan Cock-of-the-Rock	Rupicola rupicola	-	-	X
Sharpbill	Oxyruncus cristatus	X	X	X
White-tipped(Red-br.)Plantcutter	Phytotoma rutila	X	-	-
Golden-headed Manakin	Pipra erythrocephala	-	-	X
Red-headed Manakin	Pipra rubrocapilla	X	X	X
Scarlet-horned Manakin	Pipra cornuta	-	-	X
White-crowned Manakin	Pipra pipra	-	X	X
Blue-crowned Manakin	Pipra coronata	-	-	X
White-fronted Manakin	Pipra serena	-	-	X
Opal-crowned Manakin	Pipra iris	-	-	X
Golden-crowned Manakin	Pipra vilasboasi	Para		
Snow-capped Manakin	Pipra nattereri	-	-	X
Crimson-hooded Manakin	Pipra aureola	-	-	X
Band-tailed Manakin	Pipra fasciicauda	X	-	X
Wire-tailed Manakin	Pipra (Teleonema) filicauda	-	-	X
Helmeted Manakin	Antilophia galeata	X	-	X
Blue-backed Manakin	Chiroxiphia pareola	X	X	X
Swallow-tailed (Blue) Manakin	Chiroxiphia caudata	X	X	-
Pin-tailed Manakin	Ilicura militaris	X	X	-
White-throated Manakin	Corapipo gutturalis	-	-	X
White-bearded Manakin	Manacus manacus	X	X	X
Fiery-capped Manakin	Machaeropterus pyrocephalus	X	-	X
Striped Manakin	Machaeropterus regulus	-	X	X
Black Manakin	Xenopipo atronitens	-	-	X
Olive Manakin	Chloropipo uniformis	-	-	X
Cinnamon Manakin(Tyrant-Manakin)	Neopipo cinnamomea	-	-	X
Flame-crested (-crowned) Manakin	Heterocercus linteatus	X	-	X
Yellow-crested(-crowned) Manakin	Heterocercus flavivertex	-	-	X
Wied's Tyrant-Manakin	Neopelma aurifrons	-	X	-
Sulphur-bellied Tyrant-Manakin	Neopelma sulphureiventer	-	-	X

English Name	Scientific Name			
Saffron-crested Tyrant-Manakin	Neopelma chrysocephalum	-	-	X
Pale-bellied Tyrant-Manakin	Neopelma pallescens	X	X	X
Tiny Tyrant-Manakin	Tyranneutes virescens	-	-	X
Dwarf Tyrant-Manakin	Tyranneutes stolzmanni	-	-	X
E Black-capped Manakin (Piprites)	Piprites pileatus	X	X	-
Wing-barred Manakin (Piprites)	Piprites chloris	X	X	X
Greater Manakin (Schiffornis)	Schiffornis major	-	-	X
Greenish Manakin (Schiffornis)	Schiffornis virescens	X	X	-
Thrush-like Manakin(Schiffornis)	Schiffornis turdinus	X	X	X
Chocolate-vented Tyrant	Neoxolmis rufiventris	X	-	-
Black-and-white Monjita	Heteroxolmis (Xolmis) dominicana	X	-	-
Gray Monjita	Xolmis cinerea	X	X	X
White-rumped Monjita	Xolmis velata	X	X	X
Black-crowned Monjita	Xolmis coronata	X	-	-
White Monjita	Xolmis irupero	X	X	-
Little Ground-Tyrant	Muscisaxicola fluviatilis	-	-	X
Patagonian(Rufous-backed)Negrito	Lessonia rufa	X	-	-
Long-tailed Tyrant	Colonia colonus	X	X	X
Streamer-tailed Tyrant	Gubernetes yetapa	X	X	-
Cock-tailed Tyrant	Alectrurus tricolor	X	X	-
Strange-tailed Tyrant	Alectrurus (Yetapa) risora	X	-	-
Crested Black-Tyrant	Knipolegus lophotes	X	X	-
E Velvety Black-Tyrant	Knipolegus nigerrimus	X	X	-
White-winged Black-Tyrant	Knipolegus aterrimus	X	X	-
Riverside Tyrant	Knipolegus orenocensis	X	-	X
Rufous-tailed Tyrant	Knipolegus poecilurus	-	-	X
Blue-billed Black-Tyrant	Knipolegus cyanirostris	X	X	-
Amazonian Black-Tyrant	Knip.(Phaeotriccus) poecilocercus	X	-	X
Hudson's Black-Tyrant	Knipolegus (Phaeotriccus) hudsoni	X	-	-
Cinereous Tyrant	Knip.(Entotriccus) striaticeps	X	-	-
Spectacled Tyrant	Hymenops perspicillata	X	X	-
Shear-tailed Gray Tyrant	Muscipipra vetula	X	X	-
Pied Water-Tyrant	Fluvicola pica	-	-	X
Black-backed(Wing-barred)Water-T.	Fluvicola albiventer	X	X	X
Masked Water-Tyrant	Fluvicola nengeta	X	X	X
White-headed Marsh-Tyrant	Arundinicola (Fluvic.) leucocephala	X	X	X
Vermilion Flycatcher	Pyrocephalus rubinus	X	X	X
Drab Water-Tyrant	Ochthornis littoralis	-	-	X
Yellow-browed Tyrant	Satrapa icterophrys	X	X	-
Cattle Tyrant	Machetornis rixosus	X	X	X
Sirystes	Sirystes sibilator	X	X	X
Fork-tailed Flycatcher	Tyrannus (Muscivora) savana	X	X	X
Eastern Kingbird	Tyrannus tyrannus	X	-	X
Tropical Kingbird	Tyrannus melancholicus	X	X	X

Gray Kingbird	Tyrannus dominicensis			X
White-throated Kingbird	Tyrannus albogularis	X	X	X
Variegated Flycatcher	Empidonomus varius	X	X	X
Crowned Slaty Flycatcher	Griseotyrannus(E.)aurantioatrocristatus	X	-	X
Piratic Flycatcher	Legatus leucophaius	X	X	X
Three-striped Flycatcher	Conopias trivirgata	X	X	X
Yellow-throated Flycatcher	Conopias parva (albovittata)			X
Boat-billed Flycatcher	Megarynchus pitangua	X	X	X
Sulphur-bellied Flycatcher	Myiodynastes luteiventris			X
Streaked Flycatcher	Myiodynastes maculatus	X	X	X
Rusty-margined Flycatcher	Myiozetetes cayanensis	X	X	X
Social Flycatcher	Myiozetetes similis	X	X	X
Gray-capped Flycatcher	Myiozetetes granadensis			X
Dusky-chested Flycatcher	Myiozetetes (Tyran.) luteiventris	X	-	X
Sulphury Flycatcher	Tyrannopsis sulphurea	X	-	X
Great Kiskadee	Pitangus sulphuratus	X	X	X
Lesser Kiskadee	Philohydor (Pitangus) lictor	X	X	X
Bright-rumped Attila	Attila spadiceus	-	X	X
White-eyed (Dull-capped) Attila	Attila bolivianus	X	-	X
Gray-hooded Attila	Attila rufus	X	X	-
Citron-bellied Attila	Attila citriniventris			X
Cinnamon Attila	Attila cinnamomeus	X	-	X
Rufous-tailed Attila	Attila (Pseudattila) phoenicurus	X	X	X
Rufous Casiornis	Casiornis rufa	X	X	X
Ash-throated Casiornis	Casiornis fusca	X	X	X
Cinereous Mourner	Laniocera hypopyrra	-	X	X
Grayish Mourner	Rhytipterna simplex	X	X	X
Pale-bellied Mourner	Rhytipterna immunda			X
Short-crested Flycatcher	Myiarchus ferox	X	X	X
Brown-crested Flycatcher	Myiarchus tyrannulus	X	X	X
Swainson's Flycatcher	Myiarchus swainsoni	X	X	X
Dusky-capped (Olivaceous) Fly.	Myiarchus tuberculifer	X	X	X
Olive-sided Fly.(Boreal Pewee)	Contopus (Nuttallornis) borealis	-	X	X
Eastern Wood-Pewee	Contopus virens			X
Tropical Pewee	Contopus cinereus	X	X	X
White-throated Pewee	Contopus albogularis			X
Blackish Pewee	Contopus nigrescens			X
{Smoke-colored (Greater) Pewee}	{Contopus fumigatus}		-	{X}
Alder Flycatcher	Empidonax alnorum			X
Euler's Flycatcher	Lathrotriccus (Empidonax) euleri	X	X	X
Fuscous Flycatcher	Cnemotriccus fuscatus	X	X	-
"Audible" Fuscous-Flycatcher	Cnemotriccus "bimaculatus"	X	X	X
Ruddy-tailed Flycatcher	Terenotriccus (Myiobius) erythrurus	-	-	X
Sulphur-rumped (Whiskered)Fly.	Myiobius barbatus	X	X	X

Common Name	Scientific Name			
Black-tailed Flycatcher	Myiobius atricaudus	X	X	X
Bran-colored Flycatcher	Myiophobus fasciatus	X	X	X
Roraiman Flycatcher	Myiophobus roraimae	-	-	X
Cliff Flycatcher	Hirundinea ferruginea	-	-	X
"Swallow" Flycatcher	Hirundinea "bellicosa"	X	X	-
Amazonian Royal-Flycatcher	Onychorhynchus coronatus	-	-	X
"Atlantic" Royal-Flycatcher	Onychorhynchus "swainsoni"	X	X	-
White-crested Spadebill	Platyrinchus platyrhynchos	-	-	X
Russet-winged Spadebill	Platyrinchus leucoryphus	X	X	-
White-throated Spadebill	Platyrinchus mystaceus	X	X	X
Golden-crowned Spadebill	Platyrinchus coronatus	-	-	X
Cinnamon-crested Spadebill	Platyrinchus saturatus	-	-	X
Brownish Flycatcher (Twistwing)	Cnipodectes subbrunneus	-	-	X
Yellow-olive Flycatcher	Tolmomyias sulphurescens	X	X	X
Yellow-margined Flycatcher	Tolmomyias assimilis	X	-	X
Gray-crowned Flycatcher	Tolmomyias poliocephalus	-	X	X
Yellow-breasted Flycatcher	Tolmomyias flaviventris	X	X	X
"Upper Amazonian" Flycatcher	Tolmomyias "viridiceps"	-	-	X
Olivaceous Flatbill	Rhynchocyclus olivaceus	-	X	X
Rufous-tailed Flatbill	Ramphotrigon ruficauda	X	-	X
Dusky-tailed Flatbill	Ramphotrigon fuscicauda	-	-	X
Large-headed Flatbill	Ramphotrigon megacephala	X	X	X
Yellow-browed Tody-Flycatcher	Todirostrum chrysocrotaphum	-	-	X
Painted Tody-Flycatcher	Todirostrum pictum	-	-	X
E Yellow-lored(Gray-headed)Tody-F.	Todirostrum poliocephalum	X	X	-
Common Tody-Flycatcher	Todirostrum cinereum	X	X	X
Spotted Tody-Flycatcher	Todirostrum maculatum	X	-	X
Smoky-fronted Tody-Flycatcher	Todirostrum fumifrons	X	X	X
E Buff-cheeked Tody-Flycatcher	Todirostrum senex	Amazonas		
Ruddy Tody-Flycatcher	Todirostrum russatum	-	-	X
Ochre-faced Tody-Flycatcher	Todirostrum plumbeiceps	X	X	-
Rusty-fronted Tody-Flycatcher	Todirostrum latirostre	X	-	X
Slate-headed Tody-Flycatcher	Todirostrum sylvia	-	-	X
Black-and-white Tody-Tyrant	Poecilotriccus (Tod.) capitalis	-	-	X
"Tricolored" Tody-Tyrant	Poecilotriccus (Tod.) "tricolor"	-	-	X
Black-chested Tyrant	Taeniotriccus(Poecilotriccus) andrei	-	-	X
Snethlage's Tody-Tyrant	Hemitriccus (Snethlagea) minor	-	-	X
Zimmer's Tody-Tyrant	Hemitriccus (Id.) minimus (aenigma)	-	-	X
Stripe-necked Tody-Tyrant	Hemitr.(Idioptilon) striaticollis	X	X	X
Johannes's Tody-Tyrant	Hemitriccus(Idioptilon) iohannis	-	-	X
Pearly-vented Tody-Tyrant	Hemitricc.(I.) margaritaceiventer	X	X	X
E Pelzeln's Tody-Tyrant	Hemitriccus(Idioptilon) inornatus	-	-	X
E Buff-breasted Tody-Tyrant	Hemitriccus(Idioptilon) mirandae	-	X	-
E Kaempfer's Tody-Tyrant	Hemitriccus(Idioptilon) kaempferi	Santa Catarina		

	Common Name	Scientific Name			
	White-eyed Tody-Tyrant	Hemitriccus(Idioptilon) zosterops	-	X	X
	"White-bellied" Tody-Tyrant	Hemitriccus(Idiopt.)"griseipectus"	-	-	X
E	Eye-ringed (Olivaceous)Tody-Ty.	Hemitriccus(Idioptilon) orbitatus	X	X	-
	Boat-billed Tody-Tyrant	Hem.(Microcochlearius) josephinae	-	-	X
E	Hangnest Tody-Tyrant	Hemitriccus(Idiopt.) nidipendulus	-	X	-
E	Fork-tailed Tody(Pygmy)-Tyrant	Hemitric.(Ceratotriccus) furcatus	-	X	-
	Drab-breasted Bamboo (Pygmy)-Tyr.	Hemitriccus diops	X	X	-
E	Brown-breasted Bamboo(Pygmy)-Tyr.	Hemitriccus obsoletus	X	X	-
	Flammulated Bamboo(Pygmy)-Tyrant	Hemitriccus flammulatus	-	-	X
	Double-banded Pygmy-Tyrant	Lophotriccus vitiosus	-	-	X
	"Golden-scaled" Pygmy-Tyrant	Lophotriccus "congener"	-	-	X
	Long-crested Pygmy-Tyrant	Lophotriccus eulophotes	-	-	X
	Helmeted Pygmy-Tyrant	Lophotriccus(Colopteryx) galeatus	-	-	X
	Pale-eyed Pygmy-Tyrant	Atalotriccus pilaris	-	-	X
	Eared Pygmy-Tyrant	Myiornis auricularis	X	X	-
	Short-tailed Pygmy-Tyrant	Myiornis ecaudatus	-	-	X
	Southern Bristle-Tyrant	Phylloscartes(Pogonotriccus)eximius	X	X	-
	Chapman's Tyrann.(Bristle-Tyrant)	Phylloscartes chapmani	-	-	X
	Bay-ringed Tyrannulet	Phyll.(Leptotriccus) sylviolus	X	X	-
	Olive-green Tyrannulet	Phylloscartes virescens	-	-	X
	Mottle-cheeked Tyrannulet	Phylloscartes ventralis	X	X	-
E	Restinga Tyrannulet	Phylloscartes kronei	Sao Paulo, S.Cat.		
	Black-fronted Tyrannulet	Phylloscartes nigrifrons	-	-	X
E	Oustalet's Tyrannulet	Phylloscartes oustaleti	-	X	-
E	Serra do Mar (Ihering's)Tyr.	Phylloscartes difficilis	X	X	-
	Sao Paulo Tyrannulet	Phylloscartes paulistus	X	X	-
E	Minas Gerais Tyrannulet	Phylloscartes roquettei	Minas Gerais		
E	Alagoas (Long-tailed) Tyrannulet	Phylloscartes ceciliae	-	X	-
	Yellow Tyrannulet	Capsiempis flaveola	X	X	X
	Tawny-crowned Pygmy-Tyrant	Euscarthmus meloryphus	X	X	X
	Rufous-sided Pygmy-Tyrant	Euscarthmus rufomarginatus	X	-	X
	Crested Doradito	Pseudocolopteryx sclateri	X	X	-
	Warbling Doradito	Pseudocolopteryx flaviventris	X	-	-
	Bearded Tachuri	Polystictus pectoralis	X	-	X
E	Gray-backed Tachuri	Polystictus superciliaris	X	X	-
	Sharp-tailed Tyrant (Grass-Tyr.)	Culicivora caudacuta	X	-	-
	Many-colored Rush-Tyrant	Tachuris rubrigastra	X	-	-
	Greater Wagtail Tyrant	Stigmatura budytoides	-	X	-
	Lesser Wagtail-Tyrant	Stigmatura napensis	-	X	X
	River Tyrannulet	Serpophaga hypoleuca	-	-	X
	White-crested Tyrannulet	Serpophaga subcristata	X	X	X
	White-bellied Tyrannulet	Serpophaga munda	X	-	-
	Sooty Tyrannulet	Serpophaga nigricans	X	X	-
	Pale-tipped Tyrannulet	Inezia subflava	X	-	X

Plain Tyrannulet - - -	- Inezia inornata- - - - -	X	-	X
White-throated Tyrannulet -	- Mecocerculus leucophrys- - -	-	-	X
Yellow-bellied Elaenia - -	- Elaenia flavogaster- - -	X	X	X
Large Elaenia - - - -	- Elaenia spectabilis- - -	X	X	X
E Noronha Elaenia- - -	- Elaenia ridleyana - - -	-	-	X
White-crested Elaenia - -	- Elaenia albiceps - - -	X	X	-
Small-billed Elaenia - -	- Elaenia parvirostris - -	X	X	X
Olivaceous Elaenia - - -	- Elaenia mesoleuca - - -	X	X	-
{Slaty Elaenia}- - -	- {Elaenia strepera} - - -	-	-	{X}
Brownish Elaenia - -	- Elaenia pelzelni - - -	-	-	X
Plain-crested Elaenia - -	- Elaenia cristata - - -	X	X	X
Lesser Elaenia - - -	- Elaenia chiriquensis - -	X	X	X
Rufous-crowned Elaenia - -	- Elaenia ruficeps - - -	-	-	X
Highland Elaenia - - -	- Elaenia obscura- - -	X	X	-
{Great Elaenia}- - -	- {Elaenia dayi} - - -	-	-	{X}
Sierran Elaenia- - -	- Elaenia pallatangae- - -	-	-	X
Forest Elaenia - - -	- Myiopagis gaimardii- - -	X	X	X
Gray Elaenia - - -	- Myiopagis caniceps - -	X	X	X
Yellow-crowned Elaenia - -	- Myiopagis flavivertex - -	-	-	X
Greenish Elaenia - - -	- Myiopagis viridicata - -	X	X	X
Chaco Suiriri - - -	- Suiriri suiriri- - --	X	-	-
Campo Suiriri - - -	- Suiriri affinis- - -	X	X	X
Southern Scrub-Fly.(Short-b.Fly)	Sublegatus modestus- - -	X	X	X
"Northern" Scrub-Flycatcher-	- Sublegatus "arenarum" - -	-	-	X
Mouse-colored Tyrannulet -	- Phaeomyias murina - - -	X	X	X
Southern Beardless-Tyrannulet	- Camptostoma obsoletum -- -	X	X	X
Greenish Tyrannulet- - -	- Phyllomyias (Xanth.) virescens -	X	X	-
Reiser's Tyrannulet- - -	- Phyllomy.(Xanthomyias) reiseri -	X	-	-
Planalto Tyrannulet- - -	- Phyllomyias fasciatus - -	X	X	X
Sooty-headed Tyrannulet- -	- Phyllomyias griseiceps - - -	-	-	X
Rough-legged Tyrannulet- -	- Phy.(Acrochordopus) burmeisteri-	X	X	-
E Gray-capped Tyrannulet - -	- Phy.(Oreotriccus) griseocapilla-	-	X	-
Slender-footed Tyrannulet -	- Zimmerius(Tyranniscus) gracilipes	--	-	X
Yellow-crowned Tyrannulet -	- Tyrannulus elatus - - -	-	-	X
White-lored Tyrannulet -	- Ornithion inerme - - -	X	X	X
Sepia-capped Flycatcher- -	- Leptopogon amaurocephalus - -	X	X	X
Ochre-bellied Flycatcher -	- Mionectes(Pipromorpha) oleagineus	X	X	X
McConnell's Flycatcher - -	- Mionectes(Pipromo.) macconnelli	-	-	X
Gray-hooded Flycatcher - -	- Mionectes(Pipromor.) rufiventris	X	X	-
Ringed Antpipit- - -	- Corythopis torquata- - -	X	-	X
Southern (Delalande's) Antpipit-	Corythopis delalandi - - -	X	X	--
White-winged Swallow - -	- Tachycineta albiventer - -	X	X	X
White-rumped Swallow - -	- Tachycineta leucorrhoa - -	X	X	-
Chilean Swallow- - - -	- Tachycineta leucopyga (meyeni) -	X	-	-

Brown-chested Martin	Phaeoprogne tapera	X	X	X
Purple Martin	Progne subis	X	X	X
Gray-breasted Martin	Progne chalybea	X	X	X
Southern Martin	Progne modesta (elegans)	-	-	X
Blue-and-white Swallow	Notiochelidon cyanoleuca	X	X	X
White-banded Swallow	Atticora fasciata	-	-	X
Black-collared Swallow	Atticora melanoleuca	X	-	X
White-thighed Swallow	Neochelidon tibialis	-	X	X
Tawny-headed Swallow	Alopochelidon (Stelgid.) fucata	X	X	X
Southern Rough-winged Swallow	Stelgidopteryx ruficollis	X	X	X
Bank Swallow (Sand Martin)	Riparia riparia	X	X	X
Barn Swallow	Hirundo rustica	X	X	X
Cliff Swallow	Hirundo (Petrochelidon) pyrrhonota	X	X	X
Azure Jay	Cyanocorax caeruleus	X	-	-
Purplish Jay	Cyanocorax cyanomelas	X	-	X
Violaceous Jay	Cyanocorax violaceus	-	-	X
Curl-crested Jay	Cyanocorax cristatellus	X	X	X
Azure-naped Jay	Cyanocorax heilprini	-	-	X
Cayenne Jay	Cyanocorax cayanus	-	-	X
Plush-crested Jay	Cyanocorax chrysops	X	X	-
E White-naped Jay	Cyanocorax cyanopogon	X	X	X
Bicolored Wren	Campylorhynchus griseus	-	-	X
Thrush-like Wren	Campylorhynchus turdinus	-	X	X
"Plain-breasted" Wren	Campylorhynchus "unicolor"	X	-	-
Tooth-billed Wren	Odontorchilus cinereus	-	-	X
Grass (Sedge) Wren	Cistothorus platensis	X	-	X
Moustached Wren	Thryothorus genibarbis	X	X	X
Coraya Wren	Thryothorus coraya	-	-	X
Buff-breasted Wren	Thryothorus leucotis	X	X	X
Fawn-breasted Wren	Thryothorus guarayanus	X	-	X
E Long-billed Wren	Thryothorus longirostris	X	-	
E Gray Wren	Thryothorus griseus	-	-	X
House Wren	Troglodytes aedon	X	X	X
Tepui Wren	Troglodytes rufulus	-	-	X
White-breasted Wood-Wren	Henicorhina leucosticta	-	-	X
Southern Nightingale-Wren	Microcerculus marginatus	-	-	X
Flutist Wren	Microcerculus ustulatus	-	-	X
Wing-banded Wren	Microcerculus bambla	-	-	X
Musician (Organ) Wren	Cyphorhinus aradus	-	-	X
Black-capped Donacobius	Donacobius atricapillus	X	X	X
Tropical Mockingbird	Mimus gilvus	-	X	X
Chalk-browed Mockingbird	Mimus saturninus	X	X	X
White-banded Mockingbird	Mimus triurus	X	-	-
Rufous-brown Solitaire	Cichlopsis (Myadestes) leucogenys	-	X	-

Common Name	Scientific Name			
"Guyanan" Solitaire	Cichlopsis (Myadestes) "gularis"			X
Veery	Catharus fuscescens	X		X
Gray-cheeked Thrush	Catharus minimus			X
Swainson's Thrush	Catharus ustulatus		X	X
Yellow-legged Thrush	Platycichla flavipes	X	X	X
Pale-eyed Thrush	Platycichla leucops			X
Eastern Slaty Thrush	Turdus subalaris	X	X	
Black-hooded Thrush	Turdus olivater			X
Rufous-bellied Thrush	Turdus rufiventris	X	X	
Pale-breasted Thrush	Turdus leucomelas	X	X	X
Creamy-bellied Thrush	Turdus amaurochalinus	X	X	X
Black-billed Thrush	Turdus ignobilis			X
Lawrence's Thrush	Turdus lawrencii			X
Cocoa Thrush	Turdus fumigatus	X	X	X
Hauxwell's Thrush	Turdus hauxwelli			X
Bare-eyed Thrush	Turdus nudigenis			X
White-necked Thrush	Turdus albicollis	X	X	X
Collared Gnatwren	Microbates collaris			X
Long-billed Gnatwren	Ramphocaenus melanurus	X	X	X
Tropical Gnatcatcher	Polioptila plumbea		X	X
Cream-bellied Gnatcatcher	Polioptila lactea	X		
Guianan Gnatcatcher	Polioptila guianensis			X
Masked Gnatcatcher	Polioptila dumicola	X		
Short-billed Pipit	Anthus furcatus	X		
Hellmayr's Pipit	Anthus hellmayri	X	X	
Yellowish Pipit	Anthus lutescens	X	X	X
Correndera Pipit	Anthus correndera	X		
Ochre-breasted Pipit	Anthus nattereri	X		
Rufous-browed Peppershrike	Cyclarhis gujanensis	X	X	X
Slaty-capped Shrike-Vireo	Vireolanius(Smaragdola.) leucotis			X
Red-eyed Vireo	Vireo olivaceus (chivi)	X	X	X
{Yellow-green Vireo}	{Vireo flavoviridis}			{X}
E Noronha Vireo	Vireo gracilirostris			X
Black-whiskered Vireo	Vireo altiloquus			X
Rufous-crowned Greenlet	Hylophilus poicilotis	X	X	
E Gray-eyed Greenlet	Hylophilus amaurocephalus	X	X	
Lemon-chested Greenlet	Hylophilus thoracicus		X	X
Gray-chested Greenlet	Hylophilus semicinereus			X
Ashy-headed Greenlet	Hylophilus pectoralis	X		X
Tepui Greenlet	Hylophilus sclateri			X
Buff-cheeked (-chested)Greenlet	Hylophilus muscicapinus			X
Brown-headed Greenlet	Hylophilus brunneiceps			X
Dusky-capped Greenlet	Hylophilus hypoxanthus			X
Tawny-crowned Greenlet	Hylophilus ochraceiceps			X

Common Name	Scientific Name			
"Red-fronted" Greenlet	Hylophilus "rubrifrons"	-	-	X
Shiny (Common) Cowbird	Molothrus bonariensis	X	X	X
Screaming Cowbird	Molothrus rufoaxillaris	X	-	-
Bay-winged Cowbird	Molothrus badius	X	X	-
Giant Cowbird	Scaphidura oryzivora	X	X	X
Band-tailed Oropendola	Ocyalus (Psarocolius) latirostris	-	-	X
Casqued Oropendola	Psarocolius (Clypicterus) oseryi	-	-	X
Crested Oropendola	Psarocolius decumanus	X	X	X
Green Oropendola	Psarocolius viridis	-	-	X
Russet-backed Oropendola	Psarocolius angustifrons	-	-	X
Para (Amazonian) Oropendola	Psar.(Gymnostinops) bifasciatus	-	-	X
"Olive" Oropendola	Psar.(Gymnostinops) "yuracares"	-	-	X
Yellow-rumped Cacique	Cacicus cela	X	X	X
Red-rumped Cacique	Cacicus haemorrhous	X	X	X
Golden-winged Cacique	Cacicus chrysopterus	X	X	-
Solitary (Solitary Black)Cacique	Cacicus solitarius	X	-	X
Carib Grackle	Quiscalus lugubris	-	-	X
Golden-tufted(Tepui Mountain)Gra.	Macroagelaius imthurni	-	-	X
Velvet-fronted Grackle	Lampropsar tanagrinus	-	-	X
E Forbes's Blackbird	Curaeus forbesi	-	X	-
Chopi Blackbird	Gnorimopsar chopi	X	X	X
Yellow-winged Blackbird	Agelaius thilius	X	-	-
Chestnut-capped Blackbird	Agelaius ruficapillus	X	X	X
Unicolored Blackbird	Agelaius cyanopus	X	X	X
Yellow-hooded Blackbird	Agelaius icterocephalus	-	-	X
Saffron-cowled Blackbird	Agelaius (Xanthopsar) flavus	X	-	-
Epaulet Oriole	Icterus cayanensis	X	X	X
Moriche Oriole	Icterus chrysocephalus	-	-	X
Troupial (Campo Oriole)	Icterus icterus jamacaii	X	X	X
"Orange-backed" Troupial	Icterus "croconotus"	X	-	X
Yellow Oriole	Icterus nigrogularis	-	-	X
Oriole Blackbird	Gymnomystax mexicanus	-	-	X
Scarlet-headed Blackbird	Amblyramphus holosericeus	X	-	-
Yellow-rumped Marshbird	Pseudoleistes guirahuro	X	X	-
Brown-and-yellow Marshbird	Pseudoleistes virescens	X	-	-
Red-breasted Blackbird	Sturnella (Leistes) militaris	-	-	X
White-browed Blackbird	Sturnella (Leistes) superciliaris	X	X	X
Pampas(Lesser Red-br.)Meadowlark	Sturnella defilippi	X	-	-
Eastern Meadowlark	Sturnella magna	-	-	X
Bobolink	Dolichonyx oryzivorus	X	X	X
Prothonotary Warbler	Protonotaria citrea	X	-	-
{Tennessee Warbler}	{Vermivora peregrina}	-	-	{X}
Tropical Parula	Parula pitiayumi	X	X	X
Yellow Warbler	Dendroica petechia	-	-	X

Common Name	Scientific Name			
Cerulean Warbler - - -	- Dendroica cerulea - - -	-	X	-
Blackburnian Warbler - -	- Dendroica fusca- - - -	-	X	X
{Bay-breasted Warbler} - -	- {Dendroica castanea} - -	-	-	{X}
Blackpoll Warbler - - -	- Dendroica striata - - -	-	X	X
Northern Waterthrush - -	- Seiurus noveboracensis - - -	Para		
Connecticut Warbler- - -	- Oporornis (Geothlypis) agilis -	Amazonas & Mato		
{Mourning Warbler} - - -	- {Oporornis (Geoth.) philadelphia}	-	-	{X}
Masked Yellowthroat- - -	- Geothlypis aequinoctialis - -	X	X	X
Rose-breasted Chat - - -	- Granatellus pelzelni - - -	X	-	X
Canada Warbler - - -	- Wilsonia canadensis- - -	-	-	X
American Redstart - - -	- Setophaga ruticilla- - -	-	-	X
Slate-throated Redstart- -	- Myioborus miniatus - - -	-	-	X
Tepui Redstart - - -	- Myioborus castaneocapillus - -	-	-	X
Flavescent Warbler - - -	- Basileuterus flaveolus - -	X	X	-
Two-banded Warbler - -	- Basileuterus bivittatus- -	-	-	X
Golden-crowned Warbler -	- Basileuterus culicivorus - -	X	X	X
White-bellied Warbler - -	- Basileuterus hypoleucus- -	X	X	-
E White-striped Warbler - -	- Basileuterus leucophrys- -	X	-	-
White-rimmed (-browed) Warbler	- Basileuterus leucoblepharus- -	X	X	-
Neotropical River Warbler -	- Basileuterus rivularis - -	X	X	X
Buff-rumped Warbler- - -	- Basileuterus fulvicauda- -	-	-	X
Bananaquit - - - -	- Coereba flaveola - - -	X	X	X
Chestnut-vented Conebill -	- Conirostrum speciosum - -	X	X	X
Bicolored Conebill - -	- Conirostrum bicolor- - -	-	X	X
Pearly-breasted Conebill -	- Conirostrum margaritae - -	-	-	X
Scaled Flower-piercer - -	- Diglossa duidae- - - -	-	-	X
Greater Flower-piercer -	- Diglossa major - - - -	-	-	X
Short-billed Honeycreeper -	- Cyanerpes nitidus - - -	-	-	X
Purple Honeycreeper- - -	- Cyanerpes caeruleus- - -	-	-	X
Red-legged Honeycreeper-	- Cyanerpes cyaneus - - -	X	X	X
Green Honeycreeper - -	- Chlorophanes spiza - - -	-	X	X
Blue Dacnis - - -	- Dacnis cayana - - - -	X	X	X
E Black-legged Dacnis- - -	- Dacnis nigripes- - - -	-	X	-
Black-faced Dacnis - -	- Dacnis lineata - - - -	-	-	X
Yellow-bellied Dacnis -	- Dacnis flaviventer - - -	-	-	X
White-bellied Dacnis - -	- Dacnis albiventris - - -	-	-	X
Swallow-Tanager- - -	- Tersina viridis- - - -	X	X	X
Blue-naped Chlorophonia- -	- Chlorophonia cyanea- - -	X	X	X
Golden-rumped(Blue-hooded)Eupho.	Euphonia cyanocephala (aureata)-	X	X	X
Orange-bellied Euphonia- -	- Euphonia xanthogaster - -	X	X	X
White-vented Euphonia - -	- Euphonia minuta- - - -	-	-	X
Finsch's Euphonia - -	- Euphonia finschi - - -	-	-	X
Purple-throated Euphonia -	- Euphonia chlorotica- - -	X	X	X
Thick-billed Euphonia - -	- Euphonia laniirostris - -	X	-	X

"Black-tailed" Euphonia-	-	- Euphonia "melanura"-	-	-	-	-	X
Violaceous Euphonia-	-	- Eupholia violacea -	-	-	-	X X X
Rufous-bellied Euphonia-	-	- Euphonia rufiventris	-	-	-	- X
Golden-sided Euphonia	-	- Euphonia cayennensis	-	-	-	- X
Chestnut-bellied Euphonia	-	- Euphonia pectoralis-	-	-	X X	-
{Bronze-green Euphonia}-	-	- {Euphonia mesochrysa}	-	-	-	{X}
White-lored(Golden-bellied)Euph.	Euphonia chrysopasta	-	-	-	- X
Plumbeous Euphonia -	-	- Euphonia plumbea	-	-	-	- X
Green-chinned(-throated)Euphonia	Euphonia chalybea	-	-	-	X X	-
Fawn-breasted Tanager	-	- Pipraeidea melanonota	-	-	X X X
Opal-rumped Tanager-	-	- Tangara velia	-	-	-	- X
"Silvery-breasted" Tanager -	- Tangara "cyanomelaena" -	-	-	X	-
Opal-crowned Tanager	-	- Tangara callophrys -	-	-	- X
Paradise Tanager	-	- Tangara chilensis -	-	-	- X
E Seven-colored Tanager	-	- Tangara fastuosa	-	-	-	X -
Green-headed Tanager	-	- Tangara seledon-	-	-	X X -
Red-necked Tanager -	-	- Tangara cyanocephala	-	-	X X -
E Gilt-edged Tanager -	-	- Tangara cyanoventris	-	-	X X -
E Brassy-breasted Tanager-	-	- Tangara desmaresti -	-	-	X -
Green-and-gold Tanager -	- Tangara schrankii	-	-	-	X
Spotted Tanager-	-	- Tangara punctata	-	-	-	X
Speckled Tanager	-	- Tangara guttata-	-	-	-	X
Dotted Tanager -	-	- Tangara varia -	-	-	-	X
Yellow-bellied Tanager -	- Tangara xanthogastra	-	-	-	X
Blue-necked Tanager-	-	- Tangara cyanicollis-	-	-	-	X
Masked Tanager -	-	- Tangara nigrocincta-	-	-	-	X
Turquoise Tanager -	-	- Tangara mexicana	-	-	X - X
"White-bellied" Tanager-	-	- Tangara "brasiliensis" -	-	X	-
Bay-headed Tanager -	-	- Tangara gyrola -	-	-	- X
Chestnut-backed Tanager-	-	- Tangara preciosa	-	-	X - -
E Black-backed Tanager	-	- Tangara peruviana	-	-	X -
Burnished-buff Tanager -	- Tangara cayana -	-	-	X X X
Black-headed Tanager	-	- Tangara cyanoptera -	-	-	- X
Diademed Tanager	-	- Stephanophorus diadematus	-	X X -
Blue-gray Tanager -	-	- Thraupis episcopus -	-	-	- X
Sayaca Tanager -	-	- Thraupis sayaca-	-	-	X X X
E Azure-shouldered Tanager	-	- Thraupis cyanoptera-	-	X X -
E Golden-chevroned Tanager	-	- Thraupis ornata-	-	-	X X -
Palm Tanager	-	- Thraupis palmarum -	-	-	X X X
Blue-and-yellow Tanager-	-	- Thraupis bonariensis	-	-	X - -
Brazilian Tanager -	-	- Ramphocelus bresilius	-	-	-	X -
Silver-beaked Tanager -	- Ramphocelus carbo -	-	-	X X X
Masked Crimson Tanager -	- Ramphocelus nigrogularis	-	-	-	X
Lowland Hepatic-Tanager-	-	- Piranga flava -	-	-	X X X

"Highland" Hepatic-Tanager	- Piranga "lutea"-			X
Summer Tanager	- Piranga rubra			X
Scarlet Tanager-	- Piranga olivacea			X
White-winged Tanager	- Piranga leucoptera			X
Blue-backed Tanager-	- Cyanicterus cyanicterus-			X
E Olive-green Tanager-	- Orthogonys chloricterus-		X	-
Red-crowned Ant-Tanager-	- Habia rubica	X	X	X
Fulvous Shrike-Tanager	- Lanio fulvus			X
White-winged Shrike-Tanager-	- Lanio versicolor			X
White-lined Tanager-	- Tachyphonus rufus	X	X	X
Ruby-crowned Tanager	- Tachyphonus coronatus	X	X	-
Flame-crested Tanager	- Tachyphonus cristatus	X	X	X
"Natterer's" Tanager	- Tachyphonus "nattereri"-	X	-	-
Fulvous-crested Tanager-	- Tachyphonus surinamus	-	-	X
Red-shouldered Tanager	- Tachyphonus phoenicius	X	-	X
Yellow-crested Tanager	- Tachyphonus rufiventer			
White-shouldered Tanager	- Tachyphonus luctuosus	X	-	X
Olive-backed Tanager	- Mitrospingus oleagineus-	-	-	X
Gray-headed Tanager-	- Eucometis (Trichothr.) penicillata	X	-	X
Black-goggled Tanager	- Trichothraupis melanops-	X	X	-
White-rumped Tanager	- Cypsnagra hirundinacea	X	-	X
Chestnut-headed Tanager-	- Pyrrhocoma ruficeps-	X	X	-
E Cherry-throated Tanager-	- Nemosia rourei	Minas G. & E.Sant
Hooded Tanager	- Nemosia pileata-	X	X	X
E Rufous-headed Tanager-	- Hemithraupis ruficapilla	X	X	-
Guira Tanager	- Hemithraupis guira	X	X	X
Yellow-backed Tanager-	- Hemithraupis flavicollis	-	X	X
Orange-headed Tanager-	- Thlypopsis sordida	X	X	X
E Scarlet-throated Tanager	- Compsothraupis(Sericos.)loricata	X	X	-
White-banded Tanager	- Neothraupis fasciata	X	-	X
Black-and-white Tanager-	- Conothraupis speculigera	Amazonas
E Cone-billed Tanager-	- Conothraupis mesoleuca	Mato Grosso
E Brown Tanager	- Orchesticus abeillei	-	X	-
Red-billed Pied Tanager-	- Lamprospiza melanoleuca-	-	-	X
Magpie Tanager	- Cissopis leveriana	X	X	X
E Cinnamon Tanager	- Schistochlamys ruficapillus-	X	X	-
Black-faced Tanager-	- Schistochlamys melanopis-	X	X	X
Buff-throated Saltator	- Saltator maximus	X	X	X
Grayish Saltator	- Saltator coerulescens	X	X	X
Green-winged Saltator	- Saltator similis	X	X	-
Golden-billed Saltator	- Saltator aurantiirostris	X	-	-
Thick-billed Saltator	- Saltator maxillosus-	X	X	-
Black-throated Saltator-	- Saltator atricollis-	X	X	X
Yellow-green (Green) Grosbeak	- Caryothraustes canadensis	-	X	X

	Species	Scientific name					
	Yellow-shouldered Grosbeak	Caryothraustes humeralis	-	-	-	-	X
	Red-and-black Grosbeak	Periporphyrus erythromelas	-	-	-	-	X
	Slate-colored (Slaty) Grosbeak	Pitylus grossus	-	-	-	-	X
	Black-throated Grosbeak	Pitylus fuliginosus	-	-	-	X	X	-
	Yellow Cardinal	Gubernatrix cristata	-	-	-	X	-	-
	Red-crested (Brazilian) Cardinal	Paroaria coronata	-	-	-	X	-	-
E	Red-cowled Cardinal	Paroaria dominicana	-	-	-	-	X	X
	Red-capped Cardinal	Paroaria gularis	-	-	-	-	-	X
E	Crimson-fronted Cardinal	Paroaria baeri	-	-	-	X	-	-
	Yellow-billed Cardinal	Paroaria capitata	-	-	-	X	-	-
	Black-backed Grosbeak	Pheucticus aureoventris	-	-	-	X	-	X
	Blue-black Grosbeak	Cyanocompsa cyanoides	-	-	-	X	-	X
	Ultramarine Grosbeak	Cyanocompsa brissonii (cyanea)	-	-	X	X	-
	Indigo (Glaucous-blue) Grosbeak	Cyanoloxia glaucocaerulea	-	-	X	-	-
	Dickcissel	Spiza americana	-	-	-	-	-	X
	Blue Finch	Porphyrospiza caerulescens	-	-	X	X	X
	Blue-black Grassquit	Volatinia jacarina	-	-	-	X	X	X
	Sooty Grassquit	Tiaris fuliginosa	-	-	-	X	X	X
	White-naped Seedeater	Dolospingus fringilloides	-	-	-	-	X
	Buffy-fronted Seedeater	Sporophila frontalis	-	-	-	X	X	-
	Temminck's Seedeater	Sporophila falcirostris	-	-	-	X	X	-
	Slate-colored Seedeater	Sporophila schistacea	-	-	-	-	X
	Gray Seedeater	Sporophila intermedia	-	-	-	-	X
	Plumbeous Seedeater	Sporophila plumbea	-	-	-	X	X	X
	Wing-barred Seedeater	Sporophila americana	-	-	-	-	X
	Rusty-collared Seedeater	Sporophila collaris	-	-	-	X	X	X
	Lined Seedeater	Sporophila lineola	-	-	-	X	X	X
	Lesson's Seedeater	Sporophila bouvronides	-	-	-	-	X
	Black-and-white Seedeater	Sporophila luctuosa	-	-	-	-	X
	Yellow-bellied Seedeater	Sporophila nigricollis	-	-	-	X	X	X
E	Dubois's Seedeater	Sporophila ardesiaca	-	-	-	X	X	-
E	Hooded Seedeater	Sporophila melanops	-	-	-	Goias
	Double-collared Seedeater	Sporophila caerulescens	-	-	-	X	X	X
E	White-throated Seedeater	Sporophila albogularis	-	-	-	-	X	-
	White-bellied Seedeater	Sporophila leucoptera	-	-	-	X	X	X
	Black-and-tawny Seedeater	Sporophila nigrorufa	-	-	-	X	-	X
	Capped Seedeater	Sporophila bouvreuil	-	-	-	X	X	X
	"Rio de Janeiro" Seedeater	Sporophila "crypta"	-	-	-	X
	Ruddy-breasted Seedeater	Sporophila minuta	-	-	-	-	X
	Tawny-bellied Seedeater	Sporophila hypoxantha	-	-	-	X	-	X
	Dark-throated Seedeater	Sporophila ruficollis	-	-	-	X	-	X
	Marsh Seedeater	Sporophila palustris	-	-	-	X	-	-
	Chestnut-bellied Seedeater	Sporophila castaneiventris	-	-	-	-	X
	Rufous-rumped Seedeater	Sporophila hypochroma	-	-	-	X	-	X

		Col1	Col2	Col3
Chestnut Seedeater - - -	- Sporophila cinnamomea - - -	X	-	-
E Black-bellied Seedeater- -	- Sporophila melanogaster- - -	X	-	-
Large-billed Seed-Finch- -	- Oryzoborus crassirostris - - -	-	-	X
Great-billed Seed-Finch- -	- Oryzoborus maximiliani - - -	X	X	X
Lesser(Chestnut-bell.)Seed-Finch	Oryzoborus angolensis - - -	X	X	X
Blackish-blue Seedeater- -	- Amaurospiza moesta - - -	X	X	-
Paramo Seedeater - - -	- Catamenia homochroa- - - -	-	-	X
Stripe-tailed Yellow-Finch -	- Sicalis citrina- - - -	X	X	X
Orange-fronted Yellow-Finch-	- Sicalis columbiana - - -	X	X	X
Saffron Finch - - - -	- Sicalis flaveola - - -	X	X	X
Grassland Yellow-Finch - -	- Sicalis luteola- - - -	-	-	X
Misto (Southern) Yellow-Finch	- Sicalis luteiventris - - -	X	X	-
Common Duica-Finch - - -	- Diuca diuca- - - - -	X	-	-
{Slaty Finch} - - - -	- {Haplospiza rustica} - - -	-	-	{X}
Uniform Finch - - - -	- Haplospiza unicolor- - -	X	X	-
Coal-crested Finch - - -	- Charitospiza eucosma - -	X	X	X
Red Pileated-Finch(Red-cr.Finch)	Coryphospingus cucullatus - -	X	X	X
Gray Pileated-Finch(Pileated F.)	Coryphospingus pileatus- -	X	X	-
Tepui Brush-Finch - - -	- Atlapetes personatus - . - -	-	-	X
Pectoral Sparrow - - -	- Arremon taciturnus - - -	X	X	X
"Semipectoral" Sparrow - -	- Arremon "semitorquatus"- - -	-	X	-
Saffron-billed Sparrow - -	- Arremon flavirostris - - -	X	-	-
"Gray-backed" Sparrow - -	- Arremon "polionotus" - -	X	-	-
Black-striped Sparrow - -	- Arremonops conirostris - - -	-	-	X
Grassland Sparrow - - -	- Ammodramus (Myospiza) humeralis-	X	X	X
Yellow-browed Sparrow - -	- Ammodramus (Myospiza) aurifrons-	X	-	X
Rufous-collared Sparrow- -	- Zonotrichia capensis - -	X	X	X
Wedge-tailed Grass-Finch -	- Emberizoides herbicola - - -	X	X	X
Lesser(Gray-cheeked) Grass-Finch	Emberizoides ypiranganus - -	X	-	-
Black-masked Finch - - -	- Coryphaspiza melanotis - - -	X	-	X
Long-tailed Reed-Finch - -	- Donacospiza albifrons - -	X	X	-
E Bay-chested Warbling-Finch -	- Poospiza thoracica - - -	X	X	-
Cinereous(Gray-and-white)Warb-F.	Poospiza cinerea - - -	X	X	-
Black-capped Warbling-Finch-	- Poospiza melanoleuca - - -	X	-	-
Black-and-rufous Warbling-Finch-	Poospiza nigrorufa - - -	X	-	-
Red-rumped(Buff-throated)W.-Fin.	Poospiza lateralis - - -	-	X	-
"Gray-throated" Warbling-Finch -	Poospiza "cabansi" - - -	X	-	-
Great Pampa-Finch - - -	- Embernagra platensis - -	X	X	-
E Buff (Pale)-throated Pampa-Finch	Embernagra longicauda - -	X	X	-
Yellow-faced Siskin- - -	- Carduelis (Spinus) yarrellii -	-	X	-
Hooded Siskin - - - -	- Carduelis (Spinus) magellanica -	X	X	X
{Olivaceous Siskin}- - -	- {Carduelis olivacea} - -	-	-	{X}
House Sparrow - - - -	- Passer domesticus - - -	X	X	X
Common Waxbill - - - -	- Estrilda astrild - - -	X	X	X

-234-

Appendix A

The following list consists of species that have at one time been considered to occur within Brazil but have since been withdrawn. Where possible, a coded reference gives the source of this decision for withdrawal.

Black-capped Tinamou
Crypturellus atrocapillus
No confirmed sightings, though is likely to occur in Acre.

Band-rumped Storm-Petrel
Oceanodroma castro
Previous records for Brazil now reidentified as Leach's Storm-Petrels. (T3)

American White Pelican
Pelecanus erythrorhynchos
One, Rio de Janeiro, considered an escape but this is a migrant species. (S16)

Cape Gannet
Morus capensis
With the addition of Australasian Gannet to the Brazilian avifauna, the sight record of Cape Gannet for Rio Grande do Sul (1982) becomes suspect, though it was apparently photographed. (T10)

Red-backed Hawk
Buteo polyosoma
A sight record of a melanistic bird for Rio das Mortes by Pinto and Camargo is considered unconfirmed.

Gyr Falcon
Falco rusticous
A 17th century record in mid Atlantic. (S16)

Common Ringed Plover
Charadrius hiaticula
Shot individual in Maranhão now reidentified as Semi-palmated Plover.

Yellow-sided Parakeet
Pyrrhura hypoxantha
This species is now considered to be an aberrant Green-cheeked Parakeet. (F12)

White-throated Quail-Dove
Geotrygon frenata
Shown in Dunning's *South American Land Birds* as occurring within Brazil but corrected in Dunning's *South American Birds*.

Oilbird
Steatornis caripensis
No specific records within Brazil, though may well occur.

Roraima Nightjar
Caprimulgus whitelyi
No specific records within Brazil, though may well occur.

Sooty Barbthroat
Threnetes niger
Doubtfully a distinct species. Probably best treated as a colour phase of Bronze-tailed Barbthroat. (H7)

Bronze Barbthroat
Threnetes cristinae
A colour variant of Bronze-tailed Barbthroat. (H7)

Black Barbthroat
Threnetes grzmeki
A variant of Rufous-breasted Hermit. (H7)

Green Violetear *Colibri thalassinus*	Listed in Hilty & Brown's *Birds of Colombia* as occurring within Brazil but no specific record has come to light.
Natterer's Emerald *Ptochoptera iolaima*	Probably a hybrid. (S9)
Gould's Woodnymph *Eucephala smaragdocaerulea*	Hybrid. (S9)
Berlioz's Woodnymph *Augasma cyaneoberyllina*	Hybrid. (S9)
Emerald Woodnymph *Augasma smaragdinea*	Hybrid. (S9)
Flame-rumped Sapphire *Hylocharis pyropygia*	Hybrid. (G6) Considered a valid species by Sibley and Monroe. (S9) Five specimens from Bahia.
Short-billed Emerald *Amazilia brevirostris*	A form of Versicolored Emerald which can appear quite distinctive between Espírito Santo and coastal São Paulo, but merges into a more normal Versicolored Emerald plumage the farther inland.
Coppery-chested Jacamar *Galbula pastazae*	Misidentified Bluish-fronted Jacamar. (H1)
Guianan (Arrowhead) Piculet *Picumnus minutissimus*	Not found in Brazil. Previous records are considered here to represent White-wedged Piculet.
Golden-breasted Woodpecker *Colaptes melanolaimus*	The true form of this is not found in Brazil, though intermediaries between this and Green-barred Woodpecker do occur in south Rio Grande do Sul.
Red-fronted Woodpecker *Melanerpes rubrifrons*	Colour variable of Yellow-tufted Woodpecker which can be distinctive in some areas. (S8)
Slaty Spinetail *Synallaxis brachyura*	Misidentified Cinereous-breasted Spinetail. (V1)
Neblina Foliage-gleaner *Philydor hylobius*	Synonym of White-throated Foliage-gleaner. (S9)
Jet Antbird *Cercomacra nigricans*	Misidentification of a poorly preserved specimen which had been shot in Roraima and almost certainly represents *Cercomacra carbonaria*. (F4)

Red-banded Fruiteater *Pipreola whitelyi*	No definite records within Brazil, though suspected for north Roraima and two unconfirmed records for the Manaus area. (W30)
Exquisite Manakin *Pipra exquisita*	This variable form of Blue-crowned Manakin, can be very distinctive but hybridizes widely with the nominate form. (S9)
Sick`s Manakin *Pipra obscura*	Misidentified female and immature male specimens of Golden-crowned Manakin. (C8)
Smoke-colored Pewee Contopus fumigatus	No verified records within Brazil, though has appeared on several Brazilian lists.
Scale-crested Pygmy-Tyrant *Lophotriccus pileatus*	Misidentified Golden-scaled Pygmy-Tyrant and Snethlage`s Tody-Tyrant. (S35)
Bananal Tyrannulet *Serpophaga araguayae*	Misidentified Gray Elaenia. (S9)
Slaty-capped Flycatcher *Leptopogon superciliaris*	Ridgely intimates that the specimen for Brazil was actually a misidentified Black-fronted Tyrannulet.
Pale-vented Thrush *Turdus obsoletus*	Considered here to be split from Hauxwell`s Thrush and therefore not found within Brazil. Hauxwell`s Thrush is considered here to inhabit western Amazonia whilst, its cousin, Cocoa Thrush occurs farther east. (R11)

Appendix B

The following may help to clarify the use of certain names that appear on the checklists. In a few cases I have given a suggested vernacular English name where previously only a latin subspecies name existed.

Plain-breasted Piculet
Picumnus castelnau

On my site list, shown as occurring only close to the Brazilian border but has been recorded away from the site, on the upper Rio Solimões, Amazonas.

Rusty-neced Piculet
Picumnus fuscus

On my site list is shown as occurring close to the Brazilian border, but has occurred before in northwest Mato Grosso.

Banded Piculet
Picumnus a. transfasciatus

Being such a striking form of *P. aurifrons* this subspecies deserves to be recognised as distinctive. (S8) The name "Banded Piculet" seems appropriate.

Lower Amazonian Piculet
Picumnus c. macconnelli

In appearance this northern form of White-barred Piculet is quite distinctive and deserves special mention. (S8) The use of "LowerAmazonian Piculet" seems appropriate on distributional grounds.

Narrow-billed Woodcreeper
Lepidocolaptes angustirostris

I have been unable to ascertain the precise distribution of the various races. Birds in the far south at Uruguaiana, Rio Grande do Sul, are heavily streaked below, whilst those at Santarém, in the far north of their range, are clear below.

Orinocan Woodcreeper
Xiphocolaptes p. orenocensis

This is the race of Strong-billed Woodcreeper found widely in north Brazil (S9), though other races occur farther north, *e.g. X.p.neblinae* and *X.p.tenebrosus* on Cerro de la Neblina and Roraima respectively (M6).

Cipó Canastero
Asthenes luizae

On distributional grounds alone, this would appear to be a separate species. (C8), (F7), (P13) & (V2)

Plumbeous Antvireo
Dysithamnus plumbeus

Previously, Plumbeous Antshrike *Thamnomanes Plumbeous*, but now placed as *Dysithamnus* (S9) and split from the Andean race, becoming an endemic to southeast Brazil. (C7)

Salvadori's Antwren
Myrmotherula minor

Considered here to be an endemic to southeast Brazil (C7). Amazonian records are doubtful and may well refer to Ihering's Antwren. (R3)

Bititinga Antwren
Myrmotherula snowi

Split from *Myrmotherula unicolor*. (C8)

Lower Amazonian Antbird *Cercomacra laeta*	Until a full paper is written, this species is tentatively given this English name for the sake of convenience.
	This species has been split from Dusky Antbird and occurs generally on the south side of the Amazon but the two overlap on the north side at Manaus and in Amapá. R. Bierregaard, M.Cohn-Haft & D.F. Stotz, in press. (W25)
Such`s Antthrush *Chamaeza meruloides*	Confusion has always surrounded this species which has only recently been recognised as a distinct species. Precise distribution of the three *chamaeza* birds in southeast Brazil now becomes vague. Once everyone realises that three species are involved the true picture of their distribution will become apparent. (W27)
Brazilian Antthrush *Chamaeza ruficauda*	Here, this species has been split from the northern form of Rufous-tailed Antthrush, found in Colombia and Venezuela. That form may be called Scalloped or Schwartz`s Antthrush. The southeastern form being the

nominate race should retain the name Rufous-tailed Antthrush. However, since the other form is not located in Brazil, I have used the name Brazilian Antthrush to emphasise that this species should be considered different to the previous use of the name (Rufous-tailed Antthrush) which encompassed both forms. (W27)

Solitary Flycatcher *Myiodynastes solitarius*	Although a valid race of Streaked Flycatcher, it is unclear as to which records refer to which race. Streaked Flycatcher *M. maculatus* migrates south during its winter, whilst Solitary Flycatcher *M.m. solitarius* migrates north during its winter. (S9)
Audible Fuscous-Flycatcher *Cnemotriccus f. bimaculatus*	The nominate, eastern form of *C. fuscatus* is distinct and possibly separate from races that occur throughout most of South America (B7). Being the nominate race the name Fuscous Flycatcher should still apply for that form. An alternative name used here, though admittedly not ideal, to encompass the other races is "Audible Fuscous-Flycatcher".
Upper Amazonian Flycatcher *Tolmomyias f. viridiceps*	Although there are several forms of Yellow-breasted Flycatcher, these can perhaps fall into two distinct groups which may represent separate species (B4). The name "Upper Amazonian Flycatcher" is used here as a viable, though tentative form.

Restinga Tyrannulet *Phylloscartes kronei*	Recent split from *P. ventralis*, proposed by Ed. Willis. The Restinga Tyrannulet is confined to the coastal area from Registro (southern São Paulo) south to Joinville (northeastern Santa Catarina). Mottle-cheeked Tyrannulet is absent from this sandy coastal restinga region. (W28)
Gray-eyed Greenlet *Hylophilus amaurocephalus*	Proposed split from *H. poicilotis*, by Ed. Willis. The two species differ in eye colour, head pattern, beak length and song. Generally speaking, the Rufous-crowned Greenlet is found in the south and along the eastern seaboard as far north as Espirito Santo, while the Gray-eyed Greenlet is found in the inland northeast, normally in a drier habitat. The overlap zone is in southern interior São Paulo. (W26)
Casqued Oropendola *Psarocolius oseryi*	On my site lists this species is shown as occurring close to the Brazilian border. However, it is included on the Brazilian avifauna, having been observed by Andrew Whittaker at Eirunepé, Rio Juruá, Amazonas (1992). (W7)
Rio de Janeiro Seedeater *Sporophila b. crypta*	This distinctive race of Capped Seedeater deserves specific mention. On distributional grounds the name "Rio de Janeiro Seedeater" seems appropriate.
Semipectoral Sparrow *Arremon t. semitorquatus*	This distinctive race of Pectoral Sparrow deserves special mention and the name "Semipectoral Sparrow" appears appropriate.
Gray-backed Sparrow *Arremon f. polionotus*	Two colour forms of Saffron-billed Sparrow occur within Brazil, "Gray-backed Sparrow" seems an appropriate name for the western race which may deserve full status. (S21)

Literature consulted:

A1   Albuquerque, J.L.B.   1985.   *Notes on distribution of some Brazilian raptors.* Bull. B.O.C. 105 (3).

A2   Albuquerque, J.L.B.   1986.   *Conservation & Status of Raptors in Southern Brazil.* Birds of Prey Bull No. 3.

A3   Alden, P. & J.Gooders.   1981.   *Finding Birds Around the World.*

A4   Allen, J.A.   1891- 1893.   *On a collection of birds from Chapada, Matto Grosso, Brasil, made by Mr Herbert Smith.* Bull. Am. Mus. Nat. Hist. Vol 3 - Vol 5.

A5   Altman, A. & B. Swift.1986. *Checklist of the Birds of South America.*

A6   Alvarenga, H.M.F.   1990.   *Novos registros e expansões geográficas de aves no leste do estado de São Paulo.* Ararajuba 1.

A7   Anderson, W.   1987.   *Bird sightings at various locations in Brazil between 1981 & 87.* (Unpub.)

A8   Antas, P. de T. Z.   1989.   *Lista das aves observadas no Parque Nacional de Brasilia.* (Unpub.)

A9   Antas, P.deT.Z., A. Fillippini & S.M. e Azevedo Jr. 1990. *Novos registros de aves para o Brasil.* Resumos, VI Encontro Nacional de Anilhadores de Aves, Pelotas.

A10   Antas, P.deT.Z. 1990.   *Novos registros para a avifauna do Rio Grande do Sul.* Resumos, VI Encontro Nacional de Anilhadores de Aves, Pelotas.

B1   Baker, G.   1989.   *Southern Brazil.* (Unpub.)

B2   Barrowclough, G. F. & P. Escalante-Pliego. 1990. *Notes on the birds of the Sierra de Unturán, southern Venezuela.* Bull. B.O.C. 110 (4).

B3   Bates. J.M. & M. C. Garvin, D. C. & C. G. Schmitt. 1989. *Notes on bird distribution in northeastern Dpto Santa Cruz, Bolivia, with 15 species new to Bolivia.* Bull. B.O.C. 109 (4).

B4   Bates, J.M., T.A. Parker III., A.P. Capparella and T.J. Davis.   1992. *Observations on the campo, cerrado and forest avifaunas of eastern Dpto. Santa Cruz, Bolivia, including 21 species new to the country.*   Bull. B.O.C. 112 (2).

B5   Bege, L.A.R. & B.T. Pauli. 1990. *Two birds new to the Brazilian Avifauna.* Bull. B.O.C. 110 (2).

B6   Belton, W.   1984.   *Birds of Rio Grande do Sul, Brazil. Part I.* Bull.Am.Mus.Nat.Hist. 178 (4).

B7   Belton, W.   1985.   *Birds of Rio Grande do Sul, Brazil. Part II.* Bull.Am.Mus.Nat.Hist. 180 (1).

B8   Brazil, M. & D.R.Waugh. 1986. *Personal sightings in Brazil.* (Unpub.)

B9   Byrne, R.W.   1988.   *Brazil.* (Unpub.)

C1   Carnevalli, N. & G.T de Mattos. - *Lista de Aves do Parque Florestal do Rio Doce.* (Unpub.)

C2   Chapman F. M. 1931   *Upper Zonal Bird-life of Mts Roraima & Duida.* Bull. Am. Mus. Nat. Hist. Vol 63.

C3   Cintra, R. - Personal Sightings in Brazil. (Unpub.)

C4   Clube de Observadores de Aves 1989. *Listagem da Serra do Caraça.*(Unpub.)

C5   Collar, N.J., M da Fonseca, L.A.P. Gonzaga, P.J.Jones, J.F.Paceco, D.A.Scott, P. Sergio.(A Compilation of reports)   - *Aves Observadas na Reserva Florestal da C.V.R.D. - Linhares.* (Unpub.)

C6   Collar, N.J.,L.A.P. Gonzaga & D.A. Scott. *1987. The Status and Birds of Some Forest Fragments in Eastern Brazil.* (Unpub.)

C7   Collar, N.J. & P. Andrew.   1988. *Birds to Watch. The I.C.B.P. World Checklist of Threatened Birds.*

C8 Collar, N.J., L.A.P. Gonzaga, N. Krabbe, A. Madroño Nieto, L.G. Naranjo, T.A. Parker III., and D.C. Wege.1992. *Threatened birds of the Americas: The I.C.B.P. Red Data Book (Third edition, part 2).*

C9 Coopmans, P. 1991. *Bird sightings in Brazil.* (Unpub.)

C10 Cory, C.B., C.E. Hellmayr & B. Conover. 1918-1949. *Catalogue of Birds of the Americas.* Field Mus.Nat.Hist.XIII, 14 volumes.

D1 Davis, T.J. & J.P. O'Neill. 1986. *A new species of antwren (Formicariidae: Herpsilochmus) from Peru, with comments on the systematics of other members of the genus.* Wilson Bull. Vol 98. No.3.

D2 Deignan, H.G. 1961. *Type specimens of birds in the United States National Museum.* Smithsonian Institution Bull.221.

D3 Diamond, A.W. & T.E.Lovejoy. 1985. *Conservation of Tropical Forest Birds.*

D4 Dickerman, R.W. & W.H. Phelps Jr. 1982. *An annotated list of the birds of Cerro Urutaní on the border of Estado Bolívar, Venezuela, and Territorio Roraima, Brasil.* Am. Mus. Novitates:Bull. Am. Mus. Nat. Hist. June 82. No. 2732.

D5 Dickerman, R. W. 1988. *A review of the Least Nighthawk (Chordeiles pusillus).* Bull. B.O.C. 108 (3).

D6 Dobbs, G. *et al.*1989. *Brazil* (Unpub.)

D7 Dubs, B. 1983. *Vögel des südlichen Mato Grosso.*

D8 Dubs, B. 1992. *Birds of Southwestern Brazil.*

F1 Fairbank, R. & N. Preston. 1988.*Personal notes of birds seen in Brazil.* (Unpub.)

F2 Filippini, A. 1989 *Birds of Fernando de Noronha.* (Unpub.).

F3 Finch, D. - *Birds seen on the WINGS Tours to Brazil 1987 - 1990.* (Unpub.)

F4 Fitzpatrick, J.W. & D.E. Willard. 1990. *Cercomacra manu, a New Species of Antbird from Southwestern Amazonia.* Auk Vol 107. No 2.

F5 Fonseca, P.S.M da & J.F. Pacheco. 1989. *Aves Observadas na Pousada do Vale dos Veados (Parque Nacional da Bocaina).* (Unpub.).

F6 Forrester, B.C. 1987. *Brazil: July-August.* (Unpub.)

F7 Forrester, B.C. 1988. *Brazil II* (Unpub.)

F8 Forrester, B.C. 1989. *Brazil III* (Unpub.)

F9 Forrester, B.C. 1990. *Brazil IV* (Unpub.)

F10 Forrester, B.C. 1991. *Brazil V* (Unpub.)

F11 Forrester, B.C. 1992. *Brazil VI* (Unpub.)

F12 Forshaw, J.M. & W.T. Cooper. 1978 (1989) *Parrots of the World.*

F13 Friedmann, H. 1948. *Birds collected by the National Geographic Society's Expeditions to northern Brazil and southern Venezuela.* Proc. U.S. Nat. Mus. 97.

G1 Gardner, N & D. 1990. *Birding Trip to Brazil.* (Unpub.)

G2 Gonzaga, L.A.P. 1988. *A new antwren (Mymotherula) from South-eastern Brazil.* Bull. B.O.C. 108(3)

G3 Gonzaga, L.A.P. - *Birds seen at Nova Lombardia Biological Reserve.* (Unpub.)

G4 Gonzaga, L.A.P. & J.F.Pacheco. 1990. *Two new subspecies of Formicivora serrana (Hellmayr) from southeastern Brazil, and notes on the type locality of Formicivora deluzae Ménétriés.* Bull. B.O.C. 110 (4).

G5 Goodwin, D. - *Personal notes on birds seen in Brazil.* (Unpub.)

G6 Grantsau, R. 1988. *Os Beija-flores do Brasil.*

G7 Grantsau, R. e H.F.de A. Camargo. 1989. *Nova Espécie Brasileira de Amazona (Aves, Psittacidae).* Rev. Brasil. Biol.,49 (4).

G8 Graves, G.R. & R.L. Zusi. 1990. *Avian body weights from lower Rio Xingu, Brazil.* Bull. B.O.C. 110 (1).
G9 Greensmith, A. 1979. *Personal notes in Brazil.* (Unpub.)
G10 Griscom, L. & J.C. Greenway. 1941. *Birds of Lower Amazonia.* Bull Mus Comp Zoöl. 88 (3).
G11 Groen, F. van, 1986. *"Excursion Report 13", Birds of Vaupés (Amazonia, Colombia).*
G12 Gyldenstolpe. N. 1945. *The bird fauna of Rio Juruá in Western Brazil.* Kungl. Svenska. Vetenska.Handl. Ser.3. Band 22 (3).
G13 Gyldenstolpe. N. 1945. *A contribution to the ornithology of Northern Bolivia.* Kungl. Svenska. Vetenska. Handl. Ser.3. Band 23 (1).
G14 Gyldenstolpe, N. 1951. *The ornithology of the Rio Purús region in Western Brazil.* Ark. Zool. Ser. 2. Band 2 (1).
H1 Haffer, J. 1974. *Avian Speciation in Tropical South America with a systematic survey of the toucans (Ramphastidae) and jacamars (Galbulidae).* Publ. Nuttall Orn. Club No 14. Cambridge Mass.
H2 Hardy, J.W., B. Coffey Jr., & G. Reynard. 1990. *Voices of the New World Nightbirds* (cassette tape). Florida State Museum.
H3 Hellmayr, C.E. 1910. *The Birds of the Rio Madeira.* Novit. Zool XVII (3).
H4 Hellmayr, C.E. 1929. *A contribution to the ornithology of northeastern Brazil.* Field Mus.Nat.Hist. Zool.Ser 12. No 18.
H5 Hilty, S.L. & W.L.Brown. 1986. *A guide to the birds of Colombia.*
H6 Hinkelmann, C. 1988. *On the identity of Phaethornis maranhaoensis Grantsau 1968 (Trochilidae).* B.O.C. 108(1)
H7 Hinkelmann, C. 1988. *Comments on recently described new species of hermit hummingbirds.* B.O.C. 108 (4).
H8 Holt, E. G. 1928. *An ornithological Survey of the Serra do Itatiaya, Brazil.* Bull. Am. Mus. Nat. Hist. 57 (5).
I1 I.B.D.F.- *Lista preliminar de aves do Parque Nacional do Araguaia.*
I2 Ihering, H. von 1900. *Aves observadas em Cantagallo e Nova Friburgo.* Rev. Mus. Paul.IV.
I3 Ireland, C. 1987. *Personal notes on Brazil.* (Unpub.)
I4 Ireland, C. & T. Ford. 1988. *Personal notes on Brazil.* (Unpub.)
I5 Isler, M. L. & P.R. 1987. *The Tanagers.*
I6 Isler, M. L. & P.R. 1989. *Personal sightings in Brazil.* (Unpub.)
I7 I.U.C.N. 1982. *IUCN Directory of Neotropical Protected Areas .*
J1 Johns, A 1985. *Effects of Habitat Disturbance on Rainforest Wildlife in Brazilian Amazonia. 1984/85.* Project U.S. 302 - W.W.F.
J2 Joseph, L. 1988. *Range Extension of the Red-fan Parrot Deroptyus accipitrinus in Amazonian Brazil.* B.O.C. Vol 108 (3).
J3 Juniper, T. 1990. *A Very Singular Bird.* BBC. Wildlife. Vol.8. No 10.
K1 Kelsey, M. *et al.*1988. *Birds of Amaca Yacú National Park, Colombia.*(Unpub.)
K2 Kessler,M. 1989. *Personal sightings in Brazil (Unpublished).*
K3 King, W.B. 1981. *Endangered Birds of the World: The I.C.B.P. Bird Red Data Book.*
K4 Klein, B. *et al.* 1988. *The Nesting and Feeding Behaviour of the Ornate Hawk-Eagle near Manaus, Brazil.* Condor Vol 90 (1).

K5   Krabbe, N.   -   *A Collection of Birds from Southeastern Brazil 1825-1855. Based on a collection made by P.W. Lund & J. Reinhardt.*

K6   Krannitz, P. 1983. *List of birds seen at Reserva Biológica do Trombetas.* (Unpub.)

L1   Lamm, D. W. 1948. *Notes on the birds of the states of Pernambuco & Paraiba, Brazil.* Auk, 65.

L2   Lanyon, S.M., D.F. Stotz & D.E. Willard. 1990. *Clytoctantes atrogularis, A New Species of Antbird from Western Brazil.* Wilson Bull.102 (4).

L3   Lanyon, W.E. & C.H. Fry 1973. *Range and affinity of the Pale-bellied Mourner (Rhytipterna immunda).* Auk 90.

L4   Lanyon, W.E. 1978. *Revision of the Myiarchus Flycatchers of South America.* Bull.Am.Mus.Nat.Hist. Vol161 Art4.

L5   Lovejoy, T.E. 1974. *Bird diversity and abundance in Amazon forest communities.* Living Bird (13).

M1   Macdonald, D. 1989. *Bird sites in Brazil.* (Unpub.)

M2   Massie, D & N. 1992. *Brasil Species - a trip report.* (Unpub.)

M3   Mayr, E. & W.H. Phelps Jr. 1967. *The origin of the bird fauna of the South Venezuelan highlands.* Bull. Am. Mus. Nat. Hist. Vol 136 Art 5.

M4   Mayr, E. & F. Vuilleumier. 1983. *New species of birds described from 1966 to 1975.* Jo.Orn. 124.

M5   Meyer de Schauensee, R.   1966. *The species of birds of South America and their distribution.*

M6   Meyer de Schauensee, R., & W.H. Phelps, Jr.   1978. *A guide to the birds of Venezuela.*

M7   Meyer de Schauensee, R. 1970 (1990). *A guide to the birds of South America.*

M8   Midgley, M & M.C.   1990. *Bird sites of Brazil.* (Unpub.)

M9   Mitchell, M.H. 1957. *Observations on Birds of Southeastern Brazil.*

M10   Mølgaard, E. & J.Meedom.   1991. *Birdwatching in Brazil.* (Unpub.)

M11   Morrison R & R. Ross, 1989. *Atlas of nearctic shorebirds on the coast of South America.* Canadian Wildlife Service Special Publication.

M12   Moskovits D.K., J.W. Fitzpatrick & D.E. Williard. 1985 . *Lista preliminar das aves da Estação Ecológica de Maracá, Território de Roraima, Brasil, e áreas adjacentes.* Pap. Avuls.,Zool.(São Paulo) 36 (6).

M13   Moss, D.   1977. *Species seen on the 1976 / 77 University of Edinburgh Expedition to Brazil (to study termite mounds).* (Unpub.)

M14   Munn, C.A. & J.B. Thomsen.   1987. *Report on Carajas Region* (Unpub.).

N1   Nacinovic, J.B. & D.M. Teixeira. 1989. *As aves de Fernando de Noronha: Uma lista sistematica anotada.* Rev.Brasil.Biol. 49 (3).

N2   Naumburg,   E.M.B.   1930. *The Birds of the Matto Grosso, Brazil.* Bull. Am. Mus. Nat. Hist. 60.

N3   Naumburg, E.M.B.   1935-1939. *Studies of Birds from eastern Brazil and Paraguay, based on a collection made by Emil Kaempfer.* Bull. Amer. Mus. Nat. Hist. Vols. 68 (6), 74 (3) & 76 (6).

N4   Negret, A., R. Soares, J. Taylor, R.Cavalcanti & C. Johnson. 1984. *Aves da região geopolitica do Distrito Federal* (S.E.M.A. publication).

N5   Nicolle, S.   1991. *Personal Sightings in Brazil.* (Unpub.).

N6   Nisbet, I.C.T.   1984. *Migration and wintering quarters of North American Roseate Terns as shown by banding recoveries.* Journ. Field Orn. Vol 55. No 1.

N7   Novaes, F.C.   1947. *Notas sôbre os Conopophagidae do Museu Nacional (Passeriformes, Aves).* Summa Brasiliensis Biologiae. Ano II, Vol 1.

N8   Novaes, F.C.   1952. *Algumas adendas a orntologia de Goiás, Brasil.* Bol. Mus. Nac. (new Series) No.117.

N9   Novaes, F.C.   1952. *Resultados ornitológicos da "Expedição João Alberto" à Ilha da Trindade.* Rev. Brasil. Biol. 12 (2).

N10   Novaes, F.C.   1957. *Contribuição à ornitologia do noroeste do Acre.* Rev. Brasil. Biol. 9.

N11   Novaes, F.C. 1965. *Notas sôbre algumas aves da serra Parima, Território de Roraima (Brasil).* Bol. Mus. Par.Emi. Goeldi. Zool. No.54.

N12   Novaes, F.C.   1967. *Sobre algumas aves pouco conhecidas na Amazônica Brasileira.* Bol.Mus.Par.Emi.Goeldi.Zool. No 64.

N13   Novaes, F.C.   1969. *Análise ecológica de uma avifauna da região do Rio Acara, estado do Pará .* Bol. Mus. Par. Emi. Goeldi. Zool. No.69.

N14   Novaes, F.C. 1970. *Distribuição ecológica e abundancia das aves em um trecho da mata do baixo Rio Guamá (estado do Pará).* Bol.Mus.Par.Emi.Goeldi.Zool. No. 71.

N15   Novaes, F.C.& T. Pimentel. 1973. *Observações sobre a avifauna dos campos de Bragança, estado do Pará.* Separata do o Museu Goeldi no ano do Sequicetenario. Publicações Avulsas 20.

N16   Novaes, F.C. 1974 & 78. *Ornitologia de território do Amapá I & II.* Publ. Avul. Mus. Goeldi 25: 1-121 & 29: 1-75.

N17   Novaes, F.C.   1981. *Sobre algumas aves do litoral do estado do Pará.* Sociedade Sul-Riograndense de Ornitologia II.

O1   Oniki, Y. & E.O. Willis.   1972. *Studies of ant-following birds north of the eastern Amazon.* Acta Amazônica Vol 2. No.2.

O2   Oniki, Y. & E.O.Willis. 1982-82. *Breeding records of birds from Manaus.* Rev.Brasil.Biol.42 (4) & 43 (1).

O3   Oniki, Y. & E.O.Willis. 1983-84. *A study of breeding birds of the Belém area, Brazil.* I-VII. Ciene. Cult. (São Paulo).

04   Oren, D.C.   1982. *A avifauna do arquipélago de Fernando de Noronha.* Bol. Mus. Par. Emilio Goeldi. Zool. No.118.

05   Oren, D.C.   1984. *Resultados de uma nova expedição zoológica a Fernando de Noronha.* Bol.Mus.Par.Emilio Goeldi. Zool. 1 (1).

06   Oren, D.C. & F.C. Novaes. 1985. *A new subspecies of White Bellbird, Procnias alba (Hermann) from Southeastern Amazonia.* B.O.C. 105 (1).

07   Oren, D.C. & J.M.C. da Silva.   1987. *Cherrie`s Spinetail (Synallaxis cherriei gyldenstolpe) in Carajas and Gorotire, Pará, Brasil.* Bol. Mus. Para. Emilio Goeldi, Ser. Zool. 3 (1).

P1   Pacheco, F.   1988. *Black-hooded Antwren Formicivora [Myrmotherula] erythronotos re-discovered in Brazil.* Bull. B.O.C. 108 (4).

P2   Padua, M.T.J, et al.   1982. *Plano do sistema de unidades de conservação do Brasil.* (Produced jointly by MdA / IBDF / FBCN.).

P3   Parrrish, C.   -   *Leticia, Dept Amazonas, Colombia.* (Unpub.)

P4   Parker, T.A., III.   1985. *A preliminary list of the birds and mammals of Parque Nacional da Amazônia, Pará, Brazil.* (Unpublished).

P5   Parker, T.A., III & J.V. Remsen. 1987. *Fifty-two Amazonian bird species new to Bolivia.* Bull B.O.C. 107 (3).

P6   Parker, T.A.,III. 1989. *Birds of Three Brazilian National Parks and Three Brazilian Biological Reserves.* (Unpub.)
P7   Parker, T.A.,III. 1991. *Personal notes on some birds in Brazil.* (Unpub.)
P8   Parker, T.A., III. & O. Rocha O. 1991. *Notes on the status & behaviour of the Rustynecked Piculet Picumnus fuscus.* Bull. B.O.C. 111 (2).
P9   Parker, T.A.,III, A.Castillo U., M. Gell-Mann & O. Rocha O. 1991. *Records of new and unusual birds from northern Bolivia.* Bull. B.O.C. 111 (3)
P10   Pearman, M. 1987. *Species seen around Tabatinga and Benjamin Constant.* (Unpub.)
P11   Pearman, M. 1988. *Annotated Checklist of Bird Observations in Brazil.* (Unpub.)
P12   Pearman, M. 1989. *Personal sightings in Brazil* (Unpub.)
P13   Pearman, M. 1990. *Behaviour and vocalizations of an undescribed canastero Asthenes sp. from Brazil.* Bull B.O.C. 110 (3)
P14   Pelzeln, A.von., 1868. *Zur Ornithologie Brasiliens.Resultate von Johann Natterers Reisen in den Jahren 1817 bis 1835.*
P15   Peres, C.A.   - *Conservation of Western Amazonian Primate Communities. Report to the W.W.F.-U.S.* (Unpub).
P16   Peres. C.A.   1990. *A Harpy Eagle successfully captures an adult male red howler monkey.* Wilson Bull 102 (3).
P17   Peres. C.A. & A. Whittaker. 1991. *Annotated checklist of the bird species of the upper Rio Urucu, Amazonas, Brasil.* Bull. B.O.C. 111 (3).
P18   Peres, C.A. & B. Santos. - *Effects of selective understory cutting on forest bird communities in southern Bahia. (Unpub.)*
P19   Peters, J.L.   1931- 1986. *Check-list of the Birds of the World.*   Cambridge, Mass: Harvard University Press.
P20   Petry, M.V., G.A. Bencke & G.N. Klein. 1991.   *First record of the Shy Albatross Diomedea cauta for the Brazilian coast.* Bull. B.O.C. 111 (4).
P21   Phelps, W.H. 1939. *Geographical status of the birds collected at Mount Roraima.* Bol. Soc. Venez. Cien. Nat. 36.
P22   Phelps, W.H.   1962. *Cuarentinueve aves nuevas para la avifauna brasileña del Cerro Uei-tepui (Cerro del Sol).*Bol. Soc. Venez. Cien. Nat. 101.
P23   Phelps, W.H. & W.H. Phelps, Jr. 1965. *Lista de las aves del Cerro de la Neblina, Venezuela et notas sobre su descumbrimiento y ascenso.* Boletín Sociedad Venezolana Ciencias Naturales. Vol 26. No.109.
P24   Pierpont, N. & J.W. Fitzpatrick. 1983. *Specific status & behaviour of Cymbilaimus sanctaemariae, the Bamboo Antshrike, from southwestern Amazonia.*   Auk 100.
P25   Pineschi, R.B. 1990.   *Aves como dispersores de sete espécies de Rapanea (Myrsinaceae) no maciço do Itatiaia, estados do Rio de Janeiro e Minas Gerais.* Ararajuba 1.
P26   Pinto, O.M. de O.   1935.   *Aves da Bahia.* Rev. Mus. Paulista. Vol 19.
P27   Pinto, O.M. de O.   1937.   *Nova contribuição á ornitologia amazônica.* Rev. Mus. Paulista. Vol 23.
P28   Pinto, O.M. de O.1938. *Catálogo das aves do Brasil (1a parte).* Rev. Mus. Paulista 17 (2): 641-708 1944. *Catálogo das aves do Brasil (2a parte)* São Paulo: Secretario Agricultura.
P29   Pinto, O.M. de O.   1947   *Contribuição á ornitologia do baixo Amazonas.* Arq Zool (São Paulo) Vol 5 Art 6.

P30 Pinto, O.M.de O. 1948. *Notas e impressoes naturalísticas de uma viagem fluvial a Cuiabá.* Bol.Mus.Par.Emi.Goeldi. Vol X.

P31 Pinto, O.M. de O. 1949. *Esboço monográfico dos Columbidae Brasileiros.* Arq Zool (Sao Paulo) Vol 7 Art 3.

P32 Pinto, O.M. de O. 1952. *Sumula histórica e sistemática da ornitologia de Minas Gerais.* Arq. Zool.(Sao Paulo). Vol 8.

P33 Pinto, O.M. de O. 1953. *Sobre a coleção Carlos Estevão de peles, ninhos e ovos de aves de Belém (Pará).* Pap. Avuls. Zool. (São Paulo). Vol 11.

P34 Pinto, O.M. de O. 1954. *Parque Nacional do Itatiaia.* Ministerio da Agricultura Boletim No 3.

P35 Pinto,O.M. de O. & E.A. de Camargo. 1954. *Resultados ornitológicos de uma expedição ao Territorio do Acre pelo Departamento de Zoologia.* Pap. Avuls. Zool. (Sao Paulo) 11.

P36 Pinto, O.M de O. 1954. *Resultados ornitológicos de duas viagems científicas ao estado de Alagoas.* Pap. Avuls. Zool.(Sao Paulo) 12: 1-98.

P37 Pinto, O.M. de O. 1964. *Ornitologia Brasiliense.*

P38 Pope, N.J.N. 1989. *Brazil.* (Unpub.)

R1 Remsen, J. V. Jr. 1978. *Birding the Amazon at Leticia.* Birding Vol.10, No. 6.

R2 Remsen, J.V. Jr., C.G. & D.C Schmitt. 1988. *Natural history notes on some poorly known Bolivian birds. Part 3.* Le Gerfaut. Vol 78 (4).

R3 Remsen, J.V. Jr. & M.A. Traylor, Jr. 1989. *An Annotated List of the Birds of Bolivia.*

R4 Remsen. J.V. Jr. & T.A.Parker III. 1990. *Seasonal distribution of the Azure Gallinule (Porphyrula flavirostris)with comments on vagrancy in Rails and Gallinules.* Wilson Bull 102 (3).

R5 Resende, S.de M.L. 1985. *Species seen at Barra Lagoa do Peixe, Rio Grande do Sul -* Project 3416 W.W.F.

R6 Richards, G. 1989. *Bird List for Brazilian Sites.*(Unpub.)

R7 Ridgely, R.S. 1981. *List of species seen at Araguaia/Bananal region Sept/Oct 1979-81.* (Unpub.).

R8 Ridgely, R.S. 1983. *Hyacinth Macaw and Brazil's Pantanal.* Birding Vol 15. No 4/5.

R9 Ridgely, R.S. & T.S.Schulenberg.1987.*Amazonian Brazil Tour for VENT.* (Unpub.)

R10 Ridgely, R.S. 1988. *Birds seen on Southeastern Brazil Tour.* (Unpub.).

R11 Ridgely, R.S. & G.Tudor 1989. *The Birds of South America (Vol. 1. The Oscine Passerines).*

R12 Ridgely, R.S. 1990. *Fazenda Rancho Grande Area, Rondônia.* (Unpub.).

R13 Ridgely, R.S., V. Emanuel & A. Whittaker. 1990. *Brazil Specialties Tour, Oct 1990.* (Unpub.)

R14 Riker, C.B. 1890-91. *A list of birds observed at Santarém, Brazil.* Auk 7 & 8.

R15 Roever, J. de 1978. *Birds of Rondônia and adjacent Bolivia.* (Unpub.)

R16 Roever, J. de & C. de Roever. 1988. *Sites in Brazil.* (Unpub.)

R17 Rowlett, R.A. 1984. *Species seen on VENT tour of Central Brazil Sept/Oct 1984.* (Unpub.)

R18 Rumbolt, M. - *Lista de las Aves del Parque National Iguazú.*

R19 Ruschi, A. 1979 - 82. *Aves do Brasil.*

R20 Ruschi, A. 1982. *Beija-Flores do Estado do Espírito Santo.*

S1 Schofield, R. 1990. *Birdquest Bird Tour List.* (Unpub.)

S2   Schulenberg, T.S. & D.F. Stotz. 1991. *The Taxonomic Status of Myrmeciza stictothorax (Todd)*. Auk 108 No.3.
S3   Schneider, A . & H. Sick. 1962. *Sôbre a distribuição de algumas aves do sudeste do Brasil segundo coleções do Museu Nacional*. Bol.Mus. Nac. Nova Ser. Zool. 239.
S4   Scott, D.A. & M. de L. Brooke. 1985. *The endangered avifauna of southeastern Brazil: A report on the BOU/WWF expeditions of 1980/81 and 1981/82*. In *Conservation of Tropical Forest Birds* by A.W. Diamond & T.E. Lovejoy.
S5   Scott, D.A. & M. Carbonell.   1986.   *A Directory of Neotropical Wetlands*.
S6   Scott, D.A.   -   *Personal observations in Brazil*. (Unpub.)
S7   Short, L.L.   1975. *A zoogeographic analysis of the South American chaco Avifauna*. Bull. Amer. Mus. Nat. Hist. 154 (3).
S8   Short, L.L.   1982. *Woodpeckers of the World*.
S9   Sibley, C.G. & B.L. Munroe,Jr. 1990. *Distribution and Taxonomy of Birds of the World*.
S10   Sick, H. 1960. *The Honeycreeper Dacnis albiventris in Brazil*. Condor Vol 62.
S11   Sick, H. 1968. *As Aves do Rio de Janeiro (Guanabara)*. Arquivos do Museu Nacional, Rio de Janeiro. Vol 53.
S12   Sick, H. & R.S. Ridgely. 1977. *Species seen at Monte Pascoal National Park*. (Unpub.)
S13   Sick, H. 1979. *Notes on Some Brazilian Birds*. B.O.C. Vol 99 (4).
S14   Sick, H. & D.M. Teixeira.   1979. *Notas sobre aves brasileiras raras ou ameaçadas de extinção*. Publ. Avuls. Mus. Nat. Rio de J. 62.
S15   Sick, H. & D.M. Teixeira. 1980. *Discovery of the home of the Indigo Macaw in Brazil*. Amer. Birds. Vol 34, No 2.
S16   Sick, H.1984. *Ornitologia Brasileira*.
S17   Sick, H.1990. *Notes on the taxonomy of Brazilian parrots*.   Ararajuba 1.
S18   Silva, J.M.C. da 1987. *Estudos ornitológicos na estação ecológica de Maracá - primeiro relatorio de pesquisas*. Report for Royal Geographical Society.
S19   Silva, J.M.C. da, et al. 1988. *Aves da Estação Ecológica de Maracá*. (Unpub.)
S20   Silva, J.M.C. da & D. C. Oren. 1990. *Introduced & invading birds in Belém, Brazil*. Wilson Bull 102. (2).
S21   Silva, J.M.C. da 1991. *Geographical variation in the Saffron-billed Sparrow Arremon flavirostris*. Bull. B.O.C. 111 (3).
S22   Simon, H.   1989.   *Relação de passaros observados recentemente em Itatiaia*. (Unpub.)
S23   Small, A.   1986.   *Species seen at Emas National Park*. (Unpub.)
S24   Snethlage, E. 1914. *Catálogo das aves amazônicas*. Bol. Mus. Par. Emi. Goeldi. 8.
S25   Snow, D.W. 1982. *The Cotingas*.
S26   Stager, K.E. 1961.*The Machris Brazilian Expedition*. Cont. in Science No. 41.
S27   Stemple, D.   -   *Brazilian Bird List*. (Unpub.)
S28   Sterry, P.   1989 *Systematic list of species seen around Manaus*.(Unpub.)
S29   Sterry, P.   1991. *The Birds of Tefe*. (Unpub.)
S30   Stone, W. & H.R. Roberts.   1934. *Zoological results of the Matto Grosso Expedition to Brazil in 1931: III Birds*. Proc. of the Acad. of Nat. Sc. of Philadelphia. Vol 86.
S31   Storer, R.W.   1981. *The Rufous-faced Crake (Laterallus xenopterus) and its Paraguayan Congeners*. Wilson Bull. 93 (2).

S32   Stotz, D.F.   -   *Personal notes in Brazil.* (Unpub.)

S33   Stotz, D.F.   1987. *List of birds seen on Ilha Maracá for Royal Geographical Society.* (Unpub.)

S34   Stotz, D.F. 1990.*The taxonomic status of Phyllomyias reiseri.* Bull.B.O.C. 110 (4).

S35   Stotz, D.F.   1990. *Corrections and Additions to the Brazilian Avifauna.* Condor 92.

S36   Stotz, D.F., R.O. Bierregaard, M. Cohn-Haft, P. Petermann, J. Smith, A. Whittaker and S.V. Wilson.   1992. *The Status of North American Migrants in Central Amazonian Brazil.*   Condor 94.

S37   Straube, F.C.   1990. *Conservação de aves no litoral-sul do estado do Paraná (Brasil).* Arq. Biol. Tecnol. 33(1).

S38   Studer, A. & J. Vielliard.1985.   *Lista preliminar das espécies de aves de Quebrângulo e Palmeira dos Indios (Alagoas, Brasil).* (Unpub.)

T1   Teixeira, D.M & L.P.Gonzaga. 1983. *Um novo Furnariidae do nordeste do Brasil: Philydor novaesi sp. nov.* Bol.Mus. Par.Emi.Goeldi.Zool. No 124.

T2   Teixeira, D.M & A. Negret. 1984.*The Dwarf Tinamou of Central Brazil.* Auk 101.

T3   Teixeira, D.M., J.B.Nacinovic & R. Novelli. 1985. *Notes on some Brazilian Seabirds (1).* Bull. B.O.C. 105(2).

T4   Teixeira, D.M, *et al.*   1985. *Bird species observed in northeastern Brazil:* Project 3226 for W.W.F.

T5   Teixeira, D.M. D.C. Oren & R.Best. 1986. *Notes on Brazilian seabirds.(2).* Bull. B.O.C.106 (2).

T6   Teixeira, D.M. & J.B. Nacinovic & M. S.Tavares. 1986. *Notes on some birds of northeastern Brazil.* Bull. B.O.C. 106 (2).

T7   Teixeira, D.M.   1987. *A new Tyrannulet (Phylloscartes) from northeastern Brazil.* Bull.B.O.C.107 (1).

T8   Teixeira, D.M., J.B. Nacinovic & F.B. Pontual. 1987. *Notes on some birds of northeastern Brazil (2).* Bull. B.O.C. 107 (4).

T9   Teireira,D.M., J.B. Nacinovic & G.Luigi.   1988. *Notes on some birds of northeastern Brazil (3).* Bull. B.O.C. 108 (2).

T10   Teixeira, D.M., J.B. Nacinovic, I. Schloemp & E. Kischlat. 1988. *Notes on some Brazilian seabirds (3).* Bull. B.O.C. 108 (3).

T11   Teixeira, D.M., J.B.Nacinovic & G. Luigi. 1989. *Notes on some birds of north-eastern Brazil (4).* Bull. B.O.C. 109 (3).

T12   Teixeira, D.M.   -   *Personal sightings in Brazil.* (Unpub.)

T13   Thomson, A.L. 1965. *The Transequatorial Migration of the Manx Shearwater (Puffin des Anglais).* L'Oiseau 35.

T14   Traylor,M.A,Jr. 1982.*Notes on Tyrant Flycatchers (Aves:Tyrannidae)* Fieldiana Zoology. New Series No 13.

T15   Trent, D.   -   *Personal sightings in Brazil.* (Unpub.)

V1   Vaurie, C.   1980. *Taxonomy and Geographical Distribution of the Furnariidae.* Bull. Am. Mus. Nat. Hist. Vol 166. Art 1.

V2   Vielliard, J.   1990. *Uma nova espécie de Asthenes da serra do Cipó, Minas Gerais, Brasil.* Ararajuba 1.

V3   Vooren, C.M. & A.C. Fernandes. 1989. *Guia de Albatrozes e Petreis do Sul do Brasil.*

V4   Vooren, C.M. & A.Chiaradia. 1990. *Seasonal Abundance and Behaviour of Coastal Birds on Cassino Beach, Brazil.* Ornitologia Neotropical (1).

V5   Vuilleumier, F. 1965. *Relationships and evolution within the Cracidae (Aves, Galliformes). Bull. Mus. Comp. Zool. Harv. 134 (1).*

W1   Watson, G.E. *et al.* 1971. *Birds of the antarctic and subantarctic. Folio 14.* Am. Geog. Society.

W2   Weinberg, L.F.   1984. *Aves do Pantanal do Mato Grosso do Sul.*

W3   Whitney, B.M.& R.H.Barth.   1986 *Birds recorded at sites in Eastern Brazil.* (Unpub.)

W4   Whitney, B.M. 1988. *Birds recorded at sites in Eastern Brazil.* (Unpub.)

W5   Whitney B.M. & J.Rowlett   1989. *Birds recorded on the Field Guides` Southeastern Brazil Tour.* (Unpub.)

W6   Whittaker, A. & M. Cohn-Haft. 1987. *Report on Project Maracá. For Royal Geographical Society.* (Unpub.)

W7   Whittaker,A. - *Personal sightings in Brazil.* (Unpub.)

W8   Willard, D.E. *et al.* 1991. The Birds of Cerro de la Neblina, Territorio Federal Amazonas, Venezuela. Fieldiana Zoology New Series, No.65.

W9   Williams, A.J. 1984.   *Breeding distribution, numbers & conservation of tropical seabirds on oceanic islands in the South Atlantic Ocean.* (In *Status & Conservation of the World's Seabirds*). ICBP Technical Pub No 2.

W10   Willis, D.   1986-87.   *Personal notes on Brazil.* (Unpub.)

W11   Willis, D.   1988. *Personal notes on birds around Leticia, Colombia.* (Unpub.)

W12   Willis, D.   1989. *Parque Nacional Amaca Yacú Area, Colombia.* (Unpub.)

W13   Willis, D.   1990. *Personal notes on birds in Brazil.* (Unpub.)

W14   Willis, D.   1991. *Personal sightings in Brazil.* (Unpub.)

W15   Willis, E.O.   1969. *On the behavior of five species of Rhegmatorhina antfollowing antbirds of the Amazon Basin.* Wilson Bull. 81(4).

W16   Willis, E.O.   1977.   *Lista preliminar das aves da parte noroeste e áreas vizinhas da Reserva Ducke, Amazonas, Brasil.* Revista Brasil Biol. 37 (3).

W17   Willis, E.O.   1979.   *Behavior and ecology of two forms of White-chinned Woodcreepers (Dendrocincla merula, Dendrocolaptidae) in Amazonia.* Papéis Avulsos Zool., São Paulo, 33 (2).

W18   Willis, E.O. & Y.Oniki.   1981.   *Levantamento preliminar de aves em treze áreas do estado de São Paulo.* Rev. Brasil Biol. 41 (1).

W19   Willis, E.O. & Y. Oniki. 1981. *Notes on the Slender Antbird.* Wilson Bull. 93 (1).

W20   Willis, E.O. & Y. Oniki. 1985. *Bird specimens new for the state of São Paulo.* Rev. Brasil. Biol. 45.

W21   Willis, E.O.   1985.   *Aves encontradas na Fazenda Capricornio, Ubatuba, S.P. (agosto 84 e janeiro 85).* (Unpub.)

W22   Willis, E.O.   1988. *Drymophila rubricollis (Bertoni,1901) is a Valid Species (Aves, Formicariidae)* Rev. Bras. Biol. 48 (3).

W23   Willis, E.O.   1988. *A hunting technique of the Black-and-white Hawk-Eagle.* Wilson Bull. 100. (4).

W24   Willis E.O. & Y. Oniki.1990. *Levantamento preliminar das aves de inverno em dez áreas do sudoeste de Mato Grosso, Brasil.* Ararajuba 1.

W25   Willis, E.O. & Y. Oniki. 1991. *Nomes Gerais para as Aves Brasileiras.*

W26   Willis, E.O.   1991.   *Sibling species of Greenlets (Vireonidae) in southern Brazil.* Wilson Bull.103 (4).

W27   Willis, E.O.   1992.   *Three Chamaeza Antthrushes in Eastern Brazil (Formicariidae).* Condor 94.

W28   Willis, E.O. & Y. Oniki.   1992.   *A new Phylloscartes (Tyrannidae) from south-eastern Brazil.*   Bull. B.O.C. 112 (3).
W29   Winkel, Dr. W.   *et al.*   1986.   *Naturkundliche Studienreise "Brasilien"* (Unpub.)
W30   W.W.F. 1989.   *Birds of Manaus vicinity.* Biological Dynamics of Forest Fragments (formerly Minimum Critical Size of Ecosystems).   (Unpub.)
W31   W.W.F. 1989.   *Checklist of the birds at the W.W.F. Reserve, north of Manaus.* (Unpub.)
Y1   Yamashita, C. 1987.   *Field observations and comments on the Indigo Macaw (Anodorhynchus leari), a highly endangered species from northeastern Brazil.* Wilson Bull. 99 (2).
Z1   Zimmer, J.T.   1929.   *The Birds of the Neotropical Genus Deconychura.* Field Mus. Nat. Hist. Zool. Ser.Vol XVII, No 1.
Z2   Zimmer, J.T.   1935.   *Studies of Peruvian birds XVII.* Am. Mus. Novit. No 785.
Z3   Zimmer, J.T.   1955.   *Further notes on Tyrant Flycatchers (Tyrannidae).* Am. Mus. Nov. No.1749.

King Vulture

Useful Addresses

Director de Parques Nacionais,
I.B.A.M.A.
Av. L4 Norte - Setor Areas Isoladas Norte,
Brasilia - DF - 70620.

I.B.A.M.A. - Pará,
Av. Cons. Furtada, 1303 ,
Batista Campos,
Belém - PA - 66010.

I.B.A.M.A. - Minas Gerais,
Av. Contorno 8121,
Belo Horizonte - MG - 30.110.

I.B.A.M.A. - Paraná,
Rua Brigadeiro Franco, 1733,
Caixa Postal 691,
Curitiba - PR - 80420.

I.B.A.M.A. - Goiás,
Rua 299 nº 95 - Setor Universitário,
Goiânia - GO - 74210.

I.B.A.M.A. - Amapá,
R. Hamilton Silva, 1570,
Santa Rita,
Macapá - AP - 68900.

I.B.A.M.A. - Rio Grande do Sul,
Rua Miguel Teixeira, 126 - Cidade Baixa,
Porto Alegre - RS - 90050.

I.B.A.M.A. - Rondônia,
Av. Pinheiro Machado, 1523,
Porto Velho - RO - 789000.

I.B.A.M.A. - Rio de Janeiro,
Pca. XV nº 2, 6º andar, sala 409 - Centro,
Rio de Janeiro - RJ - 20010.

I.B.A.M.A. - Bahia,
Av. Juracy Magalhães Jr., 608,
Rio Vermelho,
Salvador - BA - 44050.

I.B.A.M.A. - São Paulo,
Al. Tietê, 637,
C.P. 7134,
São Paulo - SP - 01312.

I.B.A.M.A. - Espírito Santo,
Av. Marechal Mascarenhas de Morais, 2487,
Vitória. ES.

Dr. Pinheiro,
Instituto Evandro Chagas,
Av. Barroso,
Belém - PA- 66040.

Dr. Skaff,
Museu Paraense Emilio Goeldi,
Av. Independencia,
Caixa Postal 399,
Belém - PA- 66040.

Doug Trent,
Focus Tours,
North American Sales Office,   or Rua Alagoas 1460 / S503
14821 Hillside Lane,   Savassi,
Burnsville, M.N. 55337. USA.   Belo Horizonte - MG- 30130.

Director de Parques e Reservas,
I.E.F. (Instituto Estadual de Florestas),
Rua Paracatu 304 (10º Andar),
Belo Horizonte -MG- 30180.

C.E.M.I.G. - Reserva de Peti,
Sr. Antonio Procopio Sampaio Rezende,
Depto. de Programas e Ações Ambientais,
Av. Barbacena, 1200 - 20º andar,
Belo Horizonte - MG - 30190.

Paulo de Tarso Zuquim Antas,
C.E.M.A.V.E.,
Centro de Estudos de Migrações de Aves,
Sain - Av. L.4 Norte,
Brasilia - DF - 70,800.

Christoph Hrdina,
André Safari & Tours, Ltda.
Shis-Sul. Q17,
Bloco B,
Sala 203/4,
Lago Sul, CEP 7020,
Brasilia - DF- 71619.

Dr. Jacques Vielliard,
Departamento de Zoologia,
UNICAMP,
Caixa Postal 6109,
Campinas -SP- 13081.

Dr. Ted Moulton,
Director of C.E.P.A.R.N.I.C.,
Caixa Postal 43,
Cananéia - SP- 11990.

I.N.P.A. - (Instituto Nacional dePesquisas da Amazônia),
Estrada V8 do Aleixo,
Caixa Postal 478,
Manaus - AM- 69011.

Roger Hutchison,
Biological Dynamics of Forest Fragments,
Rua Rio Purus 28,
Manaus -AM- 69011.

Andrew Whittaker,
Ornithologist & Naturalist Guide,
Estrada do Aleixo,
Conj-Acariquara Sul,
Rua Samaumas 214,
Manaus -AM-69085.

WINGS, Inc,
PO. Box 31930,
Tucson, AZ 85751, USA.

Field Guides, Inc,
PO. Box 160723,
Austin, Texas 78746, USA.

WildWings,
International House, Bank Road,
Kingswood, Bristol BS15 2LX, UK.

Foreign Birdwatching Reports & Information Service,
5 Stanway Close,
Blackpole,
Worcester, WR4 9XL. UK.

WildSounds,
Dept 5 / 11, PO. Box 9,
Holt, Norfolk NR25 7AW, UK.

Localiza National,
Av. Bernardo Monteiro 1563,
Belo Horizonte- MG-30150.